T0301022

Neurobiology of Grooming Behavior

Grooming is among the most evolutionary ancient and highly represented behavior in many animal species. It represents a significant proportion of an animal's total activity and between 30% and 50% of their total awake time. Recent research has demonstrated that grooming is regulated by specific brain circuits and is sensitive to stress, as well as to pharmacological compounds and genetic manipulation, making it ideal for modeling affective disorders that arise as a function of stressful environments, such as stress and post-traumatic stress disorder. Over a series of 12 chapters that introduce and explain the field of grooming research and its significance for the human and animal brain, this book covers the breadth of grooming animal models while simultaneously providing depth in introducing the concepts and translational approaches to grooming research. The book is written primarily for graduates and researchers within the neuroscientific community.

ALLAN V. KALUEFF is Assistant Professor in the Department of Physiology and Biophysics at Georgetown University Medical Center and Department of Pharmacology at Tulane Medical School. He publishes actively on models of drug–drug and drug–receptor interactions; theories of brain disorders and their therapy; and the complex interplay between cognitive, motivational, and genetic bases of animal behavior.

JUSTIN L. LAPORTE worked at the National Institute of Mental Health in Bethesda, MA, USA. His research employs behavioral pharmacology and molecular genetics approaches to elucidate the pathogenetic mechanisms of psychiatric disorders such as depression, anxiety, and obsessive–compulsive disorder, with a specific focus on the role of serotonin transporter.

CARISA L. BERGNER is a researcher in the Department of Physiology and Biophysics at Georgetown University Medical Center. Her research involves mouse and zebrafish models of stress and depression.

Neurobiology of Grooming Behavior

Edited by

ALLAN V. KALUEFF

Georgetown University Medical Center,
Washington, DC, USA

JUSTIN L. LAPORTE

National Institute of Mental Health,
Bethesda, MD, USA

CARISA L. BERGNER

Georgetown University Medical Center,
Washington, DC, USA

CAMBRIDGE
UNIVERSITY PRESS

University Printing House, Cambridge CB2 8BS, United Kingdom

One Liberty Plaza, 20th Floor, New York, NY 10006, USA

477 Williamstown Road, Port Melbourne, VIC 3207, Australia

314-321, 3rd Floor, Plot 3, Splendor Forum, Jasola District Centre, New Delhi - 110025, India

103 Penang Road, #05-06/07, Visioncrest Commercial, Singapore 238467

Cambridge University Press is part of the University of Cambridge.

It furthers the University's mission by disseminating knowledge in the pursuit of education, learning and research at the highest international levels of excellence.

www.cambridge.org
Information on this title: www.cambridge.org/9780521116381

First published 2010

A catalogue record for this publication is available from the British Library

ISBN 978-0-521-11638-1 Hardback

Contents

Contributors

Marie-Claude Audet
Institute of Neuroscience, Carleton University, Ottawa, ON, Canada.

Carisa L. Bergner
Department of Physiology and Biophysics, Georgetown University Medical Center, Washington, DC, USA.

Wah Chin Boon
Howard Florey Institute, Parkville, Victoria, Australia.

Sam R. Chamberlain
Department of Psychiatry, University of Cambridge, Addenbrooke's Hospital, Cambridge, UK.

Howard Casey Cromwell
Department of Psychology and The J. P. Scott Center for Neuroscience, Mind and Behavior, Bowling Green State University, Bowling Green, OH, USA.

Brett D. Dufour
Department of Animal Sciences, Purdue University, West Lafayette, IN, USA.

Rupert J. Egan
Department of Physiology and Biophysics, Georgetown University Medical Center, Washington, DC, USA.

John C. Fentress
Department of Psychology/Neuroscience, Dalhousie University, Halifax, Nova Scotia, Canada.

Michael H. Ferkin
Department of Biology, Ellington Hall, University of Memphis, Memphis, TN, USA.

Naomi A. Fineberg
National OCD Treatment Service, Hertfordshire Partnership NHS Foundation Trust, Queen Elizabeth II Hospital, Welwyn Garden City, Herts, UK.

Joseph P. Garner
Department of Animal Sciences, Purdue University, West Lafayette, IN, USA.

Sonia Goulet
Centre de recherche Université Laval Robert-Giffard, Québec, Qué., Canada.

Peter C. Hart
Department of Physiology and Biophysics, Georgetown University Medical Center, Washington, DC, USA.

Rachel A. Hill
Mental Health Institute of Victoria, Parkville, Victoria, Australia.

Lara J. Hoppe
ACSENT Laboratory, Department of Psychology, University of Cape Town and Medical Research Council of South Africa, University of Cape Town, Department of Psychiatry, Groote Schuur Hospital, South Africa.

Jonathan Ipser
University of Cape Town, Department of Psychiatry, Groote Schuur Hospital, South Africa.

Allan V. Kalueff
Department of Physiology and Biophysics, Georgetown University Medical Center, Washington, DC, and Department of Pharmacology, Tulane University Medical School, New Orleans, LA, USA.

Daesoo Kim
Korea Advanced Institute of Science and Technology, Daejeon, Korea.

Hae-Young Koh
Center for Neural Science, Korea Institute of Science and Technology, Seoul, Korea.

Robert Lalonde
University of Montreal/St-Luc, Neuroscience Research Unit, Montreal, Canada.

Justin L. LaPorte
Department of Physiology and Biophysics, Georgetown University Medical Center, Washington, DC, USA.

Stuart T. Leonard
Department of Pharmacology, Louisiana State University Health Sciences Center, New Orleans, LA, USA.

Christine Lochner
MRC Unit on Anxiety and Stress Disorders, Department of Psychiatry, University of Stellenbosch, South Africa.

Sergio M. Pellis
Department of Neuroscience, University of Lethbridge, Lethbridge, AB, Canada.

Vivien C. Pellis
Department of Neuroscience, University of Lethbridge, Lethbridge, AB, Canada.

Hee-Sup Shin
Center for Neural Science, Korea Institute of Science and Technology, Seoul, Korea.

Srinivas Singisetti
National OCD Treatment Service, Hertfordshire Partnership NHS Foundation Trust, Queen Elizabeth II Hospital, Welwyn Garden City, Herts, UK.

Amanda N. Smolinsky
Department of Physiology and Biophysics, Georgetown University Medical Center, Washington, DC, USA.

Dan J. Stein
University of Cape Town, Department of Psychiatry, Groote Schuur Hospital, South Africa.

M. Frances Stilwell
Stilwell Studio, Corvallis, OR, USA.

C. Strazielle
University of Nancy I, INSERM U724, Electron Microscopy Service, Faculty of Medicine, Vandoeuvre-les-Nancy, France.

Kevin G. F. Thomas
ACSENT Laboratory, Department of Psychology, University of Cape Town, Rondebosch, South Africa.

Preface

Grooming and related behaviors

Behavioral and pharmacological research continues to play a crucial role in modern neuroscience, often spearheading new and innovative techniques and models that further our understanding of the intricate workings of the nervous system. This is particularly evident in the arena of mental health research where, with the help of animal models and novel genetic or pharmacological treatments, new insights and theories are evolving to conceptualize more accurately common brain disorders such as anxiety, depression, obsessive–compulsive disorder (OCD), and schizophrenia.

These advances are allowing for an unprecedented examination of the heritable and environmental factors that contribute to disease pathogenesis. However, although there has been marked progress, the biological substrates of many of these disorders remain unclear. To establish a more concrete understanding of these disorders, a careful dissection of experimental phenotypes must be pursued. In this way, every aspect of behavior is a potentially fruitful source of experimental data that can provide clues to the contributing mechanisms.

One important example of such a behavior is grooming. Grooming is a very highly represented behavior in many animals, comprising a large proportion of their waking time. It serves an incredibly diverse range of purposes in the life of the animal from chemocommunication to basic hygiene. It is a natural behavior, yet it can be induced as part of an experimental procedure and has been shown to be sensitive to stress and bidirectionally sensitive to anxiolytic and anxiogenic drugs in rodents, making it an ideal focal point for high-throughput behavioral studies.

In the clinic, this behavior plays a similarly important role. The complexity of human grooming behavior, of course, goes beyond simple measures of hygiene and mate attraction. However, it has proven to be an invaluable tool in clinical diagnostic assessment. Abnormalities in human self-grooming are often a component of disorder symptomology, and there are a variety of grooming disorders, such as trichotillomania, that involve behaviors like biting the nails or picking at the skin, which are indicative of a larger mental health issue.

The span of grooming behavior will be explored here in the following chapters, providing an updated summary of current research on this important topic. The opening chapter of the book will give a concise introduction, providing a comprehensive summary of past research on grooming behavior. It includes an exposition of the theoretical justifications in grooming studies, and how these are evolving to give researchers an accurate view of this quickly developing field.

The causes of grooming behaviors and environmental factors that affect them are a very active area of research. Continuing on this theme, another chapter specifically explores grooming as a response to olfactory stimuli. Here, the authors detail their research on grooming phenotypes that differentially self-groom when they come into contact with different conspecifics, and discuss the implications of this research for viewing grooming as a vehicle for targeted social communication. The olfactory component of grooming also plays a key role in animal reproductive behaviors, and the chapter examines this as a way that mate eligibility is conveyed.

Along with olfactory communication, grooming serves many other practical roles in animal behavior, and is often accompanied by behavior-associated hair loss known as barbering. These are complex, ethologically rich behaviors with sensitivity to alterations in activity and microstructure. Another chapter elucidates the utility of grooming analysis for assessing stress in individual animals, testing of psychotropic drugs, phenotyping mutant or transgenic animals, as well as selecting proper strains for experimental modeling of affective disorders. The chapter will also discuss ethologically based approaches to the assessment of animal grooming and barbering activity, present examples of genetic variation leading to altered grooming and barbering phenotypes in rodents, and summarize the growing value of these two phenotypes for translational neurobehavioral research.

In addition to barbering, there are additional forms of social grooming behavior that entail physical contact between conspecifics. Social grooming, as well as rough-and-tumble play, caressing, and hand-shaking, are all touching behaviors that are an integral part of social communication. One chapter will review the neurochemical pathways that regulate these behaviors, as well as the touch-induced changes in mood across mammals. Touching (as an important form of communication) will be discussed, with a focus on the subcortically regulated emotional state of the interactants and the cortically mediated modulation of the touching behavior that allows animals to use physical contact in a more strategic manner.

As mentioned before, grooming is also highly sensitive to pharmacological treatments, and is increasingly recognized as a reliable marker of stress-related disturbances in animal models of neuropsychiatric disorders. One chapter details the past and current literature examining induced grooming in phencyclidine-treated rats. Relevant data about the effects of this drug on grooming patterning and hygiene efficiency, and how this related to both normal and abnormal stress responses will be presented, followed by a discussion on how the qualitative observations from this study may be beneficial in identifying hygienic and stress-related irregularities in animal models.

Grooming has been shown to be regulated by a variety of factors, one of them being hormonal levels. The effect of estrogen on this behavior, as both the estrogen-synthesizing enzyme and estrogen receptors are present in the brain and have roles in several neural circuits, will be examined in another chapter. The research presented will show how the distribution of estrogen-sensitive cells in the brain corresponds to regions that control sexual differentiation, masculine and feminine sexual behaviors, aggressive behaviors as well as grooming, and will review the observations from both animal models and clinical patients indicating that estrogen has a modulatory effect on grooming and related behaviors.

The absence of grooming behaviors in an animal can be as important and significant to the translational validity of models as robustly grooming phenotypes. There are symptoms of social withdrawal that occur in many neuropsychiatric disorders, including some of the most common such as depression. Another chapter will introduce this issue, highlighting social withdrawal phenotypes like barbering behaviors. The research will focus on these aspects in a potential mouse model of schizophrenia, a knockout for the PLCβ1 gene. The authors will explain how abnormal phospholipid metabolism has been implicated in the pathogenesis of schizophrenia, and how phospholipase C (PLC) β1 has been shown to be reduced in specific brain areas of patients with schizophrenia. This chapter will also discuss the interesting array of grooming-related phenotypes and the possible signaling mechanisms that may be affected in this model, as well as its applicability as a model of schizophrenia.

In line with the animal models, another chapter will discuss the effect of brain lesions on grooming behaviors and how new information on neurobehavioral pathways can be derived using these methods. The chapter will detail research on electrical stimulation of the midline cerebellum and striatum in rats, and lesioning methods using either surgery or genetic mutations that indicate that the cerebellum, basal ganglia, and neocortical brain regions contribute to grooming behaviors. It will also discuss $Grid2^{Lc}$ mutant mice with selective cerebellar atrophy, and $Girk2^{Wv}$ mutants with combined cerebellar and substantia nigra atrophy that display different effects on grooming. Their results implicating cerebello–neocortical pathways in the completion of grooming chains, and a striato–pallido–neocortical pathway in the serial ordering of grooming chains, will also be presented in this chapter.

The role of the striatum in grooming behaviors will be specifically explored in another chapter. The research presented here will review the evidence for the role of the striatum in implementing the fixed action pattern of the grooming chain in the rat. In addition, the support for the involvement of the dorsolateral striatal subregion involved in the production of this movement sequence and the general functional significance of implementation by striatal circuitry is discussed. The authors also introduce the possibility that the general nature of striatal function of "implementing" chains of information crosses different functional boundaries between movement and reward information.

Taking a broader look at grooming-related behaviors, the next chapter will focus on the circuits surrounding barbering behavior. Many aspects still remain unclear to researchers and even the very existence of the behavior seems to be a paradox to some. Barbering behavior will be explained from an ethological perspective in an effort to resolve the barbering paradox by asking how and why barbering behavior occurs. The authors will also discuss the phylogenetic underpinnings of barbering, by comparing and contrasting the occurrence of hair-plucking behavior. The chapter reviews the developmental processes that underlie barbering behavior, with specific attention to known risk factors, learning, the laboratory environment, and transgenic mice. The authors will also review the behavioral mechanisms, eliciting stimuli, and physiological mechanisms that might mediate barbering, and outline the role of cortico–striatal circuitry in abnormal repetitive behavior in general, how it can be used to delineate disorders, and insights it provides into barbering.

Potentially maladaptive behaviors such as barbering are not uncommon in both laboratory rodents and clinical psychiatric patients alike. Human grooming disorders, such as trichotillomania, nail biting, and skin picking, are possibly linked to conditions like OCD. However, within current psychiatric classification systems trichotillomania is currently conceptualized as an impulse control disorder and nail biting and skin picking are not yet included in the official nomenclature. One clinical chapter outlines the debate over whether grooming disorders should form a separate category, or whether they should be classified as OCD spectrum disorders, impulse control disorders, or as body-focused repetitive behaviors. This chapter will also discuss these diagnostic and taxonomic issues particularly as they pertain to clinical practice.

Keeping with the theme on trichotillomania, another clinical chapter will review new advances in genetic, family, neurocognitive, neuroimaging, and neuropharmacological studies on this impulse control disorder. A particular focus will be given to new research that shows interesting similarities between trichotillomania, other impulse control disorders, and OCD, while also revealing important differences in some endophenotypic measures. The chapter will also discuss neural abnormalities in the amygdalo–hippocampal formation and frontal–subcortical circuits are discussed and how animal models of these disorders may prove to be a fruitful avenue for future research.

In conclusion, the book conveys the message that grooming is an ethologically relevant, robustly observed behavior that can be dissected with high-throughput phenotyping techniques. By understanding the behavior, the researchers can now maximize the translational significance of the data and get closer to new diagnostic tools, as well as treatments and preventions. These chapters represent a wide and thorough perspective into a very important behavior that is proving indispensible to translational mental health research. The authors have been selected from an international conglomeration of top experts in their respective fields, and have contributed data that are driving today's research. The themes range from pharmacological to genetic to behavioral, covering a wide spectrum of basic and clinical neuroscientific disciplines. However, this book has been designed to serve as a useful source of literature for both the introductory student, as well as for the experienced researcher.

The editors would like to thank the National Alliance for Research on Schizophrenia and Depression (NARSAD), the world's leading charity dedicated to mental health research, for their generous support of this work. We hope that this book can help further the goal of discovering preventions and cures for neuropsychiatric illnesses through innovative translational scientific research.

1

Grooming, sequencing, and beyond: how it all began

M. FRANCES STILWELL AND JOHN C. FENTRESS

Summary

This chapter has taken two distinctive but complementary approaches to mouse grooming. The first is based upon Frances Stilwell's intuitive perceptions in the 1970s of previously unappreciated order in grooming sequences. An important principle here is that early stages in research depend upon sensitivity to what our animals can show us. Premature narrowing of observational perspective can limit the richness of analytical questions that are initially hidden from view. In the second part of the chapter, John Fentress outlines some of the richness of subsequent research that sensitive descriptions have led up to. Mouse grooming has led to a host of studies in behavioral genetics, development, brain mechanisms, and motivational models including stress.

Introduction

This chapter is intentionally divided into two parts. The first part, by Frances Stilwell, outlines the discovery of rules underlying order in the rich patterning of mouse grooming. As Stilwell discovered in the early 1970s, there is indeed syntax, perhaps even a grammar, in these rodent movements. One of the lessons here is to look closely at rules of order in seemingly inconsequential action patterns of the animals around us. They are rich in their structure. Mouse grooming has led to a number of important insights about brain and behavior. Furthermore, Stilwell's comments are not only refreshingly personal, but also important

Neurobiology of Grooming Behavior, eds. Allan V. Kalueff, Justin L. LaPorte, and Carisa L. Bergner. Published by Cambridge University Press.

as a picture of how research sometimes actually progresses. This reminds us of the insights early ethologists, such as N. Tinbergen, came up with by just watching. Tinbergen's point was well made: sometimes we just need to open our eyes and have our minds ready to receive what nature offers.

In the second part of the chapter, John Fentress takes Stilwell's insights on the discovery of the order in mouse grooming and outlines how it opened up areas of research that were never before appreciated. Others, including the authors of this volume, have taken this richness in movement and shown conclusively how it opens up a wide range of brain/behavior issues that extend far beyond mice, or grooming. It is a rich story that continues to invite explorations from many channels.

Finally, there is an important footnote this chapter offers. The original *Nature* article on mouse grooming "grammars" listed John Fentress as first author (Fentress and Stilwell 1973). There is something uncomfortable, even wrong, if the impression was that Fentress led the discovery of order in mouse grooming. It was Frances Stilwell who discovered the order while in the lab of John Fentress. It is important to make that point of the story clear.

How the study of mouse face-grooming sequences began (M. Frances Stilwell)

Wherein the beauty of the behavior is honored, the value of the nonconscious is confirmed, and a record is clarified.

My job at the University of Oregon Chemistry Department lab ended abruptly three weeks after my arrival from Ohio in September, 1969. It never crossed my mind to go back to Ohio. Instead, I responded by driving from Eugene to British Columbia with drawings from college botany classes to show professors and doctors, hoping to sell myself as a scientific illustrator. Just before Christmas, I took my drawings to a veterinarian's office in nearby Springfield. It just so happened I was taking a course entitled introductory biology, in order to satisfy the state of Oregon's very specific teaching requirements.

The vet said, "Why don't you go to the University? A professor there has wolves and he may want you to illustrate them." I said, "Is his name Fentress?" He said, "Yes!" I said, "I'm taking a course from him and doing very well!" When I approached Dr. John Fentress about illustrating his projects, I introduced myself by saying that I was in his lecture class. He said, "Oh, yeah? What did you think of my test? Some of the students were complaining about it. They said there were trick questions."

In my opinion, the questions were not tricky although they did exact quite a bit of the focused attention required for a Graduate Records Exam to answer them. I equivocated, "Oh it was fine." He smiled, mumbled, and said, "Tsk. Tsk. Students these days." Then, when he looked up my score on the test, his tone and attitude changed. In a lower voice, sounding as if speaking just between you and me, he said, "As a matter of fact I am waiting to hear about a grant. Why don't you come back after the first of the year. I do need clerical help. How do you feel about working with mice?"

After that conversation, on New Year's Day, the *Eugene Register Guard* newspaper reported on its front page that an unnamed university professor in the Departments of Biology and Psychology had just been awarded a huge federal grant. The next day I called Dr. Fentress. "Could that professor be you?" I asked. "Yes," he said, "why don't you come in tomorrow?"

Dr. Fentress was interested in the role of sensory feedback in mouse behavior. His approach in studying this was to interrupt the route for the information from a mouse's front paw back to the brain, or to "deafferent" the limb, by surgery, and then note the behaviors affected. A postdoctoral staff member, Dr. Maria Rosdolsky, MD, had previously performed the surgeries for him. She had also devised simple tests to determine any effects on their activities. Dr. Fentress wanted me to continue her surgeries and testing work with the mice as well as perform clerical duties as an educational project aide.

On my second day at the lab, which was Dr. Rosdolsky's last day, she showed me the procedures, emphasizing how important it is not to tear the motor nerve or damage a muscle. One of the mice she had worked on carried its desensitized arm flat on its chest, as useless as a polio victim's, which was the result of such a tearing mishap. On previously deafferented mice, and control mice, she demonstrated the battery of tests she'd devised noting whether the deafferented arm adducted or abducted, rolling these two words off her tongue as if they were pablum. I could see there was much ahead for me to learn. These words meant move the limb toward the body ("adduct") or away from it ("abduct"). Her list included a test for response from her pinching the mouse's paw. Another was to hang the mouse by its tail to see if it would grab the wire grill on its cage top with its deafferented arm. A third specified dumping the unsuspecting mouse in a pan of water to watch how it would swim. At some point during the testing protocol a mouse began flapping its arms in all directions. It appeared to launch into a juggling act without a ball. And I said, "What's it doing?" She said, "It is grooming."

The following day when I came in, I lightly chloroformed a mouse, as Maria had shown me, so it would lie cooperatively on its back during the injection of anesthetic. Then I injected the mouse below its rib cage. I can still feel the resistance to the needle prick, which was too low. The mouse squeaked loudly in protest, through its chloroform, with a sound worse than a cat whose tail has been stepped on. The mouse was so vulnerable and trusting in my care. Hearing its cry, I began to cry. Mixed in with my crying for the mouse was the realization that I'd botched my first surgery. Dr. Fentress, who was talking with a graduate student, John Mates, across the room, interrupted their conversation to ask, "What's wrong?" "It squeaked," I said. "Why don't you go home for the day?" he suggested.

This experience resulted in a change in my mouse responsibilities, which sent me down a different route. That change ultimately produced a much more interesting result both for animal behavior and for myself. I don't recall whether that mouse died, but I never attempted any more surgeries. With Dr. Fentress' agreement, I was now charged just with testing the control mice and the mice whose arms Maria had already deafferented. I particularly remember the swimming tests. I would dump each mouse in the water on one side of a plastic pan and watch it swim frantically across the eight inches to the other side. The idea was to note each mouse's limb activity to check for two responses, apparent through the clear plastic walls. Firstly, to see if it used its experimental paw to swim as a normal mouse would, and secondly, to see whether it reached out that paw to touch the home-free side, its goal.

So on my fourth day of work, I began the battery of tests. I looked at both desensitized and control mice. Once each mouse had reached the opposite side of its water-filled cage, I lifted it out and let it recover on the counter top before its next test. I recorded my results. I would then wait patiently while the mouse shook off all its water droplets and calmed down from its experience. Once that test was done, I lifted each one back into its proper home cage or proceeded to the next test. I only vaguely watched each mouse as it sat on the counter top between tests, but it was always in my field of view. Though I faced my subject, my attention, and probably my eyeballs, drifted away from the mouse and back.

Between the fourth and the tenth mouse something about the mouse being tested made me focus on it. The mouse had inexplicably grabbed my attention. Even then I focused for less than an instant. It was the strangest passing of information from mouse to human! I said to myself reflexively, in an offhand way, "Oh, yeah, that's what they do." I have described it later as having my body pick up the information through its pores and that only later did the information rise up into my brain. Actually, my mouth knew it before my brain did.

I was talking to myself not to anyone else. It wasn't until my ears heard the content of my phrase that I was aware of what I knew. There was order. Another graduate student, Doug McDonald, heard me from his nearby desk. "What was that?" he said. "When the mice groom," (I'd heard this word from Maria, here I was referring to grooming of their heads) "they do it in a particular way each time." John Mates came across the room from his desk. "What was that?" he asked. After I told him, he and Doug looked at each other. One of the graduate students said, "This would fit very well into John's research," referring to our supervisor

Dr. John Fentress. I did not know what they meant by that statement, but when I heard it, I realized there was something here that could make me very valuable at my new job, so I set out to identify this order in the grooming.

Obviously, if I were to identify what this order was, I would have to describe what the mice were doing. I decided that the first approach would be to count the arm movements. The next time a mouse groomed after swimming, I tried. But the action was much too fast. It was not possible to keep up, much less to be able to speak the count out loud with the resources at hand.

I said to myself, "I need slow-motion photography." When Dr. Fentress came back to the lab, I told him what I'd noticed. He procured a slow-motion movie camera and a stop-action projector for me to use. From then on my focus at the job, when clerical activity was not required, was to film the mice, to attempt to track down this illusive order in the movements, and to document my work.

There was no question about its being there, the only question was "What is it?" This was uncharted territory for me. I was not conscious of any traditional scientific approach used to accomplish this analysis. I pretty much went as if I were a hound dog sniffing a scent or as if I were untangling a ball of yarn. Since in science, the approach is to test hypotheses, to try to eliminate them not to prove them, one might be tempted to say the notion of order was hypothesis – but I had no question about the order's existence. I just wasn't aware of what it was. I knew that the order would eventually pop out of my chronology of what turned out to be the two seconds or more of face-grooming activity. I felt everyone believed I was going to find that order so they all patiently awaited my result. Occasionally I got inklings that perhaps Dr. Fentress might not be as convinced as the others that I was going to find order or even that it was there in the first place. But, I encouraged myself, "I will find it and when I do will he ever be surprised!"

Step 1. I set up an old, cracked glass aquarium on a lab desk, and cut up a plastic cage. Then I fabricated a mouse-size plastic cage with one side open at the top and one side open where it could be taped to the inside of the aquarium. I faced the camera through the aquarium glass into the plastic cage. Then I would drop either a control or desensitized mouse by its tail into the cage, focus the camera and, with my finger on the shutter, wait and watch. Very soon, in its new surroundings the mouse would begin to groom. It took me a few tries to learn when to start the camera whirring so as to catch the whole or most of the routine. Sometimes the mouse would not be facing the camera when I started so I turned it around with a pencil while it was still grooming. I realized later that in only a certain part of the face grooming, which I called "single-stroking," could I do this without interrupting the grooming activity.

Step 2. To analyze the film frame by frame, I set up the projector in the lab to describe the action. I had to get information out there on paper to see what the order was. I remember the first grooming activity I tried to describe. My inadequate words almost immediately gave way to what came more naturally to me, and I resorted to making drawings of the arm movements along the head. This generally meant tracing the paws' trajectory with respect to the snout. As I analyzed the film, I became aware of categories of movements, which I referred to as strokes. My first such drawing of a stroke was of a half-moon trajectory. I saw that all the movements fell out quite naturally into strokes of repeated forms, which invited my naming them, which simplified my note taking. I labeled the strokes as follows:

1. Parallel was a half-moon trajectory along the snout with the paws duplicating each other's form.
2. Circling reminded me of a cheerleader introducing her cheer at a high-school basketball game, by moving her forearms in an irregular-looking way.
3. Licking was the horizontal movements back and forth beneath the lower jaw. (Occasionally I would see a tongue come out. This stroke had two versions: (a) short-licking, which was slower and of short duration and (b) long-licking, which was faster and of longer duration.)
4. Overhand reminded me of an overhand smash stroke in tennis.
5. Single-stroking was a series of ten flat strokes with the arms alternating left and right perfectly as to which made the greater excursion. It was almost a staccato percussion movement as rigid as wooden soldiers.
6. Shimmy, a blur of the whole body in every frame at 32 frames per second, reminded me of a hula dancer.
7. Pause was when both paws were held still somewhere below the chin or at chest height. It still seemed to be part of the grooming activity.

I dutifully recorded frame numbers involved for all the strokes and briefly illustrated some of them. I particularly remember drawing the meandering paths of the overhands and that I included the frame number for each point along the route. Out of curiosity at the end of the analysis of my first routine, I added up the strokes, and I was gratified but astounded to realize how many there were. The mouse had performed twenty strokes in two seconds, including the ten in single-strokes. No wonder I couldn't count them! Dr. Fentress eventually found me a desk space for my analyses in a dark closet full of photographic equipment. There I could better see the contrasts in the blacks, whites, and grays when the film was projected.

Since my work with mice was second in priority to my clerical duties around the lab, I could not focus full time on answering my question, "What is the order?" However, I was never hesitant or concerned about finding it, nor did anyone rush me. Analyzing the grooming steps was time consuming. However, I loved watching the mice do their thing, especially in slow motion. It was like witnessing a ballet. I liked working with the DBAs (the little gray mice) the most, as they seemed to thrive on grooming. The overhand strokes particularly were executed with robustness, verve, sensuousness, and pride.

The whole phenomenon of my filming and studying mouse face-grooming made some other people curious. One day I became aware that one of the professors in the neurobiology group was watching me from about five feet away as I concentrated on filming a mouse. When I finished with the filming, I looked up to see what he wanted. When he kept on looking forward in a kind of daze, someone said, "He's watching Frances watch the mice."

The spirit of inquiry was alive in what we called the ethology lab, the original European term for animal behavior. I loved being a part of it. Dr. Fentress would toss out an idea and then challenge his own thinking. I felt appreciated, and thought I'd found my niche in the whole field of animal behavior, which had more engaging stories than those in my previous academic fields of concentration, botany, and biophysics. Often while pipetting in the chemistry department I had looked longingly out of the window at the veterinary office across the boulevard and wished I were working with animals. The goodness of life had come through for me.

Dr. Fentress had a way of encouraging people in their work. He suggested that I take his course on animal behavior. I bought the thick blue textbook written by his mentor Robert Hinde at Cambridge University in England and began to consider a PhD to pursue a career in ethology. Years afterwards I met a student from that class, who said Dr. Fentress was the most inspiring professor he had ever had. Professor Robert Hinde's presence was felt in the lab long before he came to the University of Oregon for an invited lecture. In a sense he was mentor to us all.

Dr. Fentress pretty much left me on my own. I filled pages and pages of long sheets of newsprint with my penciled observation notes. I planned to describe 20 sequences before presenting the findings to him. As I worked I began to develop an impression of which strokes were associated with which. Eventually I called these associations "units." I also began to be convinced of the rhythm of the order of the units as they proceeded through the "sequence."

In addition to regaling us in the laboratory about his motor scooter ride to interview Carl Jung, Dr. Fentress often repeated his version of quotes of famous people. Of relevance to what I was doing was a quote he attributed to Nikko Tinbergen, Noble Prize winner, "I let the animals ask the questions. And when they do, I listen, for they ask very good questions." The mice had indicated to me, "We groom our faces in an order, do you know what it is?" I was searching for the answer to their question.

One day Dr. Fentress stopped by the viewing closet. That was the only occasion when we discussed my grooming work until I presented him with the results. He asked, "How's the closet working out?" "Fine." I said. "Imagine that – they do twenty strokes in two seconds! No wonder I couldn't count them!" Then, by way of conversation I continued, "For some reason I know when to start the camera," which I said as if accepting one of life's wonders. He said, "Perhaps now is the time for introspection."

Introspection meant two things to me. First, was to scavenge around internally to ask, do I feel this or do I feel that? Second, was to be open to what I found. I chose to emphasize the second. Very soon, probably at my next filming of a mouse, with my mind open to possibility, I suddenly saw that the mouse wiggled its body all over fast before the start of its routine! I'd seen this motion before, mixed in with other strokes and had called it "shimmy." This shimmy was the signal at the start of their routine for me to start the camera.

For 16 months the routine of my job was to perform my clerical duties, then when a break occurred to go to the grooming work. In April, 1971 Dr. Fentress was to give a talk at the Annual Biology Colloquium at Oregon State University in Corvallis. It seemed to me that he might be interested in the results of my grooming research to include in his speech. I had 16 sequences completed, including 3 from the deafferented mice. My observations by then had developed into a logical process I believed in concerning the flow of strokes. Generally this was it: at the start, there would be a shimmy, then a variety of stroke types would occur. Then the mice began slow-licking, which suddenly shifted to fast-licking. Then single-stroking blasted forth followed by a series of overhands, which was followed by a variety of stroke types again.

I summarized the sequences on one long sheet of newsprint, by letter for each stroke with the numbers of frames spent on each. I left it on Dr. Fentress' desk one evening after he had gone home. It felt redundant to say, "These are the units in the grooming sequences," but I did. Although they were obvious to me, for emphasis I put brackets around the units and numbered them. The order within the sequences, including the composition of the units, started to be apparent to me as I was taking notes on the 16 mice. However, I don't believe I was cognizant of it until I forced myself to write the information down. The next morning I found the list of pencil markings on my desk with his comment in ink "Excellent!"

Dr. Fentress reacted further to my breakthrough by saying: "This is the first time since birdsong, such complex stereotyped behavior has been found in vertebrates. It is the only such behavior known in mammals." He heralded the discovery as being a second example of hierarchical organization in vertebrates. His word, "hierarchy," brought to my mind a royal line, which would be an elegant and appropriate connotation for these behaviors in my mice. It also brought to mind a family tree flattened on paper, which I felt really ought to be three-dimensional with lives of children going all directions. Such a concept could better be represented by a mobile, which shifts with directional breezes but remains in balance.

About then, Dr. Fentress also suggested that the discovery was a grammar of movement. I never liked that metaphor since grammar implied such rigidity to me. It seemed inappropriate for such a lovely dance as the filmed mouse grooming, performed with a looseness in order left somewhat to the discretion of the mouse performer. This grammar metaphor gave impetus to a remark by Loren Northrup, another graduate student, who, after my results were revealed, had begun studying the grooming sequences in his neurological mutant mice. His comment was, "Do you want good grammar or good grooming?"

Dr. Fentress also straightaway applied statistics to what to me were obvious results. I thought he must have been one who ascribed to Lord Kelvin's point of view, which is essentially that "nothing is known until there are numbers on it." The data did provide a good opportunity for a statistician. The statistics were consistent with my observational findings. Actually, the mice's movement reminded me of the progression of orchestral music. I described the grooming as being like watching an orchestra with players tuning their instruments before a concert. The conductor raises his baton with circling, the symphony begins with licking, which suddenly increases in intensity and passion bursting, exploding into cymbals with single-stroking and finally, with violins, the overhands draw out the final ecstasy.

Shortly after I summarized the order, I recognized it in hamsters as well as mice. Then I saw it in gerbils, then rats of course, then in a film of golden-mantled ground squirrels. Each time the results of my pure research reminded me, "There really is order in the rodent universe!" In fact the order seemed to be a marker for rodents. Then a highly strung Sminthopsis, a marsupial rodent from Australia, arrived briefly in the lab. It groomed in long bouts and under similar circumstances as the mice but its movements, which appeared much faster than in the mice, did not look as if they would be so graceful in slow motion.

After I had listed the sequences, Dr. Fentress and I talked about whether the strokes and the ordered connections between the strokes could be detected in infant mice so I began a series of films on them, too. One time Dr. Fentress was in the lab as I filmed the infant mice. Someone asked him a question. I knew from his answer what the question concerned, because he replied, "I'm impressed with her because I wouldn't be able to do what she did." At the time I wondered whether he was impressed that I could sense the order of such a lightning-fast behavior before formal analysis, or was it that I had persisted in taking all those notes, or was it that I could spot the order from the notes? Upon reflection I continue to believe that each of us has much inside sending us signals that we don't ever take advantage of. I truly believe that children musing over their pet gerbils are probably as aware of some kind of order as I was when beginning the work with the mice. But children don't often have the circumstances that would lead them to filming.

I saw the order and defined it, but it took a different kind of mind and experience to know where the order in mouse grooming fitted into the then-known body of scientific knowledge, and a different sort of personality to promote the new information. This is where our teamwork became truly effective.

During September that year, Dr. Fentress was to present a talk at the International Ethology Conference in Edinburgh, Scotland. I elected to take my vacation in Great Britain during that time, and then hear him present my work in the capital of Scotland. When Dr. Fentress gave his talk, I was proud to have the results of my research on order in face-grooming sequences presented at an international gathering. What a wonderful culmination that trip was. It was a celebration of my joy in unknotting the order I knew was there in the mouse face-grooming sequences. A year later I wrote to Nikko Tinbergen inquiring about research positions in his lab. At the request of Professor Tinbergen, Richard Dawkins, in his research group at Oxford University, answered the letter: "I think your work on mouse grammar is fascinating, and I am sorry you want to give it up! However, I am afraid that in any case we have no money at present to employ you, much as we should like to." I also wrote to Konrad Lorenz who returned a nice note saying he was "going emerit" at the end of the year and so could not promise a guest research position in his department.

The publication from Dr. Fentress' Corvallis Colloquium at Oregon State University (Fentress 1972) described the strokes and order quite adequately, but didn't have a wide readership. In those days the proceedings of an International Ethology Conference were not published. For the discovery's real debut, therefore, we wanted to publish in a very prominent journal. Because I had wanted to publish in *Nature* ever since my MS in botany–biophysics days, we agreed we would submit an article, "The grammar of a movement sequence in inbred mice," to that journal. I had learned during my graduate studies that the first author of two was considered to be the originator of the research, actually the lead contributor, and I looked forward to a first authorship in what might become my new field of animal behavior.

So we composed the first part of the article. Dr. Fentress then said, "I will add some statistics." I accepted by then that statistics are necessary to quantify and add credibility to observational data. When I saw Dr. Fentress' name listed first in our submission to *Nature*, I was stunned. If I understood the protocol correctly, readers could conclude from his being listed as the primary author that he had had my insight and had also done the follow-up work. When I asked him, Dr. Fentress explained that first authorship was decided in several ways: (1) alphabetical order; (2) who wrote the paper; (3) in whose lab the work was funded. However, if there were misconceptions about who did the pure research that resulted in this discovery, I am glad now Dr. Fentress offered me the opportunity to clarify the record with this chapter. It may not be the first time such a false impression occurred, but it might be the first time it has been corrected in the lifetimes of the people concerned. Someone said that the article was elegant, and I hoped that he also meant the mouse's behavior was elegant. But of most importance to me was the finding that truth can first be detected without use of the conscious part of the brain.

Early rodent grooming studies and consequent research (John C. Fentress)

My interest in rodent grooming came indirectly. The project I was primarily involved in concerned the responses of two vole species, *Clethrionomys britannicus* (bank voles) and *Microtus agrestis* (field voles), to a model overhead predator (Fentress 1968a, b). The animals' responses had both important similarities and differences, and could be modified within limits by altering the conditions under which they were housed.

Fascinatingly, for each species it appeared that both the duration and apparent intensity of grooming movements could be systematically enhanced as well as depressed as an after-response to the animals' initial responses to the model predator (fleeing or freezing). In brief, the responses to the model predator could be ranked in terms of their form and duration, with grooming responses during what might be called a recovery period. This appeared to provide an analogue to previous ethological studies of "displacement activities," where it had been shown that various avian species would perform activities that seemed "irrelevant" to the circumstances they "should be" attending to. Why groom after having been disturbed by a model predator?

Part of the answer was simple. Usually fleeing was followed by freezing. Subsequently the animals appeared to relax, and groomed prior to engaging in various exploratory movements. The amount of latency and amount (and apparent intensity) of grooming could be titrated as a function of the vigor of initial responses to the predator model. This led to ideas, which will not be elaborated upon here, that during what we might call moderate levels of "stress" grooming was often enhanced compared to control periods, whereas during higher levels of "stress" grooming actions were suppressed.

Subsequent descriptions of grooming broke the actions into broad categories such as licking, face grooming, body grooming, and back grooming. These components followed one another in a predictable sequence. Informal experiments allowed me to look at the rules of sequencing in more detail. For example, if water drops were placed upon the animals' backs they would attend to the dampened area but only after starting with face grooming. If a stronger stimulus, such as a drop of ether, were placed upon the animals' backs they would groom the back without the normal sequence of preceding activities. Thus, it was clear that there was a central bias in the sequencing of these broad categories of grooming, but that could be overridden if the stimulus was strong.

Other observations suggested that during rapid phases of face grooming the animals became more or less immune to disruptors that would interrupt grooming during less vigorous actions. It appeared that during rapid movements the animals shifted their bias toward central motor control as opposed to responsiveness to sensory signals that in other cases were obvious. There was, thus, a suggestive dynamic balancing between central and peripheral control of the grooming action that deserved further exploration.

I became interested in how general two processes suggested by my grooming voles might be. The first was when extraneous events would enhance as opposed to restrict grooming expression. The second was the dynamics of this apparent shift between central and peripheral control of the actions.

As a departure I took advantage of the fact that some of the voles housed in small, covered cages had developed species-specific stereotypies. The bank voles under these conditions would jump repeatedly back and forth over the glass waterspouts in their cages. The field voles would weave around the waterspouts. In each case, once these stereotypies were well established a variety of moderate strength disturbances would markedly increase the duration and vigor of the stereotyped actions. Stronger disturbances blocked the stereotypies.

I had the fortune to be spending time at the London Zoo during this period. I became fascinated by the cage stereotypies that some of the animals developed. In particular, there was a Cape hunting dog *(Lycadon pictus)* that had developed a figure-eight pacing movement. It was clear that moderate disturbances, such as the distant approach of school children, increased these figure-eight movements. When the disturbances were stronger (e.g., the children came closer and/or made more noise) the figure-eight movements were suppressed. The Cape hunting dog oriented specifically toward the children, only to resume pacing after they moved away. There seemed to be a parallel to the grooming actions I had seen in the voles.

I became interested in the types of actions that fell into the ethological category of "displacement activities." Two things seemed clear. The first was that these actions, such as preening in bird species, were forms of behavior that were well established, relatively stereotyped, and frequently repeated. The second was that these actions fit into transitions between other actions in ways that previous workers had described. The basic idea here was that different actions inhibited (blocked) one another's expression, to varying degrees. When the animals appeared to be in conflict between two primary actions (e.g., immobility and locomotion in my voles), the two primary actions canceled one another out, "releasing" the otherwise surprising occurrences of preening, grooming, etc. (Fentress 1991).

But there was one other observation that hit me. In addition to specific inhibitory relations between actions as so beautifully described by the early ethologists, there remained the possibility that different classes of action as traditionally defined could, when not too strongly activated, facilitate one another. The basic idea was that if a given form of behavior was "set" for expression, its expression could be triggered and enhanced by a variety of moderate strength events that would otherwise lead to different actions. At higher levels of activation of these alternative events, the animals would switch, blocking the expression of their previously set behavior patterns, thus indeed suggesting inhibitory connections between action classes. There appeared to be a dual threshold in these responses: at low levels different classes of behavior could share excitatory influences, whereas at higher levels of activation the responses were mutually inhibitory. It was as if the behavioral systems could have both threshold-dependent specific and non-specific properties. There was at the time a large literature in psychology and neuroscience under the heading of "arousal": i.e., broad patterns of activation that could influence a variety of forms of behavior in a positive direction. This suggested that it might be worth pursuing such broad forms of shared activation in addition to the specific control mechanisms ethologists had previously documented so beautifully.

In summary, I became interested in, and have remained interested in, two themes. The first is that depending upon the level of activation, behavioral systems can either facilitate the expression of one another or block other forms of expression. The second is that when a system is strongly activated it becomes relatively immune to disruption by extraneous events: i.e., becomes more tightly focused and biased toward central control mechanisms (Fentress 1991).

These ideas are hard to pin down, but they led to the work Frances mentions on deafferentation of mice, subsequently examined under different behavioral conditions. It also led to a number of studies, some initiated in my laboratory but importantly carried out in others, on central brain mechanisms. In particular the neostriatum (basal ganglia) appear to glue facets of behavioral expression together, while lesions of the cerebellum disrupt individual properties of motor coordination (Berridge and Fentress 1987). Wayne Aldridge and Kent Berridge have followed these and related ideas up in a series of elegant and ongoing studies (Aldridge and Berridge 1998; Aldridge *et al.* 2004; Berridge and Aldridge 2000; Berridge *et al.* 2005).

Once these ideas occupied my attention, I left more detailed descriptions of grooming action behind. In particular, I never looked in detail at the fine structure of facial grooming. It was the pioneering efforts of Frances Stilwell that made it clear that the documentation of this fine detail reveals a highly ordered hierarchical ordering of facial grooming movements, where one can both isolate components or elements of action and look at their connections in terms of higher-order units, and the sequencing of these units. Grooming actions became even richer in our appreciation of this order. Others in my laboratory, in particular Kent Berridge and Ilan Golani, took Frances' insights and showed conclusively how they could be adapted to subsequent brain studies and development (Aldridge and Berridge 1998; Aldridge *et al.* 2004; Berridge 1990; Berridge *et al.* 1987; Golani and Fentress 1985).

These subsequent studies outlined thus far were conducted in my laboratory at the University of Oregon at Eugene. At that time we were splitting our time between the mouse research, work on cortical conditioning in squirrel monkeys, and wolves. The latter had started from my time at Cambridge when I had the good fortune to make contacts with the Regents Park Zoo in London.

After several years in Oregon I was invited to apply for the position of Chair in Psychology at Dalhousie University in Nova Scotia, with a side appointment in the Department of Biology. Besides local politics (the Nixon era), one of the appeals for moving to Dalhousie was that the University offered a lovely space for our expanding wolf research. I was also given funds for an advanced mouse behavior research laboratory, along with monies to invite outstanding colleagues both for research work and as new members of our department.

Ilan Golani met me in Halifax, helped me move into my new home on the coast, and immediately offered suggestions on how we might build upon Frances' careful evaluations of face grooming in mice by looking at kinematic details of how these patterns Frances had described might emerge during development. Ilan brought with him an expertise in the Eshkol–Wachman movement notation system, which he had applied successfully in previous work that included jackals and Tasmanian devils. One of the powers of this approach is that it allows one to step back from categorical labels of movements and to see how movement emerges from different and complementary descriptive perspectives. It is tedious work.

To get mice to groom when they were newborn, we found that we could prop them up into a sitting posture with small, mirrored supports that held them upright. Without these supports gravity won the battle, and the young mice simply did not have the strength to raise themselves into a grooming posture. The first point of these observations was that young mice had recognizable preliminary face-grooming movements from shortly after birth. However, the movements were initially highly variable, and did not fall obviously into the categories that Frances had previously described in such beautiful detail.

The obvious question that followed was how the grooming elements that Frances had documented emerged. Here we found something fascinating, which appeared to have broader implications. In early face grooming the infant mice showed a rich variety of imperfectly coordinated face-grooming movements in which the paws, through highly variable trajectories, sometimes made contact with the face, sometimes missed the face, and sometimes got stuck on the face. It was grooming behavior, but not very effective grooming behavior.

In the second phase of grooming development the mice got "smart." They simplified their rich variations in head, body, and limb movements – the result of which was that they succeeded in making more or less precise movement sequences upon their faces, but in a clearly restrictive manner. In phase three, the mice re-elaborated their grooming movements, with the head, forearms, and body working together in a richly coordinated manner (Golani and Fentress 1985).

From this descriptive database several simple experiments followed. For example, in phase two of the grooming development one could (via a small anchor cemented to the head) displace the head position from the otherwise normal (expected) trajectories of the paws. We were astonished that the paws would then wave in front of the face, where it had been. There was a disconnect between head and forepaw movements. Later in development, the paws insisted upon following the position of the head, rather like people who can eat dinner, placing food securely into their mouths while talking with dinner partners on each side of the table.

Thus, there were two things going on. The first was that the grooming elements Frances had previously described were only hinted at in early development, although their precursors could be seen if the young mice were given postural support. The second was that the development of face-grooming sequences followed three predictable phases: (a) loosely structured and to a major extent ineffectual functionally, (b) highly simplified but effective (yet limited in richness) paw-to-face contacts, and (c) a re-elaboration where different body parts clearly understood what the other body parts were doing. It was at this third stage that Frances' hierarchies took clear form. The two approaches had arrived at the same conclusion.

When Kent Berridge first came into my laboratory at Dalhousie, we had many discussions concerning how we could separate properties of behavioral performance while also looking at the links between these properties. Kent came into Halifax with an expertise at looking at ingestive and related actions in rats. We decided to pursue that first. Kent, Frances, and Ilan had a deep sympathy for the need to fractionate actions in natural movement sequences, and also to look at how these actions might be connected in sequencing and timing behavior, ranging from basic tongue-licking movements to their incorporation into larger blocks that rats use when they take in various food items. It was an exciting and important advance to things I had been thinking about before.

Soon we began looking again at grooming sequences, in this case in rats. Kent took and formalized Frances' earlier descriptions of mouse grooming and established a formal syntax among these action components. We asked numerous questions among ourselves, in the lab and in less formal situations, about such issues as how dependent these mammalian movement patterns are upon sensory information, versus how independently they could be under specific environmental contexts. Kent tackled grooming, and continues to do so with Wayne Aldridge, a colleague at Michigan (Aldridge and Berridge 1998; Aldridge *et al.* 2004; Berridge 1990; Berridge and Aldridge 2000; Berridge *et al.* 2005). Together, they and other workers have shown conclusively that the striatum contributes essentially in the temporal and sequential ordering of grooming and other actions in rodents. They have also extrapolated their data to issues that often come under the headings of motivation, stress, etc. – the important aspect that has later been addressed by several other groups (Kalueff *et al.* 2007; Kalueff and Tuohimaa 2004; Whishaw *et al.* 2001).

I have had the good fortune to work with many talented investigators. One line of research that I and my colleagues explored concerns the organization of grooming in weaver, neurological mutant mice (Bolivar *et al.* 1996; Coscia and Fentress 1993). These animals have deficits both in the cerebellum and dopaminergic system. Although homozygous weaver (*wv/wv*) mice display significant alterations in grooming, these deficiencies are both context and age dependent.

There are two critical points all readers of this volume should be aware of. The first is that without careful behavioral descriptions, such as those provided by Frances Stilwell, the details of malfunction in grooming would not be apparent. The second is that in addition to looking at the details of behavioral performance it is critical to ask broader questions, such as the developmental course of these actions and the specific contexts within which the actions are performed. There remain many rich avenues to be explored. Relations between the activation and patterning of behavior remain among the offerings that need much further investigation (Fentress 1991). Rodent grooming provides an excellent model system, one that crosses many classes of behavior, and reaches out to areas that range from the fine details of motor control to areas such as development, motivation, and emotion.

Thanks to Frances' early work, rodent grooming entered the mainstream of current behavioral neuroscience. In the space available here, we can only outline some of the high spots of this pioneering work. The bottom line in the brief summary to follow is that without Frances Stilwell's initial careful and painstaking descriptions this subsequent neurobiological work would have been robbed of its richness. Ethology and neuroscience were joined, to the benefit of each.

Conclusion

The authors have intentionally avoided going into details that are available elsewhere. Rather, it seemed important to show how the initial pictures emerged from Stilwell's efforts and insights. In some ways this is "old-fashioned" science in its emphasis. There is an appreciation in other fields, such as theoretical physics, that at the bottom line science reflects the relations between observer and observed. This certainly has been true for grooming actions. It remains premature to claim that the full lessons of rodent grooming have been appreciated. On a regular basis new mysteries and new facts are being revealed. But, as the chapter began, one must be alert to this richness in behavioral expression. It is from here that the important follow-up questions emerge. We look forward to future advances, whatever directions these might take.

Acknowledgments

The editorial assistance of Be Davison Herrera is greatly appreciated.

References

Aldridge JW and Berridge KC (1998): Coding of serial order by neostriatal neurons: a "natural action" approach to movement sequence. *J Neurosci* **18**:2777–87.

Aldridge JW, Berridge KC and Rosen AR (2004): Basal ganglia neural mechanisms of natural movement sequences. *Can J Physiol Pharmacol* **82**:732–9.

Berridge KC (1990): Comparative fine structure of action: rules of form and sequence in the grooming patterns of six rodent species. *Behaviour* **113**:21–56.

Berridge KC and Aldridge JW (2000): Super-stereotypy II: enhancement of a complex movement sequence by intraventricular dopamine D1 agonists. *Synapse* **37**:205–15.

Berridge KC and Fentress JC (1987): Disruption of natural grooming chains after striatopallidal lesions. *Psychobiology* **15**:336–42.

Berridge KC, Fentress JC and Parr H (1987): Natural syntax rules control action sequence of rats. *Behav Brain Res* **23**:59–68.

Berridge KC, Aldridge JW, Houchard KR and Zhuang X (2005): Sequential super-stereotypy of an instinctive fixed action pattern in hyper-dopaminergic mutant mice: a model of obsessive compulsive disorder and Tourette's. *BMC Biol* **3**:1–16.

Bolivar VJ, Danilchuk W and Fentress JC (1996): Separation of activation and pattern in grooming development of weaver mice. *Behav Brain Res* **75**:49–58.

Coscia EM and Fentress JC (1993): Neurological dysfunction expressed in the grooming behavior of developing weaver mutant mice. *Behav Genet* **23**:533–41.

Fentress JC (1968a): Interrupted ongoing behaviour in voles (*Microtus agrestis* and *Clethrionomys britannicus*) I. *Anim Behav* **16**:135–53.

Fentress JC (1968b): Interrupted ongoing behaviour in voles (*Microtus agrestis* and *Clethrionomys britannicus*) II. *Anim Behav* **16**:154–67.

Fentress JC (1972): Development and patterning of movement sequences in inbred mice. In: Kruger J, ed., *The Biology of Behavior*. Corvallis: Oregon State University Press, pp. 83–132.

Fentress JC (1991): Analytical ethology and synthetic neuroscience. In: Bateson P, ed., *The Development and Integration of Behaviour*. Cambridge: Cambridge University Press, pp. 77–120.

Fentress JC and Stilwell FP (1973): Letter: Grammar of a movement sequence in inbred mice. *Nature* **244**:52–3.

Golani I and Fentress JC (1985): Early ontogeny of face grooming in mice. *Dev Psychobiol* **18**:529–44.

Kalueff AV and Tuohimaa P (2004): Grooming analysis algorithm for neurobehavioural stress research. *Brain Res Brain Res Protoc* **13**:151–8.

Kalueff AV, Aldridge JW, LaPorte JL, Murphy DL and Tuohimaa P (2007): Analyzing grooming microstructure in neurobehavioral experiments. *Nat Protoc* **2**:2538–44.

Whishaw IQ, Metz GA, Kolb B and Pellis SM (2001): Accelerated nervous system development contributes to behavioral efficiency in the laboratory mouse: a behavioral review and theoretical proposal. *Dev Psychobiol* **39**:151–70.

2

Self-grooming as a form of olfactory communication in meadow voles and prairie voles (*Microtus spp.*)

MICHAEL H. FERKIN AND STUART T. LEONARD

Summary

We explore the possibility that self-grooming in response to the odors or presence of another animal plays a role in olfactory communication. For some animals, the substances released by self-grooming may make groomers more easily detected, more attractive, and/or less threatening to conspecifics that are in close proximity to them. The fact that animals self-groom at different rates when they encounter different individuals suggests that they can target particular conspecifics for purposes of communicating with them. Given that voles and other animals generally spend more time grooming in response to reproductively active, opposite-sex conspecifics than to reproductively quiescent opposite-sex conspecifics, self-grooming may be involved in attracting potential mates and is associated with the behaviors that surround reproduction. Studies have shown that conditions such as endocrine state, diet, age, and familiarity and relatedness of both the groomer and the scent donor affect the amount of time that individuals self-groom when they are exposed to the odors of opposite-sex conspecifics. Consequently, self-grooming in response to the odors of opposite-sex conspecifics may be akin to scent marking in that animals are transmitting odiferous substances into the environment that honestly signal features of their quality and condition to potential mates and competitors.

Neurobiology of Grooming Behavior, eds. Allan V. Kalueff, Justin L. LaPorte, and Carisa L. Bergner. Published by Cambridge University Press.

Introduction

As many terrestrial animals move about their home ranges they are surrounded by scent marks, some are their own and some are those of conspecifics. Animals investigating these scent marks can often determine many features about the individual that deposited them such as its sex, age, reproductive condition, diet, etc. These scent marks provide a signpost or bulletin board for the transfer of information between individuals. After gathering such information about the signaler, an individual may respond to these scents by seeking the signaler, avoiding it, or altering its own behavior in some fashion. One response that an individual may choose is to self-groom, a form of olfactory communication akin to scent marking.

Self-grooming is a behavior that is ubiquitous among mammals. Indeed, many terrestrial mammals may spend between 20% and 40% of the day self-grooming, presumably to care for the outer surface of their body (Spruijt *et al.* 1992). For instance, animals may self-groom to remove ectoparasites, reduce bacterial infections associated with nursing, stimulate wound-healing, aid in thermoregulation, and spread secretions from the integument (skin) onto the fur (Borchelt 1980; Geyer and Kornet 1982; Hainsworth 1967; Harriman and Thiessen 1985; Hart and Powell 1990; Mooring *et al.* 2000; Thiessen 1977). These functions of self-grooming occur most often when individuals are not in the presence of conspecifics or their odors (Spruijt *et al.* 1992).

Arguments swirl about the reason why terrestrial mammals self-groom when they encounter a conspecific or its odor. Some authors suggest that animals self-groom to remove parasites to prevent their transmission to a nearby conspecific during an encounter (Geyer and Kornet 1982; Hart 1990; Mooring *et al.* 2000) and/or to indicate one's health (Hamilton and Zuk 1982). Self-grooming also may be a redirected or displacement behavior displayed by individuals that are torn between fleeing and fighting (Judge *et al.* 2006; Schino *et al.* 1996). Self-grooming may also occur when groomers are anxious and attempting to relieve anxiety and thus be a form of de-arousal (Kalueff *et al.* 2007; Kalueff and Tuohimaa 2005; McFarlane *et al.* 2008; Roth and Katz 1979). In other cases, the rate of self-grooming may decrease when groomers are trying to avoid the attention of conspecifics or indicate the motivational ambivalence of groomers toward opposite-sex conspecifics (Spruijt *et al.* 1992). Self-grooming in response to odors of conspecifics may also play a role in olfactory communication between groomers and their audience. For instance, animals may self-groom when they encounter the odor of conspecifics. By doing so, groomers spread odiferous substances on their bodies that may become more volatile (Bossert and Wilson 1963; Lucas *et al.* 2004), making them more easily detected by nearby conspecifics (Brockie 1976; Ferkin and

Li 2005; Ferkin *et al.* 1996; Steiner 1973, 1974; Thiessen 1977; Wiepkema 1979; Wolff *et al.* 2002). Self-grooming may also transmit information about the groomer that facilitates interactions with opposite-sex conspecifics (Ferkin 2006; Harriman and Thiessen 1985; Leonard *et al.* 2005; Leonard and Ferkin 2005; Thiessen and Harriman 1986), maintains pair bonds (Witt *et al.* 1988, 1990), or influences interactions between same-sex conspecifics (Bursten *et al.* 2000; Ferkin *et al.* 2001; Shanas and Terkel 1997).

The goal of this chapter is to discuss studies that test the hypothesis that self-grooming plays a role in olfactory communication between conspecifics. We present evidence, primarily from our work on voles, which shows that individuals spend different amounts of time self-grooming when they encounter the odors of different conspecifics. Specifically, we discuss studies that indicate that the amount of time that voles and other individuals self-groom depends on the age, reproductive state, diet, photoperiodic history, mating history, and relatedness of both the groomer and the scent donor. We use this growing body of evidence to suggest that self-grooming, akin to scent marking, allows groomers to honestly signal the features of their quality and condition to nearby conspecifics, which may facilitate or deter interactions between groomers and their audience.

Meadow voles (*Microtus pennsylvanicus*) and prairie voles (*M. ochrogaster*) are the focal animals of much of the research that we describe in this chapter. These voles are small, secretive microtine rodents that live in grasslands and open fields, respectively, in eastern and central United States and Canada. Both species build and use existing runways to move about their home ranges. Meadow voles are characterized by having a mating system that is promiscuous. Males and females mate with multiple partners (Boonstra *et al.* 1993). In addition, male and female meadow voles do not share a nest; males generally do not participate in paternal care. Males are also not territorial. They occupy large home ranges that may encompass the territories of one or more females (Madison 1980). Competition among male meadow voles for access and copulations with females is intense (Boonstra *et al.* 1993; delBarco-Trillo and Ferkin 1994). Female meadow voles compete with one another for suitable nest sites (Madison 1980). Prairie voles are considered to be a facultative monogamous species (Getz and Carter 1996); however, multiple mating may occur on occasion. Male and female prairie voles share a nest and both rear their offspring. Both males and females defend small territories that include their nest. Males and females attempt to reduce the access of same-sex to their mates (Getz and Carter 1996). In addition, both species display seasonal shifts in behavior and social organization, with more profound seasonal changes occurring among meadow voles than prairie voles (Ferkin and Seamon 1987), which we will address later in this chapter.

General description of methods of our studies

The voles used in our research and discussed in this chapter were first- and second-generation captive voles that were raised in our colony and derived from free-living voles that we live-trapped in southern Ohio and northern Kentucky. Specifically, we have examined the response of meadow voles and prairie voles to the odors of conspecifics, heterospecifics, food, and predators in a number of papers that are highlighted in this chapter. To facilitate this overview, it is important that the reader understand our basic methodology, which is similar throughout these studies. Briefly, at the beginning of each self-grooming test, we removed the subject's own cotton nesting material and then placed approximately 5 to 8 g of the scent donor's cotton material into the subject's cage. The subject was allowed to acclimate to the presence of the donor's nesting material for two minutes. After this brief acclimation period, we continuously recorded the amount of time that the subject self-groomed over the next five minutes. In voles, the general pattern of self-grooming consists of a cephalocaudal progression that begins with rhythmic movements of the paws around the mouth and face, over the ears, descending to the ventrum, flank, anogenital area, and tail (Ferkin *et al.* 1996, 2001). We recorded self-grooming when subjects rubbed, licked, or scratched any of these body areas when they were exposed to particular scent stimuli.

Briefly, our studies have indicated that voles do not differ markedly in this pattern of self-grooming (Ferkin *et al.* 1996); however, they may spend different amounts of time self-grooming particular parts of their body, depending on the identity of the donor and the groomer's condition (Vaughn, delBarco-Trillo and Ferkin unpubl. data). At the end of each trial, we removed the donor's nesting material from the subject's cage and discarded it. In each of our experiments, the experimenter was blind to the identity of the scent donors and the groomers. Depending on the experiment, we used either univariate or multivariate statistics to determine if significant differences ($p < 0.05$) existed in the amount of time that individuals spent self-grooming to particular scent stimuli. Additional information about the methods and statistical analyses can be found in the specific papers that we cite in this chapter.

Self-grooming in social and sexual contexts

Much of our work has focused on testing hypotheses to determine why mammals self-groom when they encounter conspecifics or their odors. In a series of studies, we attempted to ascertain whether self-grooming is a form of olfactory communication that provides signals associated with sexual and social interactions between conspecifics (Ferkin and Leonard 2008; Ferkin *et al.* 1996; Leonard and Ferkin 2005). For several years, we have designed and carried out experiments that measured the amount of time that meadow voles self-groomed when they were exposed to conspecific and heterospecific odors and under different conditions and contexts. In doing so, we have tested several hypotheses surrounding the role of self-grooming in a variety of studies. We will list the broader hypotheses we have tested. The first broad hypothesis is that self-grooming in response to the odor of a conspecific is a redirected behavior or displacement activity, which indicates the groomer's ambivalence to that conspecific. The second hypothesis is that self-grooming in response to the odors of conspecifics and heterospecifics is a response to anxiety. The third hypothesis is that self-grooming in response to the odors of conspecifics and heterospecifics is a form of olfactory communication that broadcasts scents, indicating the groomer's heightened interest for a particular conspecific.

Recently, three studies were carried out, each testing some or all of the above hypotheses (Ferkin 2006; Ferkin and Leonard 2005; Ferkin *et al.* 1996; Leonard and Ferkin 2005). First we tested the hypothesis that self-grooming in response to the odors of conspecifics is a redirected behavior or displacement activity for voles. If this hypothesis is supported voles would self-groom at low rates, similar to rates of self-grooming when they are exposed to odors of heterospecifics and same-sex conspecifics, indicating that self-grooming is a low-priority spontaneous behavior. However, meadow voles spent little time self-grooming in response to the odors of heterospecifics or those of same-sex conspecifics (Figure 2.1). Thus, our studies did not support the hypothesis that self-grooming in response to odors of conspecifics is a low-priority spontaneous behavior such as a redirected behavior or displacement activity. Next, we tested the hypothesis that self-grooming to odors of conspecifics and heterospecifics is a response to anxiety. For self-grooming to be a response to anxiety, voles would have to self-groom less or experience interrupted bouts of self-grooming when they encountered the odors of conspecifics and those of heterospecifics (neutrals or predators) (Fentress 1968; Flügge *et al.* 1998; Kalueff and Tuohimaa 2005). This hypothesis was also not supported by our studies (Ferkin and Leonard 2005; Ferkin *et al.* 1996). Instead, voles spent more time self-grooming when they were exposed to odors of conspecifics, particularly those of opposite-sex conspecifics, than when they were exposed to those of heterospecifics or to the odor of fresh cotton bedding (Figure 2.1). Moreover, self-grooming in response to odors of heterospecifics may not be adaptive, as it would increase the groomer's odor field, and thereby increase the likelihood of eavesdropping, competition, or its being detected by predators (Bossert and Wilson 1963; Dell'Omo and Alleva 1994; Gosling and Roberts 2001). Such a response would be selected against in free-living populations. The third hypothesis we tested was that self-grooming in response to the odors of conspecifics and heterospecifics is a form of olfactory

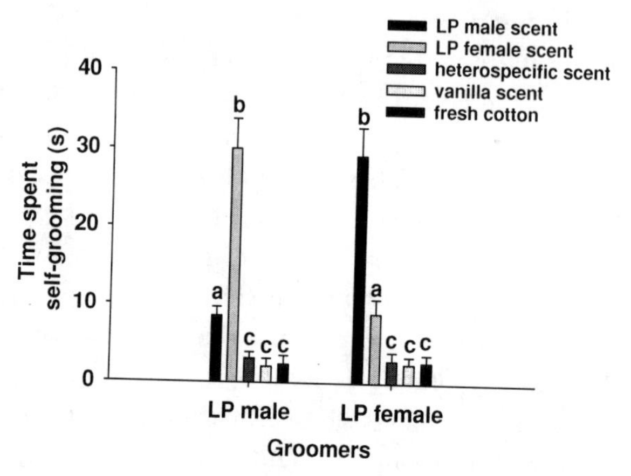

Figure 2.1 The time spent by long-photoperiod (LP) male and female meadow voles self-grooming when they were exposed for five minutes to eight grams of cotton bedding scented by an LP male conspecific, an LP female conspecific, an opposite-sex heterospecific (house mouse), vanilla, and clean cotton bedding. Histograms capped with different letters are statistically different at $p < 0.05$ (Tukey's multiple pairwise comparison, following a two-way ANOVA).

communication in which the groomer broadcasts scents that are directed toward particular conspecifics. Thus, individuals would spend different amounts of time self-grooming, depending on the odor donor. This hypothesis was supported by several studies (e.g. Ferkin and Leonard 2005; Ferkin *et al.* 1996). Firstly, we found that reproductively active voles spent more time self-grooming when they were exposed to the odors of opposite-sex conspecifics than when exposed to odors of same-sex conspecifics or those of heterospecifics (Figure 2.1). More importantly, we found that reproductively active voles self-groomed more in response to odors of opposite-sex conspecifics that were in a heightened reproductive state than to the odors of those that were not in a heightened reproductive state (Ferkin 2006; Ferkin *et al.* 1996; Leonard *et al.* 2005; Leonard and Ferkin 2005); these results are summarized below.

As we carried out additional studies on voles, it became more apparent to us that, in addition to its other functions (Fentress 1968; Kalueff and Tuohimaa 2005; Spruijt *et al.* 1992), self-grooming in response to the odor of a conspecific may provide an olfactory signal to communicate interest in particular conspecifics, which may facilitate or reduce the likelihood of interactions between them (Ferkin and Leonard 2005; Land and Seeley 2004; Lucas *et al.* 2004; Stopka and Macdonald 1998). Thus, we conducted additional studies to test in more detail those predictions that arose from the hypothesis that self-grooming in response to odors of conspecifics is a form of olfactory communication between and within the sexes.

Self-grooming and attracting the opposite sex

In this section, we highlight studies that suggest self-grooming is involved in facilitating interactions between opposite-sex conspecifics, and may be a behavior that is associated with reproduction. Studies on voles, rats, and ground squirrels demonstrate that these rodents self-groom more when they are exposed to opposite-sex conspecifics but not to same-sex conspecifics or their odors (Ferkin *et al.* 1996, 2001; Steiner 1973, 1974). In addition, male prairie voles self-groom more during encounters with their mates than during encounters with unfamiliar opposite-sex conspecifics (Witt *et al.* 1988).

Many rodents also engage in self-grooming before, during, and after courtship and copulation (Carter *et al.* 1989; Paz-y-Miño *et al.* 2002; Spruijt *et al.* 1992; Witt *et al.* 1988). In a recent study, we tested the hypothesis that the amount of time that individuals engage in self-grooming is affected by its (the subject's) reproductive condition and that of the odor donor (Ferkin 2006). If the amount of time that individuals self-groom is affected by the reproductive condition of the donor, male meadow voles should self-groom more in response to odors of a female in heightened sexual receptivity, such as females in postpartum estrus (PPE), than to odors of female conspecifics in other reproductive states and those of heterospecific females. Likewise, PPE female meadow voles should self-groom more in response to the odors of gonadally intact males as compared to those of gonadectomized (GX) males and those of heterospecific males. If self-grooming is affected by the groomer's reproductive condition, then PPE female meadow voles should spend more time self-grooming than females in other reproductive states in response to the odors of intact male conspecifics (Ferkin 2006).

Ferkin (2006) discovered that meadow voles in heightened reproductive states spent more time self-grooming when they encountered the odors of opposite-sex conspecifics than those individuals that were not in a heightened reproductive state. Specifically, PPE female voles spent more time self-grooming in response to bedding scented by GX male voles with testosterone replacement (GX + T) than did ovariectomized (OVX) females and OVX females receiving exogenous estradiol (OVX + E). In addition, GX + T male voles spent more time than GX male voles self-grooming when they were exposed to bedding scented by PPE and OVX + E female voles. Thus, these data are consistent with those from other studies, showing that the reproductive state of the groomer affects the amount of time that it self-grooms in the presence of odors of opposite-sex conspecifics (Ferkin 2006).

As a whole, these observations support and extend the hypothesis that the amount of time that meadow voles self-groom depends on their reproductive condition (Leonard *et al.* 2005; Leonard and Ferkin 2005). These findings also suggest that the groomer needs to be in a heightened reproductive state to self-groom at relatively high rates when it encounters a sexually receptive opposite-sex conspecific (Ferkin 2006; Leonard *et al.* 2005). By self-grooming at relatively high rates, male and female voles in heightened reproductive states may be more likely to attract nearby opposite-sex conspecifics. We have shown previously that self-grooming increases the attractiveness of odors produced by voles that recently self-groomed relative to the attractiveness of odors produced by voles that did not recently self-groom (Ferkin *et al.* 1996, 2001). Thus, self-grooming may allow a groomer to indicate its reproductive status to nearby conspecifics and be a form of self-advertisement (Ferkin *et al.* 1996, 2001; Wolff *et al.* 2002). However, we cannot rule out the possibility that self-grooming stimulates the groomer in some way that increases its interest in the opposite sex (Ferkin *et al.* 1996; Ferkin and Leonard 2005).

These results also support and extend the hypothesis that the amount of time that individuals self-groom depends on the reproductive condition of the odor donor (Ferkin 2006). Postpartum estrus females spent more time self-grooming when they were exposed to odors of GX + T male voles as compared to GX male voles and male mice. These findings are similar to previous studies showing that PPE voles self-groom and scent mark more in the presence of male conspecifics (Carter *et al.* 1989; Witt *et al.* 1990). Likewise, male voles self-groomed more in response to odors of PPE females than they did in response to odors of female voles in other reproductive states and to odors of female mice. Male voles also spent more time self-grooming in response to bedding scented by OVX + E female voles as compared to that of OVX female voles (Ferkin 2006). However, when exposed to bedding scented by a female house mouse that was in a heightened state of receptivity male voles spent little time self-grooming (Ferkin and Leonard 2005). Previous work shows that gonadally intact and GX + T male voles self-groom more than do GX male voles when they are exposed to odors of gonadally intact and OVX + E female voles (Ferkin *et al.* 1996; Leonard and Ferkin 2005). Voles also spend more time self-grooming in the presence of scent marks of opposite-sex conspecifics than those of same-sex conspecifics (Ferkin *et al.* 1996), and to the odors of long-photoperiod (LP) opposite-sex conspecifics than those of reproductively quiescent, short-photoperiod (SP) opposite-sex conspecifics (Leonard and Ferkin 2005). Overall, these observations indicate that male voles are responding to odors of particular conspecifics in heightened reproductive states and not simply to odors of any female in a heightened reproductive state.

The findings of Ferkin (2006) were consistent with studies showing that voles are very responsive to odors of potential mates, such as gonadally intact LP males and PPE females (Ferkin and Johnston 1995a, b). By being more responsive to odors of potential mates, self-grooming may increase the likelihood that groomers indicate their interest in them and are located by them (Ferkin 2005; Wolff *et al.* 2002). Such a speculative role in mate attraction is consistent with the natural history of meadow voles. Meadow voles are promiscuous, relatively asocial, and males and females tend to be dispersed (Boonstra *et al.* 1993; Madison 1980). Moreover, females are in PPE for only 8 to 12 hours (delBarco-Trillo and Ferkin 2007; Ferkin 2006; Ferkin and Johnston 1995b; Keller 1985). Therefore, self-grooming may allow PPE females to signal their heightened reproductive state and interest to nearby males (Ferkin 2006). In that reproductive state is an honest signal, the amount of time an individual self-grooms may be a cheat-proof signal used for mate attraction (Ferkin 2006; Gosling and Roberts 2001; Stopka and Macdonald 1998). However, the speculation about the role of self-grooming in reproduction has been questioned in a study of multiple mating in prairie voles (Wolff *et al.* 2002). In that study, a female prairie vole was housed with two males. Both males were tethered to a bar, which limited their movement about the arena and also prevented them from approaching and contacting the female and one another. Wolff and colleagues (2002) concluded that self-grooming behavior was not associated with reproduction since females spent similar amounts of time with both males, independently of how long each male self-groomed. Their conclusion is highly suspect, however, as it was not clear from the study whether the male prairie voles self-groomed in response to the other male vole, the female vole, or simply in response to being tethered. Tethering an animal is problematic and may affect the amount of time that it self-grooms (Kalueff *et al.* 2007; Kalueff and Touhimaa 2005; Thor *et al.* 1988; van Erp *et al.* 1994), independent of the animal's exposure to the scents of a conspecific.

The amount of time that individuals self-groom may also be affected by its social relationship with the scent donor. For example, self-grooming rates may differ if an individual is exposed to the scents of opposite-sex siblings or opposite-sex nonsiblings. In that self-grooming increases the attractiveness of the groomer's odors to the opposite sex (Ferkin 2006; Ferkin *et al.* 1996), individuals should not self-groom when they are exposed to scents of opposite-sex siblings. They should, however, target the scents of opposite-sex strangers as a mechanism to favor reproduction with nonrelatives (inbreeding avoidance). We addressed this question by testing the hypothesis that both meadow voles and prairie voles self-groom at different rates when exposed to the scents of their siblings as compared to those of nonsiblings (Paz-y-Miño *et al.* 2002). This prediction was studied under the context of social memory for siblings and the effects of isolation on memory for siblings. We found that within the context of sibling recognition, self-grooming behavior is directed at unfamiliar opposite-sex conspecifics, and that this response was affected by the length of isolation from their siblings. However, 20 days of isolation for male and female meadow voles and 30 days of isolation for male prairie voles were sufficient to induce these animals to spend similar amounts of time self-grooming to the scents of siblings and those of nonsiblings. Female prairie voles isolated for 30 days from siblings self-groomed more in response to the scents of nonsiblings than to those of siblings, suggesting that female prairie voles still recognize their male siblings after isolation (Paz-y-Miño *et al.* 2002).

Self-grooming and seasonality

Seasonal differences exist in the manner in which voles respond to scents of conspecifics (Ferkin and Seamon 1987). Consequently, the amount of time that voles self-groom when they encounter the scents of particular conspecifics may also vary seasonally. We carried out a study that determined whether seasonal differences exist in the amount of time meadow voles self-groom when they encounter the scents of conspecifics (Leonard and Ferkin 2005).

During the breeding season or under LP conditions, voles are attracted to opposite-sex conspecifics and produce scent marks that are also attractive to them (Ferkin and Seamon 1987). Thus, we predicted that LP male and female voles will spend more time self-grooming in response to scents of opposite-sex conspecifics than in response to those of same-sex conspecifics, and that LP males and females should self-groom more in response to scents of LP opposite-sex conspecifics than to those of SP opposite-sex conspecifics (Leonard and Ferkin 2005). This prediction is based on studies showing that LP voles are not attracted to the odors of SP voles and vice versa (Ferkin and Johnston 1995a; Ferkin and Seamon 1987). We also tested the prediction that SP voles will spend similar amounts of time self-grooming in response to the scents of opposite-sex conspecifics and those of same-sex conspecifics, independent of the scent donor's photoperiod (Leonard and Ferkin 2005). This prediction is based on experiments showing that during the winter and nonbreeding season, voles are not attracted to opposite-sex conspecifics and do not produce scent marks that are attractive to them (Ferkin and Seamon 1987).

We found that LP voles spent more time self-grooming when they encountered the odors of LP opposite-sex conspecifics than to those of LP same-sex conspecifics as well as those of SP conspecifics (Leonard and Ferkin 2005). This finding is consistent with the social biology of free-living voles. During the breeding season, voles display behaviors associated with attracting mates (Boonstra *et al.* 1993; Ferkin and Seamon 1987; Madison 1980). By spending more time self-grooming in response to LP opposite-sex conspecifics, LP groomers may produce odor(s) that increases the likelihood that they can indicate their presence in an area to nearby conspecifics (Ferkin *et al.* 1996). Such odors may allow males and females, which nest together during the breeding season (Madison 1980), to locate one another more easily, to facilitate assessment of the groomer's odor by nearby opposite-sex conspecifics, and/or to coordinate breeding (Ferkin and Leonard 2005; Wolff *et al.* 2002). During the breeding season, voles may signal their presence in an area to potential mates, which may also facilitate reproduction (Leonard and Ferkin 2005).

During the nonbreeding season, voles may signal their presence in an area to reduce agonism between same-sex conspecifics and communally nesting females (Leonard and Ferkin 2005). Short-photoperiod male voles spent more time self-grooming in response to odors of LP females. Interestingly, SP male voles do not produce odors that are attractive to opposite-sex conspecifics and are not attracted to odors produced by LP or SP conspecifics (Ferkin and Seamon 1987). Perhaps self-grooming by SP males in response to odors of LP females is similar to self-grooming by LP males in that self-grooming in response to such odors is associated with attempting to locate a potential mate. Alternatively, self-grooming by SP males may serve other functions in odor communication than it does for LP males (Leonard and Ferkin 2005). For instance, not directing behaviors towards opposite-sex conspecifics may deter or reduce reproductive behaviors of meadow voles (Ferkin and Seamon 1987; Leonard and Ferkin 2005). Self-grooming by SP females may serve functions related to indicating presence in an area to nearby females with whom they share a communal nest. In this scenario, self-grooming by SP females may be associated with the behaviors that facilitate social cohesiveness by communal nestmates. Further testing is needed to discern the communicatory role of self-grooming by SP voles.

Self-grooming and endocrine control

The effects of melatonin

Seasonal shifts in the behaviors that facilitate interactions with opposite-sex conspecifics are mediated by photoperiodically induced changes in gonadal steroid hormone concentrations. The seasonal changes in gonadal steroid titers appear to vary with the duration of the melatonin signal that an animal receives (Prendergast *et al.* 2002). Briefly, a short duration of melatonin release is characteristic of the breeding season and long daylengths. A short duration of melatonin also coincides with high circulating titers of gonadal steroids and prolactin (PRL) (Prendergast *et al.* 2002): hormones that mediate odor communication in voles (Ferkin *et al.* 1997; Leonard *et al.* 2005). In contrast, a long duration of melatonin release is characteristic of the nonbreeding season and short daylengths, and low titers of circulating gonadal steroids and PRL (Prendergast *et al.* 2002). We designed a series of experiments that assessed the effects of exogenous melatonin on the self-grooming behavior of meadow voles in response to the odors of same- and opposite-sex conspecifics. We tested the hypothesis that a constant melatonin signal (10-mm implants of melatonin for 12 weeks) was sufficient to alter the self-grooming responses of LP male and female meadow voles such that they no longer reflect those displayed by LP male and female meadow voles not receiving melatonin (Ferkin *et al.* 2007). That is, LP voles treated with exogenous melatonin will spend similar amounts of time self-grooming in response to odors of opposite-sex conspecifics and those of same-sex conspecifics shown by SP voles.

The data supported the above hypothesis. We found that 12 weeks after receiving a melatonin implant, LP meadow voles no longer spent more time self-grooming in response to bedding scented by an LP opposite-sex conspecific as compared to an LP same-sex conspecific. In contrast, LP voles not treated with melatonin spent more time self-grooming in response to scents of LP opposite-sex conspecifics than to scents of LP same-sex conspecifics (Ferkin *et al.* 2007). We also found that voles treated with 10-mm implants of melatonin for 12 weeks had lower circulating titers of T and E as compared to voles treated with empty implants for similar amounts of time; these melatonin-treated voles had titers of E and T that were similar to those of gonadally intact SP voles of similar age. Our results echo those reported in other studies, which showed that a short-day pattern of melatonin induces a decrease in gonadal steroid titers and reproductive responsiveness in LP male and female voles and deer mice, a characteristic of SP voles and deer mice (Demas *et al.* 1996, Ferkin and Kile 1996). Thus, melatonin appears to be involved in the transduction of the photoperiodic effects on circulating gonadal hormone titers, which, in turn, mediate self-grooming responses in meadow voles (Ferkin *et al.* 2007).

The effects of prolactin and testosterone

Several studies have shown that seasonal differences in the odor-related behavior of LP and SP male voles are mediated by changes in T and PRL titers (Ferkin *et al.* 1997; Leonard and Ferkin 1999). For instance, high titers of T and PRL, which are characteristic of males during the breeding season, are necessary for male meadow voles to prefer the odors of LP females to those of LP male conspecifics, and for males to prefer the odors of LP females to those of SP female conspecifics (Ferkin *et al.* 1997). Likewise, high PRL and T titers are necessary for LP male meadow voles to produce odors that are attractive to LP females (Leonard and Ferkin 1999). In contrast, reproductively quiescent male hamsters (Ebling 1977) and voles (Leonard and Ferkin 1999), which typically have low PRL titers and low T titers, no longer prefer the odors of females or produce odors that are attractive to females, independent of the female's photoperiod (LP or SP).

Interestingly, treatment with an LP-equivalent titer of T is sufficient for SP male meadow voles to respond preferentially to the odors of LP females, whereas an LP-equivalent titer of PRL is not sufficient to induce SP males to produce odors that are attractive to LP females (Ferkin *et al.* 1997; Leonard and Ferkin 1999). Thus, the endocrine control that mediates seasonal shifts in these two odor-related phenomena may also underlie seasonal changes in self-grooming behavior.

Previous studies on rats, which are not responsive to seasonal changes in day length, suggest that T and PRL are involved in the mediation of self-grooming in rodents (Drago and Bolhus 1981; Drago *et al.* 1980, 1986; Flügge *et al.* 1998; Gonzalez-Lima *et al.* 1988; Moore 1986). Thus, we sought to determine whether seasonal differences in the amount of time male meadow voles self-groom in response to odors of conspecifics are mediated by seasonal rhythms in their circulating T and PRL titers. We tested the hypothesis that high titers of both T and PRL, characteristic of LP males, are necessary for LP males and sufficient for SP males to spend more time self-grooming in response to odors of LP females as compared to those of LP male, SP male, and SP female conspecifics (Leonard *et al.* 2005). Specifically, we tested three predictions. The first prediction was that high titers of both T and PRL are necessary for LP males and sufficient for SP male voles to spend more time self-grooming in response to the odors of LP females than to those of other conspecifics. The second prediction was that high titers of T but not PRL are necessary for LP males and sufficient for SP male voles to spend more time self-grooming in response to the odors of LP females than to those of other conspecifics. The third prediction was that high titers of PRL but not T are necessary for LP males and sufficient for SP male voles to spend more time self-grooming in response to the odors of LP females than to those of other conspecifics (Leonard *et al.* 2005).

We found that GX LP males with high circulating concentrations of both PRL and T self-groomed more in response to the odors of LP females than to odors of LP males, SP females, and SP males (Leonard *et al.* 2005). This finding is in agreement with previous work showing that gonadally intact LP males, which have high circulating titers of both T and PRL, self-groom more in response to odors of LP females as compared to those of other LP and SP conspecifics (Ferkin *et al.* 1997; Leonard and Ferkin 2005). The data also show that LP males with low T titers and/or low PRL titers spent similar amounts of time self-grooming in response to the odors of all conspecifics. Taken together, the current results are consistent with the hypothesis that high titers of both PRL and T are necessary for LP male meadow voles to self-groom more in response to odors of LP females than to those of other conspecifics. It appears that the endocrine control for self-grooming in LP male voles may be similar to that for other odor-related behaviors that surround reproduction in meadow voles and other rodents. For example, high titers of PRL and T are also necessary for LP rodents to maintain their preferences for the odors of LP female conspecifics (Leonard and Ferkin 1999), to produce odors that are attractive to LP females (Ferkin *et al.* 1997), and to stimulate other aspects of reproduction (Duncan and Goldman 1984; Goldman *et al.* 1981; Miernicki *et al.* 1990; Powers *et al.* 1985). In contrast, gonadectomy and bromocriptine treatment (a dopamine agonist), lowers T and PRL titers, and abolishes or blocks these behaviors in LP male voles (Ferkin and Gorman 1992; Leonard and Ferkin 1999). For LP males, the proximate control of self-grooming and the other odor-related phenomena appears to be the same, i.e., a dependency on high titers of both PRL and T to maintain these behaviors. The neural and endocrine substrates that underlie these odor-related phenomena are sensitive to and likely depend on high circulating titers of PRL and T to maintain them. These results support and extend the hypothesis that in seasonally breeding mammals, both PRL and T support the behaviors that surround reproduction during long photoperiod (Duncan and Goldman 1984; Goldman *et al.* 1981; Miernicki *et al.* 1990; Powers *et al.* 1985).

Prolactin and testosterone titers also affected the amount of time that SP males self-groomed when they were exposed to bedding scented by conspecifics (Leonard *et al.* 2005). Short-photoperiod GX males that did not receive exogenous PRL and T self-groomed more when they encountered odors of LP females and SP males as compared to those of other conspecifics. Recent work has shown that SP males with intact gonads, which have low circulating titers of both T and PRL, also self-groomed more in response to odors of LP females and SP males as compared to those of other conspecifics (Leonard *et al.* 2005). It is interesting to note that SP males with low titers of PRL and T are not attracted to the odors of LP females or SP males and also do not produce odors that are attractive to them. Yet, SP male voles treated with exogenous PRL and T are attracted to odors of LP female voles (Leonard *et al.* 2005). However, SP males with high PRL and high T self-groomed at relatively low rates when they were exposed to odors of LP females and SP males. Taken together, these results suggest that a disassociation may exist among attractiveness of odors produced by SP voles, their preferences for the odors of conspecifics, and the amount of time they self-groom in response to the odors of conspecifics.

For SP male voles, the neural and endocrine substrates that mediate self-grooming, but not other odor-related phenomena in SP males, appear to be refractory to LP-equivalent titers of PRL and T (Leonard *et al.* 2005). It is possible that a short photoperiod may either increase the threshold for hormonal activation or reduce the responsiveness of the substrates that underlie self-grooming and some other odor-related phenomena. Perhaps in SP male voles, titers of T and PRL that are typical of LP male voles work antagonistically and inhibit self-grooming directed at the scents of LP females. It is possible that supraphysiological dosages of PRL and T may be needed to induce SP males to self-groom in a manner typical of LP males when the latter encounter the scents of female conspecifics (Leonard *et al.* 2005). These possibilities must be tempered by the fact that in our studies, SP males received chronic treatment of exogenous T and PRL, a regimen that fails to replicate the episodic release of these hormones (Prendergast *et al.* 2002). At present, it is not known if self-grooming in response to odors of opposite-sex conspecifics reflects sexual arousal of the groomer or can affect its hormonal milieu.

Much less is known about whether self-grooming affects the endocrine physiology of the groomer and its audience. By self-grooming, individuals may bring their own odiferous substances into contact with their nares or vomeronasal organ (Johnston 1998; Keverne 1999; Marchlewska-Koj *et al.* 1998). Once detected by these organs, this odor information can be relayed to different parts of the brain where it can affect the behavior and physiology of groomers. For example, self-grooming in response to the odors of opposite-sex conspecifics may stimulate the limbic system of the groomer. This may, in turn, affect that individual's hormonal milieu by triggering gonadotropin releasing hormone pulses in the hypothalamus, which stimulate the luteinizing hormone release from the anterior pituitary. Luteinizing hormone release would stimulate the release of T or E from the gonads, which may cause arousal on the part of the groomer (Beltramino and Taleisnik 1983; Dluzen *et al.* 1981; Vandenbergh 1994). High T and E titers are necessary for the maintenance and expression of the traits and odor-related behaviors that surround reproduction in many terrestrial mammals (Ferkin and Zucker 1991; McClintock 2002; Prendergast *et al.* 2002). Luteinizing hormone and T or E titers may rise even higher among individuals that self-groom at high rates in response to the odors of opposite-sex conspecifics relative to those of individuals that self-groomed at low rates in response to such odors. In this case, self-grooming in response to the odors of opposite-sex conspecifics may be self-stimulatory and increase the groomer's sexual arousal (Ferkin, Leonard, Pierce, and Hobbs unpubl. data). If self-grooming increases the likelihood that opposite-sex conspecifics attract one another and mate, it may cause a cascade of endocrine events related to sexual arousal in the groomer and its audience.

Self-grooming and age

In free-living populations, voles may live up to two years. However, most voles live between 3 and 12 months (mo) (Ferkin and Tamarin unpubl. data). This creates different age cohorts in voles, which display different responses to odors of conspecifics (Ferkin 1999). For instance, a study on meadow voles indicated that

12–13 mo old males spent more time than 2–3 mo old and 8–9 mo old males investigating the scent marks of female conspecifics (Ferkin 1999). That study also found that 8–9 mo old males spent more time than did 2–3 mo old males investigating the scent marks of female conspecifics (Ferkin 1999). In addition, 8–9 mo old female voles spent more time investigating the scent marks of 12–13 mo old males than those of 8–9 mo old and 2–3 mo old males (Ferkin 1999).

In that voles respond differently to the odors of older and younger conspecifics (Ferkin 1999), we designed a study that addressed how the role of an individual's age affects the amount of time they spend self-grooming. Specifically, we determined whether the age of the subject and the scent donor affects the amount of time an individual self-grooms in response to the odors of opposite-sex conspecifics (Ferkin and Leonard 2008). In doing so, we had two main objectives. The first objective was to test the hypothesis that the age of the groomer affects the amount of time that it spends self-grooming in response to odors of opposite-sex conspecifics. The second objective was to test the hypothesis that the age of the scent donor affects the amount of time opposite-sex conspecifics self-groom when exposed to the donor's odors.

We found that sex differences existed in the amount of time that 2–3, 8–9, and 12–13 mo old voles self-groomed in response to cotton scented by opposite-sex conspecifics (Ferkin and Leonard 2008). Specifically, 12 mo old males spent significantly more time self-grooming in response to the odors of females aged 2–3 mo and 8–9 mo than to those from 12–13 mo old females, suggesting that 12–13 mo old male voles may be more interested in, and attractive to, females than are younger males. This result is interesting in that it suggests that the amount of time that male voles self-groom in response to the odors of females increases as adult male voles grow older.

Why did male meadow voles, aged 12–13 mo, spend more time self-grooming than younger males when exposed to odors of female conspecifics? One explanation is that older male voles were more attracted to the odors of female conspecifics. An earlier study on meadow voles reported that 12–13 mo old males spent more time investigating the scent marks of female conspecifics than did 2–3 or 8–9 mo old males (Ferkin 1999). An alternative interpretation is that older males spend more time self-grooming and investigating the odors of female conspecifics because they have sensory or perceptual deficits that are not present in younger voles. Sensory deficits in older male rodents are not unusual (Sarter and Bruno 1998) and could cause a groomer to take longer to detect, interpret, and respond to olfactory cues. However, this explanation is not consistent with the fact that 12–13 mo old males could discriminate between the odors of females in different age groups by spending more time self-grooming in response to odors of females aged 2–3 or 8–9 mo than to those aged 12–13 mo (Ferkin and Leonard

2008). Perhaps older but not younger males are able to detect some feature found in the odors of older females that makes them less attractive relative to younger females. However, in a previous study, male voles spent more time investigating the odors from females aged 8–9 mo relative to those from 2–3 or 12–13 mo old females (Ferkin 1999). Alternatively, younger males may self-groom less than older males because the former may not need to self-groom as much in order to increase the likelihood of being detected and being more attractive to the opposite sex (Ferkin and Leonard 2008). Thus, 12–13 mo old males may have self-groomed more than younger males in response to female odors because the pelage and sebaceous tissues of the former required more tactile stimulation to release their odiferous substances as compared to the pelage and sebaceous tissues of younger male voles (Ebling 1977; Johnson 1977).

The data from Ferkin and Leonard (2008) also indicate that independent of their age, female voles spent more time self-grooming when they were presented with bedding scented by 12–13 mo old males than to bedding scented by younger males. A previous study reported that female voles in the same age categories spent more time investigating the odors of 12–13 mo old males than those of younger males (Ferkin 1999). An inference drawn from this and previous work suggests older male voles produce odors that are more easily detected by females than do younger males. It is possible that 12–13 mo old male voles produce substances in their odors that are not present in those from 2–3 and 8–9 mo old males (Ferkin and Leonard 2008). Perhaps this substance provides a cue to female voles indicating male longevity, thus making older males more interesting or attractive to females (Ferkin 1999). It is also possible that female voles are attracted to the odors of older male voles and subsequently mate with them to gain indirect benefits of longevity for their sons (Møller and Alatalo 1999). Overall, the study by Ferkin and Leonard (2008) supports the hypothesis that age of the groomer and the scent donor affects the amount of time that individuals self-groom in response to odors of opposite-sex conspecifics.

Self-grooming and protein content of the diet

The theory of sexual selection predicts that the expression of traits and signals can depend on an individual's condition (Zeh and Zeh 1988). Condition-dependent signals provide reliable information about the phenotypic quality of potential mates (Kodric-Brown and Brown 1984). For example, an individual's nutritional state would likely reflect its current relative quality (Jones and Wade 2002) and also affect their interactions with opposite-sex conspecifics (Ferkin *et al.* 1997; Pierce *et al.* 2005a, b). Studies on voles suggest that the protein content of their diet may affect their fitness (Bergeron and Joudoin 1989) and also affect how they respond to conspecifics. For example, voles fed a diet high in protein produced odors that were more attractive to the opposite sex than did voles fed a diet low in protein content (Ferkin *et al.* 1997). Interestingly, the protein content of the diet did not affect the preferences voles had for the odors of opposite-sex conspecifics (Pierce *et al.* 2005a). In that meadow voles inhabit transitional grasslands where patches of food can vary greatly in quality (Batzli 1985), and wandering males and territorial females (Madison 1980) can live in areas in which the amount of forage and its quality are variable (Bergeron and Joudoin 1989), the ability of male and female voles to find patches of forage high or low in protein content may affect their self-grooming behavior.

We tested the hypothesis that the amount of time that individuals self-groom to opposite-sex conspecifics is affected by the amount of protein in their diet. If so, an individual will self-groom more in response to odors of opposite-sex conspecifics that are fed a high protein diet than to individuals fed a lower protein diet. Likewise, if self-grooming is affected by the amount of protein in the diet of the groomer, then individuals fed a high protein diet will self-groom more than individuals fed a lower protein diet. We tested these predictions by measuring the amount of time that male and female meadow voles fed diets that differed in protein content self-groomed when they encountered each other's odors (Hobbs *et al.* 2008).

We discovered that female meadow voles self-groomed more to males fed a 22% protein diet than to males fed a 9% or 13% protein diet (Hobbs *et al.* 2008). Protein content of the diet of male scent donors affected the amount of time that female voles investigated the odors of males (Ferkin *et al.* 1997). In that study, male voles fed diets higher in protein produced odors that were more attractive to female voles than males fed diets lower in protein content. Interestingly, the protein content of the diet of female groomers did not affect the amount of time they spent self-grooming when they were exposed to the odors of male scent donors. Rather, female voles spent a similar amount of time spent self-grooming regardless of the amount of protein in their diet (Hobbs *et al.* 2008). This finding is not consistent with the prediction that the protein content of a female groomer's diet will affect the amount of time that she self-grooms when exposed to the scents of a male donor.

The protein content of the diet also did not affect the amount of time male meadow voles self-groomed in response to the odors of female donors. Specifically, male meadow voles, independently of the protein content of their diet, self-groomed for similar amounts of time to the bedding scented by females fed a 22%, 13%, or 9% protein diet (Hobbs *et al.* 2008). Thus, our results are not consistent with the hypothesis that the amount of time that males self-groom depends on the diet of the female donors. Similarly, the protein content of the diet of male meadow voles did not affect their odor preference for female conspecifics over that of male conspecifics (Pierce *et al.* 2005a), but did affect the attractiveness of their odors to females (Ferkin *et al.* 1997). Overall, these findings suggest that for voles the protein content of the diet affects the behaviors that indicate interest in the opposite sex (Ferkin *et al.* 1997) and those that may facilitate interactions with them, such as self-grooming (Hobbs *et al.* 2008).

Self-grooming and same-sex competition

If social self-grooming projects or produces odors that are responded to by nearby opposite-sex conspecifics, these odors may also be intercepted by nearby same-sex conspecifics. Consequently, social self-grooming may also influence the responses of nearby same-sex conspecifics to the groomer and vice versa. It is not clear, however, why individuals may self-groom when they encounter the odors of same-sex conspecifics. Two hypotheses that may account for self-grooming in response to same-sex conspecifics focus on indicating social tolerance between the groomer and nearby conspecifics. The first hypothesis is that social self-grooming is a signal of dominance to a nearby same-sex conspecific, indicating a willingness on the part of the groomer to escalate their agonistic behavior (Bursten *et al.* 2000; Silbaugh and Ewald 1987). The second hypothesis is that self-grooming is a signal of subordination to a nearby same-sex conspecific and is representative of reduced agonistic behavior by the groomer (Shanas and Terkel 1995, 1997). The rationale for these two hypotheses is similar. By self-grooming, individuals may facilitate or avoid encounters with particular same-sex conspecifics by adjusting the amount of time that they self-groom. The null hypothesis is that individuals do not adjust the amount of time they self-groom when they encounter the odors of particular same-sex conspecifics, indicating that self-grooming is not associated with social tolerance.

Interestingly, male prairie voles (Ferkin *et al.* 2001), but not male meadow voles (Ferkin *et al.* 1996), self-groom at high rates in response to male odors. One explanation may be associated with the fact that male prairie voles are philopatric, territorial, and have relatively small home ranges, which allow them to encounter the scent marks of many familiar males (McGuire *et al.* 1990). By self-grooming in response to odors of nearby or wandering males, resident prairie voles may be reaffirming their ownership of the territory, making it easier for male intruders to detect them. In contrast, male meadow voles are not philopatric and inhabit large home ranges (Madison 1980), which allow them to encounter the scent marks of many unfamiliar males. It may be too energetically costly to self-groom when meadow voles encounter the scent marks of both female and male conspecifics. It may be prudent for males to self-groom when they encounter the odors of female conspecifics and not those of male conspecifics. If a male meadow vole encounters scent marks of an unfamiliar conspecific male, he may simply ignore this scent mark and move to another location. In contrast, high rates of self-grooming by a male prairie vole in response to odors of both male and female conspecifics may be associated with the strong male–female pair bond that is characteristic of this species (Getz and Carter 1996). By self-grooming at high rates in response to male odors, male prairie voles may be attempting to deter other males from interacting with their mates by indicating their presence in the area to nearby males (Ferkin *et al.* 2001). Alternatively, the resident male prairie vole may be masking the scent of the other male with his own scent, thereby preventing a potential mate from being stimulated by another male's odor. In contrast, the low rate of self-grooming by a male meadow vole in response to male odors may be associated with tactics male voles use to compete with other males for access to female conspecifics. Male meadow voles do not form a bond with a female and do not attempt to guard her from intruders (Boonstra *et al.* 1993). Instead, males deposit scent marks near sexually receptive females. These scent marks cause males that visit these females to increase the amount of sperm that they ejaculate during coitus (delBarco-Trillo and Ferkin 2004).

Self-grooming may also facilitate a stress response from same-sex conspecifics. In this case, catecholamine and glucocorticoid titers of nearby conspecifics may rise after an individual encounters the odors of same-sex groomers (Dluzen and Ramirez 1987; Marchlewska-Koj and Zacharczuk-Kakietek 1990; Zalaquett and Thiessen 1991). However, it is not clear from the current literature whether socially dominant individuals self-groom more than subordinate individuals or vice versa (Bursten *et al.* 2000; Shanas and Terkel 1997). It is possible that socially dominant individuals would self-groom when they encounter the scents of subordinate individuals to signal their social status and announce their presence to nearby subordinates (Bursten *et al.* 2000; Payne 1977). Such an announcement may reduce agonism between the groomer and nearby conspecifics (Shanas and Terkel 1995, 1997). However, we do not know whether previous encounters with same-sex conspecifics affect the amount of time that voles self-groom when they are exposed to odors from those conspecifics. It is also possible that self-grooming allows dominant individuals to come into a heightened state of arousal (Spruijt *et al.* 1992). Further investigations are needed to answer these questions.

Self-grooming and olfactory communication

Among terrestrial animals, self-grooming is ubiquitous and may be a form of scent marking (Ferkin and Leonard 2005; Halloran and Bekoff 1995). A previous model of self-grooming suggests that it serves multiple functions, some of which are unique to each species, whereas others serve a general function across species (Spruijt *et al.* 1992). A revision of this model provides several hypotheses (Ferkin and Leonard 2005). One hypothesis is that self-grooming provides a chemical résumé of the signaler, which may include information about condition-dependent features of its identity (i.e., diet, age, relatedness, and reproductive status) and condition independent-factors such as its sex. A prediction of this hypothesis is that self-grooming provides "cheat-proof" signals. A second hypothesis is that animals match the scents produced by self-grooming with their signalers and use this information to distinguish between familiar and unfamiliar individuals, residents and intruders, dominants and subordinates, reproductively active and reproductively quiescent, opposite-sex conspecifics, adults and juveniles, related and nonrelated conspecifics, and other possible relationships (Ferkin 2006; Gosling and Roberts 2001; Leonard *et al.* 2005; Paz-y-Miño *et al.* 2002). A third hypothesis is that individuals can, by self-grooming, alter features of the odors that they produce. Predictions of this hypothesis are that by self-grooming individuals produce or release odors that are more attractive than those produced by individuals that do not self-groom (Payne 1977; Thiessen 1977). Such odors may reduce agonism between the groomer and opposite-sex conspecifics, and/or increase the odor field that surrounds the groomer, making it more likely that it will be detected by nearby opposite-sex conspecifics (Bursten *et al.* 2000; Ferkin *et al.* 1996, 2001; Shanas and Terkel 1997). The last hypothesis of our model is that self-grooming is a sexually selected trait used by individuals during same-sex competition and mate choice (Ferkin *et al.* 1996). That is, the benefits of self-grooming and providing olfactory information to conspecifics to attract mates and to compete with same-sex conspecifics outweigh the costs of being detected more easily and eavesdropping (Ferkin 2005).

In this chapter, we have highlighted studies on voles that suggest a role for self-grooming, which may be associated with facilitating interactions between opposite-sex conspecifics (Ferkin and Leonard 2005). The present findings and those of a growing literature indicate that voles and likely other mammals spend different amounts of time engaged in self-grooming when they encounter the scents of particular conspecifics, which may reflect the type or the nature of interaction that groomers would have with those conspecifics (Ferkin and Leonard 2005).

Acknowledgments

We thank Lara LaDage, Nick Hobbs, Ashlee Vaughn, Andrew Pierce, Javier delBarco-Trillo, Justin LaPorte, and Allan Kalueff for commenting on earlier drafts of this manuscript. This work was supported by National Science Foundation grants IBN 9421529 and IOB 444553 and National Institutes of Health grants AG16594-01 and HD0 49525-01 to M. H. Ferkin.

References

Batzli GO (1985): Nutrition. In: Tamarin RH, ed., *Biology of New World Microtus*, Special Publ. 8. Lawrence, KS: The American Society of Mammalogists, pp. 779–811.

Beltramino C and Taleisnik S (1983): Release of LH in the female rat by olfactory stimuli: effect of the removal of the vomeronasal organs or lesioning of the accessory olfactory bulbs. *Neuroendocrinology* **36**:53–8.

Bergeron JM and Joudoin L (1989): Patterns of resource use, food quality, and health status of voles (*Microtus pennsylvanicus*) trapped from fluctuating populations. *Oecologia* **79**:306–14.

Boonstra R, Xia X and Pavone L (1993): Mating system of the meadow vole, *Microtus pennsylvanicus*. *Behav Ecol* **4**:83–9.

Borchelt PL (1980): Care of the body surface. In: Denny RM, ed., *Comparative Psychology: An Evolutionary Analysis of Animal Behavior*. New York: Wiley Press, pp. 362–84.

Bossert WH and Wilson EO (1963): The analysis of olfactory communication among animals. *J Theoret Biol* **5**:443–69.

Brockie R (1976): Self-anointing by wild hedgehogs, *Erinaceus europaeus*. *Anim Behav* **24**:68–71.

Bursten SN, Berridge KC and Owings DH (2000): Do California ground squirrels (*Spermophilus beecheyi*) use ritualized syntactic cephalocaudal grooming as an agonistic signal? *J Comp Psychol* **114**:281–90.

Carter CS, Witt DM, Manock SR *et al.* (1989): Hormonal correlates of sexual behavior and ovulation in male-induced and postpartum estrus female prairie voles. *Physiol Behav* **46**:941–8.

delBarco-Trillo J and Ferkin MH (2004): Male mammals respond to a risk of sperm competition conveyed by odours of conspecific males. *Nature* **431**:446–9.

delBarco-Trillo J and Ferkin MH (2007): Female meadow voles, *Microtus pennsylvanicus*, experience a reduction in copulatory behavior during postpartum estrus. *Ethology* **113**:466–73.

Dell'Omo G and Alleva E (1994): Snake odors alter behavior, but not pain sensitivity in mice. *Physiol Behav* **55**:125–8.

Demas GE, Klein SL and Nelson RJ (1996): Reproductive and immune responses to photoperiod and melatonin are linked in *Peromyscus maniculatus*. *J Comp Physiol* (A) **179**:819–25.

Dluzen DE and Ramirez VD (1987): Involvement of olfactory bulb catecholamines and luteinizing hormone-releasing hormone in response to social stimuli mediating reproductive functions. *Ann N Y Acad Sci* **519**:252–68.

Dluzen DE, Ramirez VD, Carter CS and Getz LL (1981): Male vole urine changes luteinizing hormone-releasing hormone and norepinephrine in female olfactory bulb. *Science* **212**:573–5.

Drago F and Bohus B (1981): Hyperprolactinemia-induced excessive grooming in the rat: time-course and element analysis. *Behav Neural Biol* **33**:117–22.

Drago F, Canonico PL, Bitetti R and Scapagnini U (1980): Systemic and intraventricular prolactin induced excessive grooming. *Eur J Pharmacol* **65**:457–8.

Drago F, Bohus B, Bitetti R *et al.* (1986): Intracerebroventricular injection of anti-prolactin serum suppresses excessive grooming of pituitary homografted rats. *Behav Neural Biol* **46**:99–105.

Duncan MJ, Goldman BD (1984): Hormonal regulation of the annual pelage color cycle in Djungarian hamster, *Phodopus songurus*, II. Role of prolactin. *J Exp Zool* **230**:97–103.

Ebling FJ (1977): Hormonal control of mammalian skin glands. In: Muller-Schwarze D and Mozell MM, eds., *Advances in Chemical Signals in Vertebrates*. New York: Plenum Press, pp. 17–33.

Fentress JC (1968). Interrupted ongoing behaviour in two species of vole (*Microtus* and *Clethrionomys brittanicus*) II. Extended analysis of motivational variables underlying fleeing and grooming behaviour. *Anim Behav* **16**:154–67.

Ferkin MH (1999): Attractiveness of opposite-sex odor and responses to it vary with age and sex in meadow voles (*Microtus pennsylvanicus*). *J Chem Ecol* **4**:757–69.

Ferkin MH (2005): Self-grooming in meadow voles. In: Mason RT, LeMaster MP and Muller-Schwarze D, eds., *Chemical Signals in Vertebrates*, Vol. 10. New York: Plenum Press, pp. 64–9.

Ferkin MH (2006): The amount of time that a meadow vole, *Microtus pennsylvanicus*, self-grooms is affected by its reproductive state and that of the odor donor. *Behav Processes* **73**:266–71.

Ferkin MH and Gorman MR (1992): Photoperiod and testosterone control seasonal odor preferences of meadow voles, *Microtus pennsylvanicus*. *Physiol Behav* **51**:1087–91.

Ferkin MH and Johnston RE (1995a): Meadow voles, *Microtus pennsylvanicus*, use multiple sources of scent for sexual recognition. *Anim Behav* **49**:37–44.

Ferkin MH and Johnston RE (1995b): Effects of pregnancy, lactation, and postpartum oestrous on odour signals and the attraction to odours in female meadow voles, *Microtus pennsylvanicus*. *Anim Behav* **49**:1211–17.

Ferkin MH and Kile JR (1996): Melatonin treatment affects the attractiveness of the anogenital area scent in meadow voles (*Microtus pennsylvanicus*). *Horm Behav* **30**:227–35.

Ferkin MH and Leonard ST (2005): Self-grooming by rodents in social and sexual contexts. *Acta Zool Sinica* **51**:772–9.

Ferkin MH and Leonard ST (2008): Age of the subject and scent donor affects the amount of time that voles self-groom when they are exposed to odors of opposite-sex conspecifics. In: Hurst JL, Beynon RJ, Roberts SC and Wyatt TD, eds., *Chemical Signals in Vertebrates 11*. New York: Springer Press, pp. 281–9.

Ferkin MH and Li HZ (2005): A battery of olfactory-based screens for phenotyping the social and sexual behaviors of mice. *Physiol Behav* **85**:489–99.

Ferkin MH and Seamon JO (1987): Odor preference and social behavior in meadow voles, *Microtus pennsylvanicus*: seasonal differences. *Can J Zool* **65**:2931–7.

Ferkin MH and Zucker I (1991): Seasonal control of odour preferences of meadow voles (*Microtus pennsylvanicus*) by photoperiod and ovarian hormones. *J Reprod Fert* **92**:433–41

Ferkin MH, Sorokin ES and Johnston RE (1996): Self-grooming as a sexually dimorphic communicative behaviour in meadow voles, *Microtus pennsylvanicus*. *Anim Behav* **51**:801–10.

Ferkin MH, Sorokin ES and Johnston RE (1997): Effect of prolactin on the attractiveness of male odors to females: independent effects and synergism with testosterone. *Horm Behav* **31**:55–63.

Ferkin MH, Leonard ST, Heath LA and Paz-y-Miño CG (2001): Self-grooming as a tactic used by prairie voles *Microtus ochrogaster* to enhance sexual communication. *Ethology* **107**:939–49.

Ferkin MH, Leonard ST and Gilless JP (2007): Melatonin treatment affects conspecific odor preferences and self-grooming of meadow voles (*Microtus pennsylvanicus*). *Physiol Behav* **91**:255–63.

Flügge G, Kramer M, Rensing S and Fuchs E (1998): 5HT1A-receptors and behaviour under chronic stress: selective, counteraction by testosterone. *Eur J Neurosci* **10**:2685–93.

Getz LL and Carter CS (1996): Prairie vole partnerships. *Am Sci* **84**:56–62

Geyer LA and Kornet CA (1982): Auto- and allo-grooming in pine voles (*Microtus pinetorum*) and meadow voles (*Microtus pennsylvanicus*). *Physiol Behav* **28**:409–12.

Goldman BD, Matt KS, Roychoudhury P and Stetson MH (1981): Prolactin release in golden hamsters: photoperiod and gonadal influences. *Biol Reprod* **24**:287–92.

Gonzalez-Lima F, Velex D and Blanco R (1988): Antagonism of behavioral effects of bromocriptine by prolactin in female rats. *Behav Neural Biol* **49**:74–82.

Gosling LM and Roberts SC (2001): Scent marking in male mammals: cheat-proof signals to competitors and mates. *Adv Study Behav* **30**:169–217.

Hainsworth FH (1967): Saliva spreading, activity, and body temperature regulation in the rat. *Am J Physiol* **212**:1288–92.

Halloran ME and Bekoff M (1995): Cheek rubbing as grooming by Abert squirrels. *Anim Behav* **50**:987–93.

Hamilton WD and Zuk M (1982): Heritable true fitness and bright birds: a role for parasites? *Science* **218**:384–7.

Harriman AE and Thiessen DD (1985): Harderian letdown in male Mongolian gerbils (*Meriones unguiculatus*) contributes to proceptive behavior. *Horm Behav* **19**: 213–19.

Hart BL (1990): Behavioral adaptations to pathogens and parasites: five strategies. *Neurosci Biobehav Rev* **14**:273–94.

Hart BL and Powell KL (1990): Antibacterial properties of saliva: role in material grooming and in licking wounds. *Physiol Behav* **48**:383–6.

Hobbs NJ, Aven AM, Ferkin MH (2008): Self-grooming response of meadow voles to the odor of opposite-sex conspecifics in relation to the dietary protein content of both sexes. *Ethology* **114**:1210–17.

Johnson EK (1977): Seasonal changes in the skin of mammals. *Symp Zool Soc London* **39**:373–404.

Johnston RE (1998): Pheromones, the vomeronasal system, and communication: from hormonal responses to individual recognition. *Ann NY Acad Sci* **855**: 333–48.

Jones JE and Wade GN (2002): Acute fasting decreases sexual receptivity and neural estrogen receptor-alpha in female rats. *Physiol Behav* **77**:19–25.

Judge PG, Griffaton NS and Fincke AM (2006): Conflict management by hamadryas baboons (*Papio hamadryas hamadryas*) during crowding: a tension-reduction strategy. *Am J Primatol* **68**:993–1005.

Kalueff AV and Tuohimaa P (2005): The grooming analysis algorithm discriminates between different levels of anxiety in rats: potential utility for neurobehavioural stress research. *J Neurosci Methods* **143**:169–77.

Kalueff AV, Fox MA, Gallagher PS and Murphy DL (2007): Hypolocomotion, anxiety and serotonin syndrome-like behavior contribute to the complex phenotype of serotonin transporter knockout mice. *Genes Brain Behav* **6**:389–400.

Keller BL (1985): Reproductive patterns. In: Tamarin RH, ed., *Biology of New World Microtus*, Special Publ. 8. Lawrence, KS: The American Society of Mammalogists, pp. 725–78.

Keverne EB (1999): The vomeronasal organ. *Science* **286**:716–20.

Kodric-Brown A and Brown JH (1984): Truth in advertising: the kinds of traits favored by sexual selection. *Am Nat* **124**:309–23.

Land BB and Seeley TD (2004): The grooming invitation dance of the Honey Bee. *Ethology* **110**:1–10.

Leonard ST and Ferkin MH (1999): Prolactin and testosterone affect seasonal differences in male meadow vole, *Microtus pennsylvanicus*, odor preferences for female conspecifics. *Physiol Behav* **68**:139–43.

Leonard ST and Ferkin MH (2005): Seasonal differences in self-grooming in meadow voles, *Microtus pennsylvanicus*. *Acta Ethol* **8**:86–91.

Leonard ST, Alizadeh-Naderi R, Stokes K and Ferkin MH (2005): The role of prolactin and testosterone in mediating seasonal differences in the self-grooming behavior of male meadow voles, *Microtus pennsylvanicus*. *Physiol Behav* **85**:461–8.

Lucas C, Pho DB, Fresneau D and Jallon JM (2004): Hydrocarbon circulation and colonial signature in *Pachycondyla villosa*. *J Insect Physiol* **50**:595–607.

Madison DM (1980): An integrated view of the social biology of meadow voles *Microtus pennsylvanicus*. *Biologist* **62**:20–33.

Marchlewska-Koj A and Zacharczuk-Kakietek M (1990): Acute increase in plasma corticosterone level in female mice evoked by pheromones. *Physiol Behav* **48**:577–80.

Marchlewska-Koj A, Kruczek M, Olejniczak P and Pochron E (1998): Involvement of main and vomeronasal systems in modification of oestrous cycle in female laboratory mice. *Acta Theriol* **43**:235–40.

McClintock MK (2002): Pheromones, odors, and vasanas: the neuroendocrinology of social chemosignals in humans and animals. In: Pffaf D, ed., *Hormones, Brain, and Behavior*. New York: Elsevier Science, pp. 797–870.

McFarlane HG, Kusek GK, Yang M *et al.* (2008): Autism-like behavioral phenotypes in BTBR T+tf/J mice. *Genes Brain Behav* **7**:152–63.

McGuire B, Pizzuto T and Getz LL (1990): Potential for social interaction in a natural population of prairie voles (*Microtus ochrogaster*). *Can J Zool* **68**:391–8.

Miernicki M, Pospichal MW and Powers JB (1990): Short photoperiods affect male hamster sociosexual behaviors in the presence and absence of testosterone. *Physiol Behav* **47**:95–106.

Møller AP and Alatalo RV (1999): Good-genes effects in sexual selection. *Proc R Soc Lond B* **266**:85–91.

Moore CL (1986): A hormonal basis for sex differences in the self-grooming of rats. *Horm Behav* **20**:155–65.

Mooring MS, Benjamin JE, Harte CR and Herzog NB (2000): Testing the interspecific body size principle in ungulates, the smaller they come, the harder they groom. *Anim Behav* **60**:5–45.

Payne AP (1977): Pheromonal effects of Harderian gland homogenates on aggressive behaviour in the hamster. *J Endocrinol* **73**:191–2.

Paz-y-Miño CG, Leonard ST, Ferkin MH and Trimble JF (2002): Self-grooming and sibling recognition in meadow voles (*Microtus pennsylvanicus*) and prairie voles (*M. ochrogaster*). *Anim Behav* **63**:331–8.

Pierce AA, Ferkin MH and Patel NP (2005a): Protein content of the diet does not influence proceptive or receptive behavior in female meadow voles, *Microtus pennsylvanicus*. In: Mason RT, LeMaster MP and Muller-Schwarze D, eds., *Chemical Signals in Vertebrates*, Vol. 10. New York: Plenum Press, pp. 70–6.

Pierce AA, Ferkin MH and Williams TK (2005b): Food-deprivation-induced changes in sexual behavior of meadow voles, *Microtus pennsylvanicus*. *Anim Behav* **70**:339–48.

Powers JB, Bergondy ML and Matochik JA (1985): Male hamster sociosexual behaviors: effects of testosterone and its metabolites. *Physiol Behav* **35**:607–16.

Prendergast BJ, Nelson RJ and Zucker I (2002): Mammalian seasonal rhythms: behavior and neuroendocrine substrates. In: Pffaf D, ed., *Hormones, Brain and Behavior*. New York: Elsevier Press, pp. 93–156.

Roth KA and Katz RJ (1979): Stress, behavioural arousal and open-field activity: a reexamination of emotionality in the rat. *Neurosci Biobehav Rev* **3**:247–63.

Sarter M and Bruno JP (1998): Age-related changes in rodent cortical acetylcholine and cognition: main effects of age versus age as an intervening variable. *Brain Res Rev* **27**:143–56.

Schino G, Perretta G, Taglioni AM, Monaco V and Troisi A (1996): Primate displacement activities as an ethological model of anxiety. *Anxiety* **2**:186–91.

Shanas U and Terkel J (1995): Grooming expresses Harderian gland materials in the blind mole rat. *Aggr Behav* **21**:137–46.

Shanas U and Terkel J (1997): Mole-rat Harderian gland secretions inhibit aggression. *Anim Behav* **54**:1255–63.

Silbaugh JM and Ewald PW (1987): Effects of unit payoff asymmetries on aggression and dominance in meadow voles, *Microtus pennsylvanicus*. *Anim Behav* **35**:606–8.

Spruijt BM, Van Hooff JARAM and Gispen WH (1992): Ethology and neurobiology of grooming behavior. *Physiol Rev* **72**:825–52.

Steiner AL (1973): Self- and allo-grooming behavior in some ground squirrels (Sciuridae), a descriptive study. *Can J Zool* **51**:151–61.

Steiner AL (1974): Body-rubbing, marking, and other scent-related behavior in some ground squirrels (Sciuridae). *Can J Zool* **52**:889–906.

Stopka P and Macdonald DW (1998): Signal interchange during mating in the wood mouse (*Apodemus sylvaticus*) the concept of active and passive signaling. *Behaviour* **135**:231–49.

Thiessen DD (1977): Thermoenergetics and the evolution of pheromone communication. *Prog Psychobiol Physiol Psychol* **7**:91–191.

Thiessen DD and Harriman AE (1986): Harderian gland exudates in the male *Meriones unguiculatus* regulate female proceptive behavior, aggression, and investigation. *J Comp Psychol* **100**:85–7.

Thor DH, Harrison RJ and Schneider SR (1988): Sex differences in investigatory and grooming behaviors of laboratory rats (*Rattus norvegicus*) following exposure to novelty. *J Comp Psychol* **102**:188–92.

van Erp AMM, Kruk MR, Meelis W and Willekens-Bramer DC (1994): Effect of environmental stressors on time course, variability, and form of self-grooming in the rat: handling, social contact, defeat, novelty, restraint and fur moistening. *Behav Brain Res* **65**:47–55.

Vandenbergh JG (1994): Pheromones and mammalian reproduction. In: Knobil E and Neill JD, ed., *The Physiology of Reproduction*, 2nd edn. New York: Raven Press, pp. 343–59.

Wiepkema PR (1979): The social significance of self-grooming in rats. *Netherlands J Zool* **29**:622–3.

Witt DM, Carter CS, Carlstead K and Read LD (1988): Sexual and social interaction preceding and during male-induced oestrous in prairie voles, *Microtus ochrogaster*. *Anim Behav* **36**:1465–71.

Witt DM, Carter CS, Chayer R and Adams K (1990): Patterns of behavior during postpartum estrous in prairie voles. *Anim Behav* **39**:528–34.

Wolff, JO, Watson MH and Thomas SA (2002): Is self-grooming by male prairie voles a predictor of mate choice? *Ethology* **108**:169–79.

Zalaquett C and Thiessen D (1991): The effects of odors from stressed mice on conspecific behavior. *Physiol Behav* **50**:221–7.

Zeh DW and Zeh JA (1988): Condition-dependent sex ornaments and field-tests of sexual-selection theory. *Am Nat* **132**:454–9.

3

Phenotyping and genetics of rodent grooming and barbering: utility for experimental neuroscience research

CARISA L. BERGNER, AMANDA N. SMOLINSKY, BRETT D. DUFOUR, JUSTIN L. LAPORTE, PETER C. HART, RUPERT J. EGAN, AND ALLAN V. KALUEFF

Summary

Grooming and barbering (behavior-associated hair loss) are complex, ethologically rich behaviors. They are commonly observed in different animal species, and represent important phenotypes to study in experimental models utilizing rodent research. Due to sensitivity to alterations in activity and microstructure, grooming analysis has utility in the assessment of stress in individual animals, the testing of psychotropic drugs, phenotyping mutant or transgenic animals, as well as the selection of proper strains for experimental modeling of affective disorders. Similarly, barbering shows context- and strain-specific variations, and may serve as an indicator of social dominance or behavioral perseveration. While little is known about the genetics of barbering phenotypes, evaluation of this behavior has implications in neurophysiology and biological psychiatry, providing insight into trichotillomania, obsessive–compulsive disorder (OCD), aggression-related and other human brain disorders. Here, we discuss ethologically based approaches to the assessment of animal grooming and barbering activity. Additionally, we present examples of genetic variation leading to altered grooming and barbering phenotypes in rodents, and summarize the growing value of these two phenotypes for translational neurobehavioral research.

Neurobiology of Grooming Behavior, eds. Allan V. Kalueff, Justin L. LaPorte, and Carisa L. Bergner. Published by Cambridge University Press.

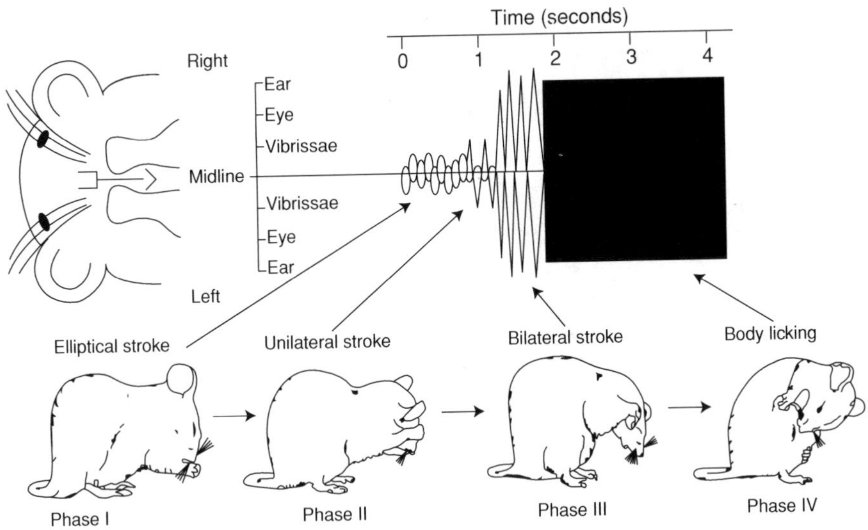

Figure 3.1 Prototypical syntactic grooming chain pattern in mice (K. Berridge, with permission). Phase I: series of ellipse-shaped strokes tightly around the nose (paw, nose grooming). Phase II: series of unilateral strokes (each made by one paw) that reach up the mystacial vibrissae to below the eye (face grooming). Phase III: series of bilateral strokes made by both paws simultaneously. Paws reach back and upwards, ascending usually high enough to pass over the ears (head grooming). Phase IV: body licking, preceded by postural cephalocaudal transition from paw/head grooming to body grooming.

Introduction

Grooming is an innate behavior shared across many animal species with remarkable homology (Fentress 1988; Sachs 1988; Spruijt *et al.* 1992). Common in laboratory and wild rodents, grooming occupies a substantial portion of their waking time, thereby representing an important phenotype to study (Bolles 1960; Hyman 2007; Kalueff *et al.* 2007a; Kalueff and Tuohimaa 2004a, 2005a). Rodent grooming is a patterned behavior, which generally proceeds in a cephalo caudal direction (Berridge *et al.* 2005; Fentress 1988). This pattern begins with paw licking, followed by washing of the nose and face, head, body, legs, and finally, the tail and genitals (Figure 3.1). Regulation of grooming behavior is mediated by multiple brain regions, especially the basal ganglia and hypothalamus (Aldridge *et al.* 2004; Berntson *et al.* 1988; Kruk *et al.* 1998; Roeling *et al.* 1993). Various endogenous and exogenous substances, such as the neuromediators dopamine, GABA (γ-aminobutyric acid) or serotonin, as well as many hormones and psychotropic drugs, have been shown to modulate grooming activity (Barros *et al.* 1994; Bertolini *et al.* 1988; Dunn 1988; Dunn *et al.* 1987; Hill *et al.* 2007; Kalueff

Table 3.1 *Examples of grooming phenotypes in different genetically modified mice (data obtained from Mouse Genome Informatics and PubMed)*

Model	Background strain	Grooming behavior	References
Engrailed 2 (En2) gene knockout mice	129S2/SvPas × C57BL/6	Increased grooming	(Cheh *et al.* 2006)
Vitamin D receptor knockout mice	129S1	Increased grooming	(Kalueff *et al.* 2006a)
Homeobox (B8) gene knockout mice	129S1/Sv × 129 × 1/SvJ	Increased grooming	(Greer and Capecchi 2002)
Paired related homeobox protein-like 1 (Prrxl1) knockout mice	129S7/SvEvBrd × C57BL/6J × CD-1	Increased grooming	(Chen *et al.* 2001)
Cholinergic receptor, nicotinic, alpha polypeptide 4 (Chrna 4) knockout mice	129S4/SvJae × C57BL/6	Decreased grooming	(Ross *et al.* 2000)
D-aspartate oxidase knockout mice	129S4/SvJae	Decreased grooming	(Huang *et al.* 2006)
Oxytocin knockout mice	129S/SvEv × C57BL/6	Decreased grooming	(DeVries *et al.* 1997)
AT rich interactive domain 5B (Arid5b) knockout mice	129S4/SvJae× BALB/c	Decreased grooming	(Lahoud *et al.* 2001)

and Tuohimaa 2005b; Kruk *et al.* 1998; Navarro *et al.* 1995; Yalcin *et al.* 2007). Genes also play an important role in the regulation of this behavior (Greer and Capecchi 2002; Welch *et al.* 2007), and various genetic manipulations in animals have been reported to produce robust grooming phenotypes (Table 3.1).

Given the importance of grooming in animal phenotypes, it is reasonable to predict that alterations in this domain would be seen in various experimental models of brain disorders. For example, as a displacement behavior, grooming is frequently displayed in animal models of stress, suggesting that it may simply be an anxiogenic response (Choleris *et al.* 2001). However, recent data show that higher stress in animals does not necessarily cultivate increased grooming activity, as it may also be increased under conditions of low stress (e.g., "comfort" grooming that occurs spontaneously as a transition between rest and activity) (Kalueff *et al.* 2007a; Kalueff and Tuohimaa 2004a, 2005a).

Table 3.2 *Methodological approaches to animal grooming phenotyping (according to Aldridge* et al. *2004; Berridge* et al. *2005; Kalueff* et al. *2007a; Kalueff and Tuohimaa 2004a, 2005a; Piato* et al. *2008)*

Global assessment
Coat state
General cumulative measures
The latency to onset, the duration and the number of grooming episodes (bouts). Temporal patterning (e.g., per-minute distribution) of grooming duration and frequency may be recorded to examine habituation of this behavior.
The following patterns can be recorded for each bout: paw licking; nose/face grooming; head washing; body and leg grooming/scratching; tail/genitals grooming.
Additional cumulative indices: the average duration of a single grooming bout, total number of transitions between grooming stages, and average number of transitions per bout.
Patterning (sequencing)
The percentages of incorrect transitions, as well as interrupted and incomplete grooming bouts.
Regional distribution of grooming
Can be assessed as directed to the following five anatomic areas: forepaws, head, body, hind legs, and tail/genitals. Rostral grooming includes forepaw (preliminary rostral grooming) and head grooming. Body, legs and tail/genital grooming can be considered as caudal grooming. Each bout can be categorized as being directed to (i) multiple regions or (ii) a single region, and the percentages of grooming bouts and of time spent grooming can be calculated for both categories.
Additional useful indices of grooming
Probability of chain initiation (frequency of chain initiation per minute of grooming time), probability of pattern completion once initiated.

Due to this complexity, grooming phenotypes must be examined both qualitatively and quantitatively. While general cumulative measures provide a gross assessment of grooming activity, its patterning and regional distribution indices are also important for comprehensive evaluation of this behavior (Table 3.2). Understanding the cumulative, patterning, and regional alterations in grooming has implications for developing improved animal models of human brain disorders (e.g., anxiety, depression, OCD, or Tourette's syndrome), behavioral phenotyping of mutant or transgenic strains, and the testing of psychotropic drugs (Berridge *et al.* 2005; Campbell *et al.* 1999; Kalueff *et al.* 2007a; Kalueff and Tuohimaa 2005b; Welch *et al.* 2007).

Like grooming, barbering (Figure 3.2) is a common phenotype in many different species. Representing a behavior-associated hair loss, it is also known in the literature as whisker-eating, whisker trimming, hair nibbling, hair pulling,

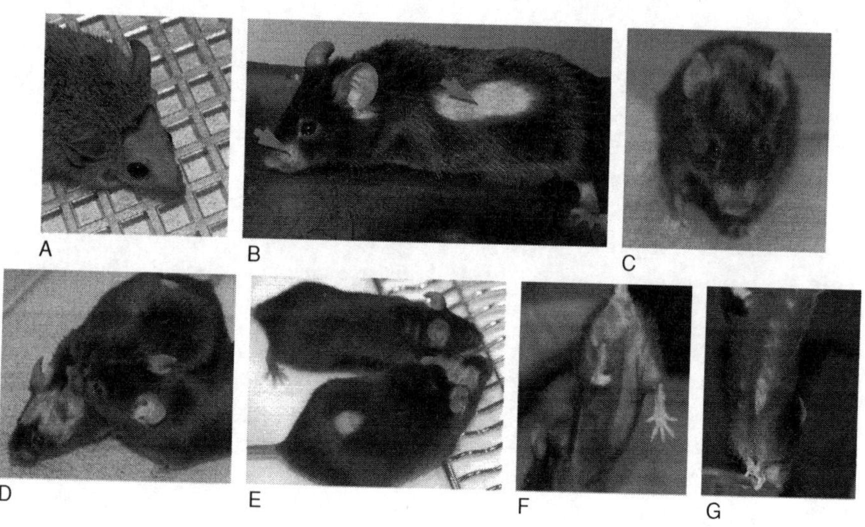

Figure 3.2 Examples of barbering phenotypes in different mice. A: hetero-barbering in 129Sv mice; B, C, D, E: hetero-barbering in C57BL/6 mice; F, G: self-barbering in C57BL/6 mice. (Photos: B. Dufour, J. Garner.)

behavior-associated alopecia areata, and the Dalila effect (Garner *et al.* 2004a; Kurien *et al.* 2005; Long 1972; Sarna *et al.* 2000). Barbering behavior is frequently seen in laboratory mice, when an individual plucks or trims fur and/or whiskers from cage-mates and/or itself, leaving idiosyncratic patches of hair loss on the nose, head, shoulders, forearms, or elsewhere (Garner *et al.* 2004b; Hill *et al.* 2007; Kalueff *et al.* 2006b; Kurien *et al.* 2005; Sarna *et al.* 2000).

There has been a growing interest in barbering phenotype recently, both as a husbandry problem (De Luca 1997; Garner *et al.* 2004a) and as a behavioral assay in biomedical research (Garner *et al.* 2004b; Hill *et al.* 2007; Kalueff *et al.* 2006b). However, relatively little is known about how or why barbering occurs. The existence of barbering behavior is a biological paradox, since it is sometimes performed without an apparent adaptive benefit for the barber, and in spite of fitness costs associated with this behavior. Answers to this paradox are the most contentious issues within the barbering literature, to which several hypotheses have been developed (Garner *et al.* 2004a; Kalueff *et al.* 2006b; Kurien *et al.* 2005; Sarna *et al.* 2000). Briefly, the *dominance hypothesis* claims that mice pluck hair in order to establish their dominance over their cage-mates (Long 1972). For example, hetero-barbering may be a dominant behavior related to social hierarchy, since in mouse groups, there is often one individual with unbarbered whiskers who appears to play a dominant role in the cage (Kalueff *et al.* 2006b; Sarna *et al.* 2000). Thus, the adaptive value of barbering may be to facilitate murine social hierarchy development and/or maintenance, which will reduce the incidence of aggression and improve the health and survival of both the barber and its cage-mates.

The *coping hypothesis* suggests that barbering may represent a form of aberrant behavior developed to cope with inadequate housing conditions. Since coping is often invoked as a functional explanation for other abnormal behaviors seen in captivity (such as stereotypy), the adaptive value of barbering may be to reduce stress in individuals that pluck, which should improve the health and survival of laboratory mice. However, while plausible, there is no evidence that the performance of barbering behavior provides any anxiolytic or stress-reducing effects for barbers. The *pathology hypothesis* explains the barbering paradox by claiming that mice pluck hair as a result of abnormal brain function, which is induced by the unnatural environment in which they develop. In contrast to the other hypotheses, it implies that barbering behavior has no adaptive value, but instead occurs as a symptom of disturbed neurophysiology (Garner *et al.* 2004a, b).

Although the exact biological reasons for barbering remain unclear, many studies indicate that there may be a strong genetic component, as barbering occurs more frequently in some mouse strains than others (Figure 3.2), and because some genetic manipulations may robustly affect barbering phenotypes (Table 3.4). Thus, the *genetic hypothesis* of barbering may also be an interesting avenue for further research in this field. Finally, it is possible that barbering represents a more *complex, multifactorial* behavioral phenomenon, and several different context-specific factors play a role in this behavior (Kalueff *et al.* 2006b).

Interestingly, while appearing similar enough, animals' barbering profiles do not always correlate with grooming phenotypes (Kalueff *et al.* 2006b; Sarna *et al.* 2000), and the two activities are likely to represent related but distinct behavioral domains. For example, in hetero-grooming, the mouse licks the body surface, often in specific regions, and sometimes even gently bites the fur without pulling any hair out. In whisker plucking, some mouthing and licking may also take place, but hairs are often plucked out in the absence of these normal grooming behaviors after the cage-mate has been pressed down (Sarna *et al.* 2000).

Overall, there are several reasons why domain-specific behavioral analyses may benefit neurobehavioral research. Firstly, like grooming, barbering is an interesting behavior per se, which plays an important role in mouse activity (Garner *et al.* 2004b; Kalueff *et al.* 2006b; Sarna *et al.* 2000). Secondly, because barbering often affects the reception of essential sensory input from the whiskers, barbered whiskers may affect all rodent behaviors, including behavioral performance in experimental tasks. Thirdly, barbering is observed more commonly in some strains than others (Carruthers *et al.* 1998; Garner *et al.* 2004a; Kalueff *et al.* 2006b; Sarna *et al.* 2000), enabling studies of different genetic contributors to the behavior.

Finally, phenotyping rodent barbering could lead to ethologically oriented experimental models of many prevalent human disorders, such as trichotillomania, OCD, and aggression (Garner *et al.* 2004b; Hill *et al.* 2007; Kalueff *et al.* 2006b; Kurien *et al.* 2005). Therefore, further in-depth ethological analyses are necessary to achieve a detailed understanding of the nature, etiology, and genetics of rodent barbering.

Given the considerable amount of time animals spend on grooming and barbering (Bolles 1960; Fentress 1988; Garner *et al.* 2004b; Sarna *et al.* 2000), these behaviors are a noteworthy subject of research in behavioral neuroscience. This chapter will provide updated information on the phenotyping and genetics of animal grooming and barbering behaviors, and how their analysis may foster further advances in translational biopsychiatry research.

Behavioral phenotyping

Animal grooming

Procedures

Coat-state assessment is a simple method to evaluate animal grooming activity (Piato *et al.* 2008; Yalcin *et al.* 2005, 2007), and may be performed in each individual rodent in eight separate body parts: head, neck, forepaws, dorsal coat, ventral coat, hind legs, tail, and genital region. For example, a score of 0 could be attributed to a coat in good form, and a score of 1 may be given to a dirty disheveled coat. The resulting score will represent the average (or the sum) of all body areas, and can also be compared across different experimental groups. Although this approach may lack some ethological sensitivity, poor coat state generally correlates with experimental depression. Indeed, chronically stressed "depressed" mice typically display poor coat status, whereas antidepressant treatments tend to reverse this phenotype (Piato *et al.* 2008; Yalcin *et al.* 2005, 2007). Therefore, coat-state assessment can be a useful tool in measuring animal brain pathology.

To induce acute stress-evoked grooming, researchers may use a brief mild stress, such as exposure to a novelty (Barros *et al.* 1994; Clement *et al.* 1994; Crusio *et al.* 1989; Crusio and van Abeelen 1987; Enginar *et al.* 2008; Kalueff and Tuohimaa 2004b). In addition, stronger stressors (e.g., a bright light, social aggression, a predator, or a predator's scent) will also generate stress-evoked grooming, which is highly relevant to emotionality and experimental modeling research.

While chronically applied mild stress may reduce animal grooming, stronger stressors (e.g., olfacto-bulbectomy or peripheral anosmia) produce pronounced activation of stereotypic grooming activity. This "pathological" grooming is generally focused on one specific area of the body (e.g., flanks), and is often accompanied by severe depression-like behaviors (such as anhedonia, hypoactivity, and aggression) (Kalueff *et al.* 2001; Makarchuk 1998, 1999; Makarchuk and Zyma 2002). In general, these observations seem to parallel clinical data showing overall increases in stereotypic behavior (grooming disorders, hair pulling) in depressed patients.

Unlike spontaneous stress-induced grooming, artificially induced grooming can be evoked by swimming or by smearing the animal with food (Audet *et al.* 2006; Burne *et al.* 2006). The splash test (in which a sucrose solution is squirted onto the dorsal region of the animal, and grooming is recorded for five minutes after solution vaporization) will also stimulate artificial grooming in rodents (Piato *et al.* 2008; Yalcin *et al.* 2005). Misting with water is another easy and reliable method to evoke artificial grooming behavior, and is widely used in neurobehavioral experiments (Audet *et al.* 2006; Hartley and Montgomery 2008). Since spontaneous and artificial grooming represent two different forms of this behavior, abnormalities in one type do not necessarily imply deficits in the other. Thus, a parallel assessment of stress-evoked and artificial grooming is necessary for an accurate characterization of animal behavioral phenotypes (Kalueff *et al.* 2005; Kalueff and Tuohimaa 2004b, 2005c).

In addition to these methods, a "smart battery" that combines several other behavioral tests may be used (Kalueff *et al.* 2008). For example, a five-minute open-field test (to assess baseline anxiety and spontaneous novelty-induced grooming) may be followed by the Porsolt's forced swim test to evaluate depression-related immobility or despair. In order to maximize the number of behavioral endpoints and domains per experiment, researchers may place animals into an observation cylinder (for five minutes) to investigate artificial, swim-induced grooming immediately following the forced swim test. Comparing the patterning and activity of the artificial post-swim grooming with the spontaneous pre-swim grooming may provide intriguing data regarding grooming phenotypes. In some instances, animals may also have a "fatigability" phenotype that should be discriminated from other grooming behaviors, as it will often be a confounding factor in such studies (Kalueff *et al.* 2008).

Behavioral analysis

Table 3.2 summarizes a systematic and high-throughput approach to analyzing mouse grooming activity and microstructure. To accurately evaluate grooming, researchers may develop a standardized scale to represent specific grooming activity and use it consistently within each laboratory. A typical scale may be as follows:

- no grooming (0)
- paw licking (1)
- nose, face, and head wash (2)
- body grooming, including body fur licking and scratching with hind paws (3)
- leg licking (4)
- tail or genital grooming (5).

However, researchers may modify this scale to suit their individual needs (e.g., by including additional strain-specific grooming behaviors of interest or by simplifying this scale for better detectibility).

A "correct" bout is cephalo caudal in direction and follows a (0–1) (1–2) (2–3) (3–4) (4–5) (5–0) pattern of transitions (Table 3.2). An "incorrect" bout can vary from the model in one of four ways:

- aborted or prematurely terminated bouts (2–0, 4–0)
- skipped transitions (1–3, 2–5)
- reversed bouts (4–3, 5–2)
- incorrectly initiated bouts (0–2, 0–5).

A "complete" bout consists of a strict (0–1–2–3–4–5–0) sequence and any other pattern is considered incomplete. Frequently, researchers will notice grooming interruptions. Any sequence that contains at least one interruption is deemed "interrupted." However, an interruption of six seconds or longer is judged to be an entirely separate bout.

Using this approach, researchers may assess the three primary ethological measures of grooming patterning: the percentage of incorrect transitions, interrupted bouts, and incomplete bouts. In addition, the duration of correct versus incorrect patterns, the number of interruptions during bouts, and the duration of complete versus incomplete bouts may be calculated. It is also useful to investigate the regional distribution of grooming patterning. For example, data may be collected based on five anatomic areas (forepaws, head, body, hind legs, and tail/genitals) or simply a rostral (forepaw and head) versus caudal (body, legs, and tail/genitals) distinction. Researchers may also classify each grooming bout as being directed to a single anatomic region or multiple regions, and calculate the percentage of grooming bouts and the percentage of time spent grooming for each category. Furthermore, the percentage of total grooming patterns, the percentage of time spent grooming, and the number of interruptions for each anatomic area may be assessed (Table 3.2).

Animal barbering

Procedures and behavioral analyses

Barbering behavior can often be observed in individual mice (such as C57BL/6 mice) that pluck whiskers from cage-mates, and can be broken into four distinct stages (see Sarna *et al.* 2000 for details):

- Hold: the barber presses down on the back and neck of its cage-mate
- Grasp: the barber grasps a single hair from the victim with its incisors
- Pluck: the barber pulls its head away from the victim, removing the hair from the root
- Manipulation: the barber often manipulates the removed hair with its paws, sometimes ingesting the hair.

The process for plucking fur has not been described in such detail, but is presumably very similar. Over time, barbers pluck hair from focused areas on the body of the recipients, leaving idiosyncratic patterns of alopecia (Garner *et al.* 2004a, b; Long 1972; Sarna *et al.* 2000; Figure 3.2). The skin of these regions is nonpruritic, since barbering per se does not involve tissue damage. Each barber typically plucks a similar, matching pattern from all accessible cage-mates, and this pattern is referred to as the barber's "cutting style" (Sarna *et al.* 2000). For example, one barber may pluck the whiskers, between the ears, and around the tail of its cage-mates, while another barber may pluck only a spot on the left flank of its cage-mates. Cutting styles also differ between strains, as some only pluck whiskers and from the face, while others pluck their idiosyncratic pattern from any area that is accessible (Figure 3.2; Garner *et al.* 2004b; Kalueff *et al.* 2006b; Sarna *et al.* 2000).

Table 3.4 summarizes some approaches to behavioral assessment of rodent barbering phenotypes. Patterns of hair loss can be drawn on a standardized mouse map. Cage-mates of individuals with no dorsal and ventral hair loss are classified as "non-barbers," whereas animals with ventral or low-forelimb hair loss can be classified as "self-barbers." Mice with cage-mates having similar patterns of alopecia on the face, whiskers, or dorsal surface only, and with the indicated mouse missing that pattern are classified as "cage-mate barbers," and those showing both cage-mate and self-barbering as "both barbers." Mice with any type of hair loss are categorized as "barbered," and mice with no hair loss as "intact" for each time-point.

On the mouse maps, both nonpruritic alopecia (hair loss without any redness, tissue damage, or scabbing) as well as pruritis must be recorded and differentiated. Typically, patterns of hair loss due to barbering have smooth and well-defined borders (Figure 3.2), and are distinct from hair loss caused by other factors. Presence

Table 3.3 *Examples of strain differences in mouse grooming behavior*

Measure	Strain ranking	References
Frequency	DBA/2J, F1 [C57BL/6J-DBA/2J] > C57BL/6J	(van Abeelen 1966)
	A/Ibg, BALB/cIbg > DBA2Ibg, C57BL/6Ibg	(Streng 1971)
	C57BL/6J > 129S1	(Kalueff and Tuohimaa 2004b)
Duration	DBA/2J > CPB-K-Nmg > C3H/St, C57BL/6J	(Crusio and van Abeelen 1986)
	C57BL/6J > FVB/N	(Mineur and Crusio 2002)
	C57BL/6J > 129S1	(Kalueff and Tuohimaa 2004b)
	BALB/c > 129S1, NMRI	(Kalueff and Tuohimaa 2004b)

or absence of pruritis can be recorded for each mouse on alopecia scoring days, and mice with skin/tissue damage consistent with *ulcerative dermatitis* must be categorized accordingly.

Genetics

Grooming behavior

Interesting data on the behavioral genetics of grooming is currently available in the literature (Tables 3.1 and 3.3). For example, increased grooming was found to be associated with the *pink-eyed dilution* (*p*) and *brown coat color* (*b*) loci on chromosomes 7 and 4, respectively (van Abeelen 1963a, b, c). The *p* locus is located close to a cluster of GABAergic genes, and because the GABAergic system regulates both grooming and emotional behaviors, it is possible that these genes play a role in grooming phenotypes. Indeed, as both *p* and *b* loci are associated with increased anxiety (Clement *et al.* 1994; Clement and Chapouthier 1998), they may modulate the interplay between grooming and anxiety at a genetic and behavioral level.

Several studies have examined strain differences in mouse grooming (Table 3.3). For example, when BALB/cIbg, C57BL/6Ibg, A/Ibg, and DBA/2Ibg were tested in the open field, their cumulative grooming scores showed a significant increase in time effect and time × strain effect (grooming increase over a test time, observed in all strains) (Streng 1971). Additionally, ABP/Le mice groomed significantly more in the open field than less anxious C57BL/6 mice, while F1 ABP/Le-C57BL/b mice groomed more than F1 C57BL/6-ABP/Le, indicating a possible maternal effect (Clement *et al.* 1994).

There were also strain differences between C57BL/6 and some 129 substrains. The 129SvEm and 129SvHsD mice showed lower light and dark grooming, but grooming scores rose for C57BL/6J mice during the dark phase (Rodgers *et al.*

Table 3.4 *Assessment of animal barbering phenotypes*

The following five-point scale can be used to assess barbering: 0 – no barbering; 1 – whisker removal or shortening; 2 – snout/face denuding; 3 – individual bald patches on head and body; 4 – multiple alopecic areas on head and/or body; 5 – severe alopecia including complete snout denuding and large alopecic areas on head and body. This scale may be modified if necessary (Kalueff *et al.* 2006b), depending on the requirements of the study, but must remain consistent within the laboratory.

Hair loss can be scored at baseline, and every two weeks thereafter through the completion of the experiment (Garner *et al.* 2004a, b). Mice can be inspected on both dorsal and ventral surfaces for hair loss. Within each pattern, the severity of hair loss can be recorded as follows: 0 – intact; 1 – slight; 2 – medium; 3 – heavy; 4 – completely nude.

Hair loss can be scored as barbering only if the hair lesion was nonpuritic, there was no scarring or scabbing around the lesion, and the animal was otherwise in good health and the fur (where present) was in good condition.

The following parameters of barbering can be assessed: the number (%) of cages in which the barbering occurred; the average severity of barbering in each cage; and the percentages of barbers and barbered animals (of total animals of each strain). Barber animals can be easily identified as the single intact mouse in the cage (see Garner *et al.* 2004b; Sarna *et al.* 2000 for details).

If necessary, self-barbering may be assessed in mice housed individually (to prevent hetero-barbering) for three to four weeks (see Kalueff *et al.* 2006b for details). Note, however, that such isolation stress may trigger animal anxiety that can further provoke stereotypic behaviors, including self-barbering.

Sometimes, excessive grooming in mice (e.g., Greer and Capecchi 2002) may lead to pronounced barbering-like alopecia (homecage observations may be needed in such cases, to distinguish between the two behaviors).

2002). In a similar study, FVB/N and C57BL/6J mice displayed comparable frequencies of grooming, but exhibited differences in duration, confirming that grooming frequency and duration may vary independently in different mouse strains (Mineur and Crusio 2002). Interestingly, grooming behavior in that study did not correlate with open-field horizontal and vertical activity (FVB/N > C57), suggesting that grooming represents a distinct dimension in the organization of rodent behavior.

Strain differences in grooming were also reported between NMRI, 129S1, and BALB/c mice (Kalueff and Tuohimaa 2005c). The NMRI mice displayed a clear tendency to earlier onset of grooming than 129S1 and BALB/c strains; however, there was no correlation between grooming activity and anxiety. Anxious strains display high (BALB/c) and low (129S1) grooming profiles, and nonanxious mice showed moderate to high (NMRI, C57BL/6) grooming profiles. Thus, overall grooming activity cannot accurately measure anxiety in mice. In contrast, studies investigating grooming microstructure did reveal significant differences between anxious and nonanxious mouse strains; anxious 129S1 mice displayed higher percentages of incorrect transitions and interrupted grooming bouts (Kalueff and Tuohimaa 2005c).

Grooming responses also vary across selectively bred mouse strains. For example, female Turku Aggressive mice spent considerably less time grooming during predatory aggression than the Turku Nonaggressive strain (Sandnabba 1995). In contrast, anxious high-thigmotaxis strain exhibited fewer grooming bouts in the open field than did less anxious, low-thigmotaxis strain (Leppanen and Ewalds-Kvist 2005; Leppanen *et al.* 2006). When these strains were cross-fostered, data revealed similar grooming activity in both the high- and low-thigmotaxis strains, indicating that both genetic and epigenetic factors influence mouse grooming.

Barbering behavior

There is limited data on the behavioral genetics of barbering, particularly on genetic mapping and strain differences. We have recently (Kalueff *et al.* 2006b, 2007b; Kalueff 2006, unpubl. data) assessed barbering in several strains and their F1 offspring, focusing on distinct domains of this behavior. These included social dominance barbering in same-sex cages (observed in C57BL/J6, A/J, 129S1, and NMRI but not BALB/c mice), barbering of males by females in breeding pairs (C57BL/J6, 129S1, and NMRI but not BALB/c mice), maternal barbering (removal of lactating dam's ventral fur by pups) (C57BL/6 and 129S1), and whisker barbering of pups by their mothers (129S1). Notably, the percentage of mice exhibiting barbered hair varies markedly from strain to strain. For example, BALB/c mice never exhibit barbering, while C57BL/6, A/J, A2G, and NMRI show frequent barbering behavior (Carruthers *et al.* 1998; Kalueff *et al.* 2006b; Sarna *et al.* 2000). Additionally, several studies have demonstrated that mice may have consistent individual (Sarna *et al.* 2000) or strain-specific (Kalueff *et al.* 2006b) "cutting styles"; see Figure 3.2 for more examples.

Rodent barbering has been shown to be associated with social dominance and low levels of aggression. For example, the strain ranking of barbering activity (NMRI, C57 > 129 >>> BALB/c) generally negatively correlated with that of aggressiveness (BALB/c >> 129, C57 >> NMRI), which suggests that barbering might emerge in rodents to minimize potential aggression (Kalueff *et al.* 2006b).

Further revealing the behavioral complexity and multifactorial nature of mouse barbering, four different outcomes have been observed following genetic and epigenetic barbering crosses. In the first case, one of the parental phenotypes can outcompete the other in a hybrid cross. For example, F1 hybrids derived from the BALB/c strain exhibit the low-barbering phenotype associated with that strain, regardless of maternal influence (Kalueff *et al.* 2007b). The second case results in a blending of the parental strain barbering phenotypes in the hybrid offspring that is independent of maternal influences, as in the different crosses (between 129S1 and C57 or NMRI) performed by Kalueff *et al.* (2006a, b, and unpubl. data). The third case results from cross-fostering experiments in which offspring take up the barbering phenotype of their foster parents. For example, nonbarbering strains raised with barbering foster parents may develop barbering behavior, although only in a small percentage of animals (Carruthers *et al.* 1998). Finally, the fourth outcome occurs when the genotype overcomes the maternal influence in cross-foster experiments, as in the same study by Carruthers *et al.* (1998) in which pups of barbering strains raised by nonbarbering foster parents continued to develop a barbering phenotype.

Assessment of the phenotypes of nonbarbering strains may be another useful approach to understanding the behavioral genetics of mouse barbering. For example, over 800 BALB/c mice did not show barbering activity (Carruthers *et al.* 1997, 1998; also see similar data in Kalueff *et al.* 2006b). Strain differences in sociability have recently been suggested as underlying factors in barbering phenotypes (Brodkin 2007). If confirmed, this interesting hypothesis may explain the low barbering activity in "autistic" mouse strains like BALB/c, as well as the high intensity of barbering in "sociable" strains, such as C57BL/6. Thus, barbering emerges as an important part of mouse social behavior, and strain differences may reflect (or underlie) different aspects and strategies of animal socialization. These variations in socialization, in turn, may confound all other behavioral domains, implying that in-depth analyses of strain barbering phenotypes may be even more significant than has been previously recognized.

Conclusion

Overall, in-depth phenotyping of animal grooming and barbering offers clear benefits for neurobehavioral research. Firstly, it allows assessment of these biologically important behaviors per se. Secondly, grooming and barbering activity may reflect strain differences in activity, anxiety, sociability, motor activity, and behavioral patterning, in addition to data from existing methods for phenotyping emotionality. Thirdly, given the sensitivity of rodent grooming and sequencing to various pharmacological and physiological manipulations, ethological analysis of grooming may be used in pharmacogenetics and neurophysiology – for example, in the dissection of brain substrates involved in behavior regulation. Fourthly, altered

Table 3.5 *Examples of barbering phenotypes in genetically modified mice (data obtained from Mouse Genome Informatics and PubMed)*

Model	Background strain	Barbering behavior	References
Phospholipase C beta1 knockout mice	F1 C57BL/6J(N8) × 129S4/SvJae(N8)	Lack of whisker trimming	(Koh *et al.* 2008)
Complexin II knockout mice	F1, F2 129Ola × C57BL6	Lack of whisker trimming	(Glynn *et al.* 2003)
Disheveled gene 1 (Dvl1) knockout mice	129S/SvEv	Lack of whisker trimming	(Lijam *et al.* 1997)
Transgenic mice overexpressing G protein-coupled receptor 85	C57BL/6	Reduced whisker trimming	(Matsumoto *et al.* 2008)
Vitamin D receptor knockout mice	129S1	Reduced whisker trimming and fur barbering	(Kalueff *et al.* 2006a)
Transcription factor USF1 knockout mice	C57BL/6	Increased whisker trimming	(Sirito *et al.* 1998)
Aromatase knockout mice	C57B6J × J129	Increased whisker trimming and fur barbering	(Hill *et al.* 2007)

grooming and barbering profiles may indicate behavioral perseverations, which may originate from an animal's natural displacement activity. Therefore, profiling both grooming and barbering phenotypes may allow researchers to indirectly assess potential strain differences in "compulsivity." Finally, comprehensive coverage of animal grooming and barbering peculiarities (Tables 3.1, 3.5) may assist researchers in correct data interpretation and in selecting appropriate animal models for their studies.

Acknowledgments

This research was supported by the NARSAD YI Award, LA BoR P-Fund, and Tulane University Provost's grants to AVK.

References

Aldridge JW, Berridge KC and Rosen AR (2004): Basal ganglia neural mechanisms of natural movement sequences. *Can J Physiol Pharmacol* **82**:732–9.

Audet MC, Goulet S and Dore FY (2006): Repeated subchronic exposure to phencyclidine elicits excessive atypical grooming in rats. *Behav Brain Res* **167**:103–10.

Barros HM, Tannhauser SL, Tannhauser MA and Tannhauser M (1994): The effects of GABAergic drugs on grooming behaviour in the open field. *Pharmacol Toxicol* **74**:339–44.

Berntson GG, Jang JF and Ronca AE (1988): Brainstem systems and grooming behaviors. *Ann N Y Acad Sci* **525**:350–62.

Berridge KC, Aldridge JW, Houchard KR and Zhuang X (2005): Sequential super-stereotypy of an instinctive fixed action pattern in hyper-dopaminergic mutant mice: a model of obsessive compulsive disorder and Tourette's. *BMC Biol* **3**:1–16.

Bertolini A, Poggioli R and Vergoni AV (1988): Cross-species comparison of the ACTH-induced behavioral syndrome. *Ann N Y Acad Sci* **525**:114–29.

Bolles RC (1960): Grooming behavior in the rat. *J Comp Physiol Psychol* **53**:306–10.

Brodkin ES (2007): BALB/c mice: low sociability and other phenotypes that may be relevant to autism. *Behav Brain Res* **176**:53–65.

Burne TH, Johnston AN, McGrath JJ and Mackay-Sim A (2006): Swimming behaviour and post-swimming activity in Vitamin D receptor knockout mice. *Brain Res Bull* **69**:74–8.

Campbell KM, de Lecea L, Severynse DM *et al.* (1999): OCD-Like behaviors caused by a neuropotentiating transgene targeted to cortical and limbic D1+ neurons. *J Neurosci* **19**:5044–53.

Carruthers EL, Halkin SL and King TR (1997): Are mouse "barbers" dominant to their cage mates? *Anim Behav Soc Abstr.*

Carruthers EL, Halkin SL and King TR (1998): Mouse barbering: investigations of genetic and experiential control. *Anim Behav Soc Abstr.*

Cheh MA, Millonig JH, Roselli LM *et al.* (2006): En2 knockout mice display neurobehavioral and neurochemical alterations relevant to autism spectrum disorder. *Brain Res* **1116**:166–76.

Chen ZF, Rebelo S, White F *et al.* (2001): The paired homeodomain protein DRG11 is required for the projection of cutaneous sensory afferent fibers to the dorsal spinal cord. *Neuron* **31**:59–73.

Choleris E, Thomas AW, Kavaliers M and Prato FS (2001): A detailed ethological analysis of the mouse open field test: effects of diazepam, chlordiazepoxide and an extremely low frequency pulsed magnetic field. *Neurosci Biobehav Rev* **25**: 235–60.

Clement Y and Chapouthier G (1998): Biological bases of anxiety. *Neurosci Biobehav Rev* **22**:623–33.

Clement Y, Adelbrecht C, Martin B and Chapouthier G (1994): Association of autosomal loci with the grooming activity in mice observed in open-field. *Life Sci* **55**: 1725–34.

Crusio WE and van Abeelen JH (1986): The genetic architecture of behavioural responses to novelty in mice. *Heredity* **56**(Pt 1):55–63.

Crusio WE and van Abeelen JH (1987): Zinc-induced peripheral anosmia and behavioral responses to novelty in mice: a quantitative-genetic analysis. *Behav Neural Biol* **48**:63–82.

Crusio WE Schwegler H, Brust I and Van Abeelen JH (1989): Genetic selection for novelty-induced rearing behavior in mice produces changes in hippocampal mossy fiber distributions. *J Neurogenet* **5**:87–93.

De Luca AM (1997): Environmental enrichment: does it reduce barbering in mice? *AWIC Newsletter* **8**:7–8.

DeVries AC, Young WS, 3rd and Nelson RJ (1997): Reduced aggressive behaviour in mice with targeted disruption of the oxytocin gene. *J Neuroendocrinol* **9**:363–8.

Dunn AJ (1988): Studies on the neurochemical mechanisms and significance of ACTH-induced grooming. *Ann N Y Acad Sci* **525**:150–68.

Dunn AJ, Berridge CW, Lai YI and Yachabach TL (1987): CRF-induced excessive grooming behavior in rats and mice. *Peptides* **8**:841–4.

Enginar N, Hatipoglu I and Firtina M (2008): Evaluation of the acute effects of amitriptyline and fluoxetine on anxiety using grooming analysis algorithm in rats. *Pharmacol Biochem Behav* **89**:450–5.

Fentress JC (1988): Expressive contexts, fine structure, and central mediation of rodent grooming. *Ann N Y Acad Sci* **525**:18–26.

Garner JP, Dufour B, Gregg LE, Weisker SM and Mench JA (2004a): Social and husbandry factors affecting the prevalence and severity of barbering ("whisker-trimming") by laboratory mice. *Appl Anim Lab Sci* **89**:263–82.

Garner JP, Weisker SM, Dufour B and Mench JA (2004b): Barbering (fur and whisker trimming) by laboratory mice as a model of human trichotillomania and obsessive–compulsive spectrum disorders. *Comp Med* **54**:216–24.

Glynn D, Bortnick RA and Morton AJ (2003): Complexin II is essential for normal neurological function in mice. *Hum Mol Genet* **12**:2431–48.

Greer JM and Capecchi MR (2002): Hoxb8 is required for normal grooming behavior in mice. *Neuron* **33**:23–34.

Hartley JE and Montgomery AM (2008): 8-OH-DPAT inhibits both prandial and waterspray-induced grooming. *J Psychopharmacol* **22**:746–52.

Hill RA, McInnes KJ, Gong EC *et al.* (2007): Estrogen deficient male mice develop compulsive behavior. *Biol Psychiatry* **61**:359–66.

Huang AS, Beigneux A, Weil ZM *et al.* (2006): D-aspartate regulates melanocortin formation and function: behavioral alterations in D-aspartate oxidase-deficient mice. *J Neurosci* **26**:2814–19.

Hyman SE (2007): Neuroscience: obsessed with grooming. *Nature* **448**:871–2.

Kalueff AV and Tuohimaa P (2004a): Grooming analysis algorithm for neurobehavioural stress research. *Brain Res Brain Res Protoc* **13**:151–8.

Kalueff AV and Tuohimaa P (2004b): Contrasting grooming phenotypes in C57Bl/6 and 129S1/SvImJ mice. *Brain Res* **1028**:75–82.

Kalueff AV and Tuohimaa P (2005a): The grooming analysis algorithm discriminates between different levels of anxiety in rats: potential utility for neurobehavioural stress research. *J Neurosci Methods* **143**:169–77.

Kalueff AV and Tuohimaa P (2005b): Mouse grooming microstructure is a reliable anxiety marker bidirectionally sensitive to GABAergic drugs. *Eur J Pharmacol* **508**:147–53.

Kalueff AV and Tuohimaa P (2005c): Contrasting grooming phenotypes in three mouse strains markedly different in anxiety and activity (129S1, BALB/c and NMRI). *Behav Brain Res* **160**:1–10.

Kalueff AV, Maisky VA, Pilyavskii AI and Makarchuk NE (2001): Persistent c-fos expression and NADPH-d reactivity in the medulla and the lumbar spinal cord in rat with short-term peripheral anosmia. *Neurosci Lett* **301**:131–4.

Kalueff AV, Lou YR, Laaksi I and Tuohimaa P (2005): Abnormal behavioral organization of grooming in mice lacking the vitamin D receptor gene. *J Neurogenet* **19**:1–24.

Kalueff AV, Keisala T, Minasyan A *et al.* (2006a): Behavioural anomalies in mice evoked by "Tokyo" disruption of the Vitamin D receptor gene. *Neurosci Res* **54**:254–60.

Kalueff AV, Minasyan A, Keisala T, Shah ZH and Tuohimaa P (2006b): Hair barbering in mice: implications for neurobehavioural research. *Behav Processes* **71**:8–15.

Kalueff AV, Aldridge JW, LaPorte JL, Murphy DL and Tuohimaa P (2007a): Analyzing grooming microstructure in neurobehavioral experiments. *Nat Protoc* **2**:2538–44.

Kalueff AV, Keisala T, Minasyan A and Tuohimaa P (2007b): Influence of paternal genotypes on F1 behaviors: lessons from several mouse strains. *Behav Brain Res* **177**:45–50.

Kalueff AV, Laporte JL, Murphy DL and Sufka K (2008): Hybridizing behavioral models: a possible solution to some problems in neurophenotyping research? *Prog Neuropsychopharmacol Biol Psychiatry* **32**:1172–8.

Koh HY, Kim D, Lee J, Lee S and Shin HS (2008): Deficits in social behavior and sensorimotor gating in mice lacking phospholipase Cbeta1. *Genes Brain Behav* **7**:120–8.

Kruk MR, Westphal KG, Van Erp AM *et al.* (1998): The hypothalamus: cross-roads of endocrine and behavioural regulation in grooming and aggression. *Neurosci Biobehav Rev* **23**:163–77.

Kurien BT, Gross T and Scofield RH (2005): Barbering in mice: a model for trichotillomania. *BMJ* **331**:1503–5.

Lahoud MH, Ristevski S, Venter DJ *et al.* (2001): Gene targeting of Desrt, a novel ARID class DNA-binding protein, causes growth retardation and abnormal development of reproductive organs. *Genome Res* **11**:1327–34.

Leppanen PK and Ewalds-Kvist SB (2005): Crossfostering in mice selectively bred for high and low levels of open-field thigmotaxis. *Scand J Psychol* **46**:21–9.

Leppanen PK, Ravaja N and Ewalds-Kvist SB (2006): Twenty-three generations of mice bidirectionally selected for open-field thigmotaxis: selection response and repeated exposure to the open field. *Behav Processes* **72**:23–31.

Lijam N, Paylor R, McDonald MP *et al.* (1997): Social interaction and sensorimotor gating abnormalities in mice lacking Dvl1. *Cell* **90**:895–905.

Long SY (1972): Hair-nibbling and whisker-trimming as indicators of social hierarchy in mice. *Anim Behav* **20**:10–12.

Makarchuk M (1999): [An electrophysiological evaluation of the role of the olfactory analyzer in brain integrative activity]. *Fiziol Zh* **45**:77–83.

Makarchuk M and Zyma IH (2002): [Effect of anosmia on sex-related differences in conditioned avoidance in rats]. *Fiziol Zh* **48**:9–15.

Makarchuk NE (1998): [The effect of anosmia on sex dimorphism in the patterns of orienting-exploratory, emotional and passive defensive behaviors in rats]. *Zh Vyssh Nerv Deiat Im I P Pavlova* **48**:997–1003.

Matsumoto M, Straub RE, Marenco S *et al.* (2008): The evolutionarily conserved G protein-coupled receptor SREB2/GPR85 influences brain size, behavior, and vulnerability to schizophrenia. *Proc Natl Acad Sci USA* **105**:6133–8.

Mineur YS and Crusio WE (2002): Behavioral and neuroanatomical characterization of FVB/N inbred mice. *Brain Res Bull* **57**:41–7.

Navarro M, Rubio P and de Fonseca FR (1995): Behavioural consequences of maternal exposure to natural cannabinoids in rats. *Psychopharmacology (Berl)* **122**:1–14.

Piato AL, Detanico BC, Jesus JF *et al.* (2008): Effects of Marapuama in the chronic mild stress model: further indication of antidepressant properties. *J Ethnopharmacol* **118**:300–4.

Rodgers RJ, Boullier E, Chatzimichalaki P, Cooper GD and Shorten A (2002): Contrasting phenotypes of C57BL/6JOlaHsd, 129S2/SvHsd and 129/SvEv mice in two exploration-based tests of anxiety-related behaviour. *Physiol Behav* **77**:301–10.

Roeling TA, Veening JG, Peters JP, Vermelis ME and Nieuwenhuys R (1993): Efferent connections of the hypothalamic "grooming area" in the rat. *Neuroscience* **56**:199–225.

Ross SA, Wong JY, Clifford JJ, *et al.* (2000): Phenotypic characterization of an alpha 4 neuronal nicotinic acetylcholine receptor subunit knock-out mouse. *J Neurosci* **20**:6431–41.

Sachs BD (1988): The development of grooming and its expression in adult animals. *Ann N Y Acad Sci* **525**:1–17.

Sandnabba NK (1995): Predatory aggression in male mice selectively bred for isolation-induced intermale aggression. *Behav Genet* **25**:361–6.

Sarna JR, Dyck RH and Whishaw IQ (2000): The Dalila effect: C57BL6 mice barber whiskers by plucking. *Behav Brain Res* **108**:39–45.

Sirito M, Lin Q, Deng JM, Behringer RR and Sawadogo M (1998): Overlapping roles and asymmetrical cross-regulation of the USF proteins in mice. *Proc Natl Acad Sci USA* **95**:3758–63.

Spruijt BM, van Hooff JA and Gispen WH (1992): Ethology and neurobiology of grooming behavior. *Physiol Rev* **72**:825–52.

Streng J (1971): Open-field behavior in four inbred mouse strains. *Can J Psychol* **25**:62–8.

van Abeelen JH (1963a): Mouse mutants studied by means of ethological methods. I. Ethogram. *Genetica* **34**:79–94.

van Abeelen JH (1963b): Mouse mutants studied by means of ethological methods. II. Mutants and methods. *Genetica* **34**:95–101.

van Abeelen JH (1963c): Mouse mutants studied by means of ethological methods. III. Results with yellow, pink-eyed dilution, brown and jerker. *Genetica* **34**:270–86.

van Abeelen JH (1966): Effects of genotype on mouse behaviour. *Anim Behav* **14**: 218–25.

Welch JM, Lu J, Rodriguiz RM *et al.* (2007): Cortico-striatal synaptic defects and OCD-like behaviours in Sapap3-mutant mice. *Nature* **448**:894–900.

Yalcin I, Aksu F and Belzung C (2005): Effects of desipramine and tramadol in a chronic mild stress model in mice are altered by yohimbine but not by pindolol. *Eur J Pharmacol* **514**:165–74.

Yalcin I, Aksu F, Bodard S, Chalon S and Belzung C (2007): Antidepressant-like effect of tramadol in the unpredictable chronic mild stress procedure: possible involvement of the noradrenergic system. *Behav Pharmacol* **18**:623–31.

4

Social play, social grooming, and the regulation of social relationships

SERGIO M. PELLIS AND VIVIEN C. PELLIS

Summary

Social grooming and rough-and-tumble play, along with caressing and hand-shaking, have something important in common, touching. Physical contact with another can be an essential ingredient of social communication – gentle touching can place the other animal at ease, whereas rough contact can do the opposite. Although the underlying neurobiology is still to be fully mapped, it does appear that there is a common set of neurochemical pathways that regulate these touch-induced changes in mood across mammals. Given its potential value in the manipulation of the affective state of social partners, it should not be surprising that touch is an important component of communication. A close analysis of the comparative and neurobiological literature on rough-and-tumble play, or play fighting, suggests that there are two levels of control over this touch-based communication. Firstly, there is the subcortically regulated emotional state of the interactants. Secondly, there is the cortically mediated modulation of the touching behavior that allows animals to use touch in a more strategic manner. How these two levels interact and what social conditions foster the need for additional cortical control over touch remains to be determined.

Introduction

A hostile donkey is rendered peaceful by the human object of its ire vigorously rubbing the base of its tail (Ewer 1967), an anxious monkey is calmed

Neurobiology of Grooming Behavior, eds. Allan V. Kalueff, Justin L. LaPorte, and Carisa L. Bergner. Published by Cambridge University Press.

down after being groomed by a social partner (Goosen 1981), and agitated rats relax after social play (Arelis 2006; Darwish *et al.* 2001). What do all these situations have in common? In each case, some form of tactile contact provides an avenue by which to change the animal's mood. It has the effect of calming and reducing stress and has been shown in many species including rats, monkeys, and people (Field 2001; McGlone *et al.* 2007; Panksepp 1998; Schino *et al.* 1988). Indeed, studies have shown that touch can facilitate cooperative behavior among people, even among those who are strangers (e.g., Bohm and Hendricks 1997; Crusco and Wetzel 1984; Fisher *et al.* 1976; Kurzban 2001). These effects appear to be mediated by the release of a cascade of neurochemicals, including neurotransmitters such as dopamine and serotonin, peptides such as oxytocin and vasopressin, and the endogenous opioids, that are associated with the brain's reward and pleasure centers (Carter and Kervene 2002; Panksepp 1998). For instance, increased positive affective states that lead to a greater sense of trust in other people have been linked to the release of oxytocin (Zak *et al.* 2005).

There are conflicting views as to which of these neurochemical systems may be most critical, or whether these systems serve different roles in the process (Dunbar 2010). Furthermore, although all mammals studied appear to use these various neurochemical systems in processes related to social bonding and social interactions, there are differences across lineages, such as an apparent increase in the importance of the role of endogenous opioids in primates (Curley and Kervene 2005). Irrespective of the exact mechanisms involved, what seems to be clear is that social touch can tap into a system that induces a positive mood.

Various species of mammals have been documented to use this capacity to modulate the emotional state of others through contact and so manipulate social situations to their best advantage. For example, chimpanzees and bonobos engage in a combination of social grooming and social play before a feeding session; those that interact the most are also the ones most likely to tolerate each other's proximity to the food (Palagi *et al.* 2004, 2006). Similarly, bonobos use sex to quell conflict (de Waal 1995) and monkeys use social grooming to decrease aggression and increase tolerance (Gumert and Ho 2008; Sparks 1967).

In a sense, animals use friendly patterns of social contact to manage each other's mood. Such affect-induction may be a property of communication in general (Owren and Rendall 1997) – if so, we should be alerted to the possible negative impact of such contact. For example, social grooming in groups of rats involves gently nibbling and licking a partner's fur, and this contact is directed over much of its body (Pellis and Pellis 1997a). This appears to induce a positive affective state. The recipient adopts a relaxed body tone, as is evidenced by its droopy eyelids. In contrast, when encountering an intruding male, a dominant resident may direct rough grooming using sharp nips to the nape of the intruder's neck (Barnett 1975). The recipient is anything but relaxed, in that its body tone is rigidly maintained, and each sharp nip is accompanied by a body jerk. In many species of primates, social play between adults promotes further friendly contact, such as huddling, grooming, or copulation (Pellis and Iwaniuk 1999, 2000), but for some, like the black howler monkey, dominant troop members initiate rough play with subordinates, seemingly to intimidate ("punish") them (Jones 1983). Clearly, the latter produces a different affective state to the former.

Tactile social contact then, is part of the communicatory tool kit available to many species, and is used to manipulate the affective state of partners and so gain an advantage over them. In some cases, the advantage could be mutual, such as when two bonobos share some food, or it could be unidirectional, such as when a dominant howler monkey punishes a subordinate. At a broad level, it would appear that there are common neural circuits that mediate these tactile-induced changes of affect (Lim and Young 2006; Panksepp 1998); however, with our current state of knowledge, it is unclear whether different ways of gaining access to this circuitry, such as by grooming or play, involve unique components of the system, or how these underlying neural mechanisms are differentially activated depending on whether a negative or positive affective state is induced. In this chapter, we will assume that such a common system for mediating touch, with its multifaceted combinations, exists, and instead, explore two questions related to the origins and functions of this system in the behavioral domain. Firstly, do play and grooming have access to this system independently, or are they themselves linked, at least with regard to their origins? Secondly, given that our personal knowledge of social play is greater than our knowledge of social grooming, we will use social play, particularly that of rats, to explore how the manipulative role of tactile communication can become increasingly sophisticated. After all, any attempt of manipulation by one animal is likely to be countered by the other, leading to an arms race in neural mechanisms of ever-growing complexity.

To groom or to play: is a choice necessary?

As already noted for bonobos and for rats, some species can obviously use both social play and social grooming in their relationships. It is possible that social play originates from social grooming, or, alternatively, that the two are both facets of the same coin – both are attempts to alter a partner's affective state, with situational factors dictating which is the most appropriate. The evidence, although limited, suggests that there are multiple routes to access the neural mechanisms that influence tactile-based affect and that different lifestyles and modes of communication provide the context for which combination is the most likely to be used by a particular species. Prosimian primates provide some clues to how these mechanisms may have changed over evolutionary time.

The prosimian primates are marked by their longish snouts and "wet-dog" noses – the big-eyed tarsiers, although grouped with the prosimians, have short snouts and monkey-like noses, and so are more properly aligned with the anthropoid primates (i.e., the human-like primates). Prosimians are thought to resemble more closely the ancestral primate state. The social systems of the nocturnal members of this group are also thought to be more like those present in ancestral primates. There are several families of nocturnal prosimians: two belong to the superfamily Lorisoidea and the others, to the superfamily Lemuroidea (Fleaggle 1988). Each family has members that exhibit high frequencies of adult play intermixed with social grooming, but, importantly, in different ways. In the family Lorisidae, males and females have separate, individual territories that are adjacent to one another, and territory holders are antagonistic to the intrusions of others. The males do, however, visit the females in the adjacent territories (Charles-Dominique and Bearder 1979). Observations of the potto, an African member of this group, both in the wild (Charles-Dominique 1977) and in captivity (Epps 1974), portray the pattern of these male–female encounters. A male will intrude into the territory of a neighboring female each night. As she becomes accustomed to his presence, she allows him to groom her; gradually, over succeeding nights, this routine becomes more prolonged and reciprocal. Eventually, grooming leads to grappling, in which both animals, typically hanging upside down from branches, grab and pull at each another. After more days of this grooming and grappling, copulation may take place. This prolonged sequence of courtship appears to function as a means of overcoming the antagonism that individuals exhibit towards intruders. During social grooming, the animals will deposit a drop of their urine on their palm and then will rub this into the area of their partner's body that is being groomed. This scent sharing, which probably also takes place during grappling, must increase the pair's familiarity with one another, and so reduce the likelihood that aggression is elicited. As Epps (1974) notes, "[g]rappling must cement the pair-bond, since all aggressive responses must be suppressed in what is, in essence, a play fight" (p. 242). Thus, grappling, a form of play fighting, is part of a complex courtship sequence. Observations on captive groups of another lorisid, the slow loris, has shown that adult–adult play fighting interactions are also frequent, and, indeed, that even when a juvenile is present, more play occurs between the males and females than with the juvenile (Ehrlich 1977; Ehrlich and Musicant 1975).

The adults of Coquerel's mouse lemur – which belongs to the family Cheirogaleidae, from the superfamily Lemuroidea – have an even more elaborate pattern of playful contact between males and females (Pagés 1978). In this species, the adults have partially overlapping home ranges, and male–female pairs spend a lot of time in these common zones. Every few days, the pair meet and engage in prolonged social contact; this involves resting side by side, mutual grooming, and play fighting in which the pair, entwined, hang suspended from a branch. Such tactile contact is maintained all year round, but occurs at its highest frequency in the breeding season, peaking just before copulation and then declining following copulation. Disturbance of this developing pair bond by other animals can disrupt mating, or even if mating occurs, a viable fetus may not be brought to fruition. A pair bond based on repeated tactile contact appears necessary for successful reproduction to take place. The social structure of this species seems to be intermediate between more solitary and more gregarious prosimians (Charles-Dominque 1978). Social association involving social grooming and other bodily contact has also been reported for other species of cheirogaleids (Hladick and Charles-Dominique 1974; Russell 1975).

The third group, also from the Lorisoidea, is the Galagidae, comprising species where the natural social unit is the family, incorporating an adult male, one or more adult females, and their offspring (Charles-Dominique 1978). These family groups congregate in communal sleeping sites during the day, and then move off, individually, to forage at night. When congregating together at the sleeping sites, the adults may engage in huddling, social grooming, and play fighting. The latter has the appearance of low-intensity aggression: they grab and pull at each other and roll around on the ground, but do not bite one another (Doyle 1974). These social interactions are dissociated from sex, with courtship and mating being relatively simple in its form. Typically, the male follows the female, sniffs and licks her anogenital area, whereupon she may run forward and then stop. At this point, he will approach her again, then sniff and mount her (Blackwell 1969; Doyle *et al.* 1967). Grooming and play fighting are not necessary precursors of mating, although a continued linkage between grooming and play is suggested by the high frequency with which play is either preceded or proceeded by social grooming (Doyle 1974). In captive groups of galagos, play fighting is relatively infrequent between adult males and females, and most play by adults involves juveniles (Doyle 1974; Ehrlich 1977; Roberts, 1971; Rosenson, 1973).

From these descriptions, two general patterns emerge. Firstly, for nocturnal solitary animals, tactile contact between the sexes, which may include play, is an essential feature of courtship. Secondly, courtship behavior of nocturnal prosimians may involve surveillance and contact with the females all year long, and so is divorced from sexual motivation. Indeed, such "courtship" mostly occurs when the male's testes are regressed. This seems true not only for the lorisoids, but also for species of the Lepilemuridae, a family of the superfamily Lemuroidea that is nocturnal and has a more solitary lifestyle compared to other lemuroids (Hladick and Charles-Dominique 1974). In particular, play fighting seems to be the most sexually emancipated form of contact in the galago, in that it no longer occurs as a precursor to mating: instead, "social play is a mechanism to enhance intra-group harmony and it may aid friendly contact between compatible strangers" (Charles-Dominique and Bearder 1979). We suggest that for some nocturnal prosimians, such as the lorisids and cheirogaleids, play fighting is intimately related to mating and intra-group (pair) friendship maintenance, but has become, in others, such as the galagids, increasingly emancipated from sexual motivation. In a study of captive groups of the greater galago and the slow loris, play fighting was more frequent in the loris than in the galago (Ehrlich 1977; Ehrlich and Musicant 1975). Relevant to the view that play fighting is more emancipated from sex in galagids than lorisids, is the finding that when the lorises were confronted with strangers, it led to avoidance, but if contact were made, they would engage in serious fighting. In galagos, in the same circumstances, however, positively affective social contact was frequent, with much of that contact involving play fighting (Newell 1971). For galagos, play fighting, even among adults, appears to have become a general form of nonagonistic social contact.

Note that when these different groups of nocturnal prosimians are considered, play fighting seems to be derived from competitive social grooming (pottos), and even when structurally distinct, high rates of one are associated with high rates of the other (galagos). Furthermore, although both forms of interaction appear, initially, to be closely linked to sex, they then seem to become increasingly emancipated from sexual behavior and more broadly used for social communication and manipulation. Indeed, in primates more generally, those with infrequent male–female association are more likely to use forms of playful contact as part of their courtship behavior (Pellis and Iwaniuk 1999). The same seems to apply to the use of social play in nonsexual contexts – friendship maintenance, dominance testing, punishment, etc. – species with more infrequent or unpredictable contact use playful contact more often (Pellis and Iwaniuk 2000).

The presence of signals that can be used at a distance, such as visual and auditory ones, mitigate the need for tactile forms of communication. Lineages of primates with large repertoires of visual and auditory signals are less likely to use tactile forms than are lineages with less well developed distance signals – Old World Monkeys use playful fighting the least and prosimians the most, with New World Monkeys and apes in between. In other lineages of mammals, there appear to be similar trade-offs between distance communication and touch-based communication. For example, species of dolphins that are more reliant on auditory signals use tactile communication less than those less reliant on the auditory channel (Paulos *et al.* 2008). In primates, for those using tactile forms of communication, there may be an additional trade-off: in those species that emphasize grooming, such as Old World Monkeys (Sussman and Garber 2004), the use of play is diminished (Pellis and Iwaniuk 2000). Nonetheless, as shown by the close inspection of the play and grooming of prosimians, it does appear that in primates the origin of these forms of tactile contact are intricately intermeshed. However, for other lineages, this may not be the case.

For rats and other murid rodents (i.e., the mouse-like rodents), play fighting is derived from courtship behavior. During play, the partners compete for access to the body targets contacted by the male during the precopulatory phase of sexual encounters (Pellis 1993; Pellis and Iwaniuk 2004; Pellis and Pellis 1998a). Rats compete for access to the nape of the neck, which is gently nuzzled if contacted (Pellis and Pellis 1987; Siviy and Panksepp 1987), as is the case for voles (Pellis *et al.* 1989). Syrian golden hamsters gently nibble the cheeks (Pellis and Pellis 1988a, b), grasshopper mice nuzzle the shoulders (Pellis *et al.* 2000), and Djungarian hamsters lick the mouth (Pellis and Pellis 1989). In all these species, social grooming involves more generalized bodily contact. A detailed developmental analysis has shown that the nape-directed nuzzling and the body-directed grooming of rats are distinct from their earliest appearance in the days preceding weaning (Pellis and Pellis 1997a). Therefore, it would seem that the origins of social grooming and social play are quite distinct in these rodents, although in some other rodents, such as marmots (e.g., Barash 1973, 1974), the two may be interlinked.

From a functional perspective, in a rat colony, playful fighting appears to be an active form of friendship maintenance (Pellis and Pellis 1991, 1992a; Pellis *et al.* 1993) and the means by which to test for dominance when a male encounters an intruder (Smith *et al.* 1999). Social grooming, in its gentle form, is also used for within colony interactions, and in its rough form (such as nipping the nape), for dealing with intruders (Barnett 1975). These uses of grooming, however, are temporally and contextually disassociated with playful fighting (Pellis and Pellis 1987, 1997a).

Much work remains to be done to trace the evolutionary history of social grooming and playful fighting across mammals, but from the examples for which we have sufficient data, it does appear that the two can either share a common origin or evolve independently of each other. Whatever the route of their emergence, both are capable of tapping into the same brain mechanisms that involve touch and can alter the affective state of a partner. Our own work on play fighting in the rat illustrates how such a use of tactile communication can involve novel and sophisticated mechanisms.

Playful manipulation

Many species of mammals are capable of using play fighting as a means of assessing and manipulating conspecifics (Pellis 2002). This use of play fighting may be relatively frequent, as in rats (Pellis *et al.* 1993) and spider monkeys (Pellis and Pellis 1997b), or relatively infrequent, as in rhesus monkeys (Brueggeman 1978). There are several contexts for which such play has been identified – for sexual relations, pair formation and courtship, nonsexual social contexts, within group friendship maintenance, within group dominance testing, and for between group dominance testing (Pellis 2002; Pellis and Iwaniuk 1999, 2000). As has already been pointed out, one reason that playful fighting can be used in these contexts is because the tactile contact involved can generate positive or negative affective states. However, there is also another reason: play fighting is inherently "fair." By fairness, we mean that when two animals are engaged in play fighting, neither partner pursues its advantage to the point where it is the only one winning all the play fights. Theoretical models show that the further from parity the win–loss ratio migrates, the greater the risk that play fighting will either cease or escalate to serious fighting (Dugatkin and Bekoff 2003). The little empirical data available suggest that the escalation of play fights to serious fighting is most likely to occur when one of the partners breaks the play fighting rules (Pellis and Pellis 1998b). Therefore, like social grooming, there is a degree of reciprocity involved. It remains to be determined if deviations away from reciprocity are aversive because of their lack of fairness (Bekoff 2001); *this would require a cognitive assessment by the animals.* Alternatively, it could be that these deviations add to the negative affect induced by the partner's behavior; *this would require an emotional assessment of the tactile stimuli.* Whatever the case, altering the fairness of play fighting is a means by which to assess and manipulate a social partner. To understand how it does so, requires a closer examination of the behavioral means by which fairness is achieved in play fighting.

For play fighting to remain play fighting, animals have to refrain from taking advantage of the situation. Since play fighting is a competitive activity that requires one to attempt to gain some advantage over another – such as when rats nuzzle the nape of a partner's neck – this requires striking a fine balance. The trick is to be sufficiently competitive so as to have a good chance of gaining the advantage, but not so much so as to capitalize on that advantage once gained. Although such a statement encapsulates the meta-rule for fairness (see above), there do appear to be many different ways in which this overall fairness can be maintained (Pellis *et al.* 2010). To appreciate the situation in play fighting, it may be useful to contrast it with the situation in serious fighting. In serious fighting, the aim is to bite, strike, or unbalance the opponent without receiving a retaliatory strike or bite (Geist 1978). To achieve this, an attacker will incorporate a defensive maneuver into its attack, as a way of mitigating the chance of a retaliatory strike from their opponent.

Among murid rodents, the most common target of attack is the opponent's rump and lower flanks. The defending animal can retaliate by biting the side of its attacker's face (Pellis 1997). For territorial animals such as rats, placing an unfamiliar male into the cage of another male can lead to the resident doing all the attacking and the intruder doing all the defending; such a situation makes it easier for the researcher to evaluate the tactics of both attack and defense (Blanchard and Blanchard 1990a). Because of its limited opportunity to flee, the intruder will adopt an upright posture and rotate so as to track the movements of the resident. In turn, the resident will maneuver to gain access to the intruder's flanks and rump. It will typically achieve this by adopting a lateral orientation to the intruder (Blanchard *et al.* 1977). As the resident lunges at the intruder's flank, it also runs the risk that the intruder will lunge and bite the side of its face, a risk that highlights the value of the lateral orientation.

As the resident approaches in the lateral orientation, it presses its flank against the intruder's exposed underbelly; if this pushing succeeds in unbalancing the intruder, there is but a moment of opportunity in which the resident can deliver a bite to the exposed flank (Pellis and Pellis 1987). However, if the intruder manages to regain its composure in time for it to lunge at the resident's head, the lateral orientation provides it with a vantage point from which to swerve laterally away, by pivoting on its hind feet (Pellis and Pellis 1992b). That is, the lateral orientation is an attack tactic with a defensive component, in that it enables the resident to press its attack, but also to mitigate the retaliation if one is forthcoming. Sometimes, the defensive component of an attack maneuver is even more strikingly evident. In one case, a pair of Syrian golden hamsters was in the upright *versus* lateral orientation. Just preceding a lunge at the intruder's flank, the resident thrust its hand into the opponent's face, blocking the intruder's capacity to launch a retaliatory bite (Pellis and Pellis 1998b). The situation is very different when rats engage in play fighting.

During play fighting, rats compete for access to the nape of the neck, which if contacted is gently nuzzled (Pellis and Pellis 1987; Siviy and Panksepp 1987). While the animals seem to compete actively for access to the nape, the tactics of attack and defense are not linked as they are during serious fighting. For instance, one of the most common defensive actions by the rat receiving an attack to the nape is for it to rotate to fully supine. While this maneuver can enable the defender to succeed in extricating its nape from the attacker, it also leaves it on its back with a reduced capacity to counterattack or to escape, especially if the attacker stands over it. The animal standing on top, however, may then behave in what seems to be a very peculiar manner. The obvious advantage of being on top is that by using its forepaws, the movements of its supine partner can be restrained and its counterattacks blocked. In order to use its forepaws effectively, and to make any necessary accompanying shifts of body weight with its upper trunk, the rat stands with its hind feet on the ground; this provides it with a solid base of support for its upper body movements. In this anchored position, the on-top animal has considerable stability and so can maintain a high degree of control over both its own movements and those of its supine partner. However, rats will, at times, stand on their supine partners with all four of their limbs. When they do this, they have less control over their own movements, and, certainly, a reduced ability to restrain their partner's movements (Foroud and Pellis 2003). Indeed, a supine rat that has an unanchored partner standing on top of it has a 70% chance of launching a successful counterattack, compared to only 30% when its partner is anchored (Pellis *et al.* 2005). Furthermore, in the juvenile period, when play fighting is most common, the rats increase their use of the unanchored on-top position (Foroud and Pellis 2002). Thus, during play fighting, rats engage in actions that can relinquish their advantage once they have gained it, or, to state it in more formal terms, during play fighting, rats do not incorporate defensive components in their attacks (Pellis and Pellis 1998b). Detailed studies of other species, such as spider monkeys, similarly reveal that during play fighting preparation to curtail retaliation is diminished (Pellis and Pellis 1997b). That is, in these species, attack and defense are coupled differently in play fighting as compared to serious fighting, in that in play fighting, it is easier for the partner to retaliate. Thus, in play fighting, unlike in serious fighting, reciprocity is built into the very organization of attack and defense.

Other species during play fighting seem to use different ways of modifying their competitiveness so as to retain the fairness that enables this behavior to remain playful. For example, in the degu, a South American rodent related to the chinchilla, defense appears to be integrated into their attacks in both playful and serious fighting. Once it has gained an advantage, an attacking degu solves the reciprocity problem by not using its advantage to overwhelm its partner (Pellis *et al.* 2010). So, in rats, whereas the playful character of the tactics performed can be detected in the way they are performed, in degus, it is not the tactics themselves that differentiate playful from serious fighting, but the consequences that follow the execution of a successful tactic. Although there are likely many other ways in which different species have solved this reciprocity problem, however it is achieved, what is crucial to all of them is that fairness is maintained (Pellis and Pellis 2009). This fairness can be exploited in order to use play fighting as a tool for social assessment and manipulation.

Navigating social relationships

Consider a young, adult male rat: after growing up in a colony, it is now ready to confront the social world of adults. In that colony are many adult male rats, one of which is an alpha male, dominant over all the others. Our youngster faces some stark choices – it can remain in its natal colony, but will likely be doomed to occupy a subordinate position; this mean a life of making do – only getting crumbs to eat and rare opportunities to mate. Alternatively, it can head off and try to break into an existing colony or set up one of its own. These options, however, are likely to involve quite severe competition from similarly ambitious rats. In either case, having the social skills to probe and test potential rivals without getting into overt fights is of considerable advantage, and adult male rats appear to use play fighting as part of their repertoire for doing so. The dominant male of a colony will periodically attack subordinate males that live in the colony, forcing many to live a furtive life of avoidance (Blanchard and Blanchard 1990b; Blanchard *et al.* 1988). Such seemingly unprovoked attacks appear to function to reinforce the dominance of the alpha male. Resident subordinates face two choices – either accept being subordinates or try to reverse the relationship. Play fighting has a role in both.

Some male rats, it appears, are content with their role as subordinates and will behave in a manner that reinforces that role. Firstly, subordinate males and females will initiate social contact with the dominant male – this can involve social grooming and social play (Adams and Boice 1983, 1989; Pellis and Pellis 1991). Secondly, when engaged in play fighting, the actions taken by the subordinate can be modified – this is where fairness fits into the strategic use of play as a social tool. However, in order to understand this use of play, we must first discuss the age-related changes in play fighting by male rats. As juveniles, both males and females are most likely to roll over onto their backs when defending against a nape attack, but as adults, males are most likely to rotate only partially around their longitudinal axis, leaving them standing on one or both hind feet. From this partially rotated position, the defender can either push and pull at its partner with its fore feet, rear upright so as to push the partner off balance, or remain in the horizontal position and hip slam its partner to do so (Pellis and Pellis 1987, 1990). If the defender does roll over to supine, the attacker is more likely to remain in this anchored position when an adult than when a juvenile (Foroud and Pellis

2002). That is, following puberty, the play fighting of male rats becomes rougher. This roughness increases the male rat's chances of gaining the advantage during play fighting.

As adults, female rats retain the juvenile-typical pattern of gentle play while all males develop the rougher form (Pellis and Pellis 1990; Smith *et al.* 1996, 1998). However, depending on their partner, males can alter the pattern of play used. When play fighting with a female or another subordinate male rat, a male will mainly use the rougher form of play fighting, but when play fighting with a dominant male, it will revert to the more gentle form of play. Now let's reconsider this from the point of view of fairness.

If a subordinate male rat plays roughly with a female that plays gently, the advantage will go to the male, but if this same rat plays with another subordinate male, that also plays roughly, both rats have an equal chance of gaining the advantage. However, when a subordinate male plays with a dominant male, the subordinate will play gently and the dominant will play roughly, which gives the dominant the advantage. Thus, the subordinate male can play in a manner that is unfair or fair, and it can strategically modify that fairness – by relinquishing the advantage to the dominant, it does, in effect, reinforce the other rat's dominance (Pellis and Pellis 1991, 1992a; Pellis *et al.* 1993; Smith *et al.* 1998). When playing roughly with females and other subordinate males, however, this same male demonstrates that it has a dominance advantage (with females) or that its partner does not (with subordinate males). But not all subordinates are the same. Some subordinates are better thought of as dominant wannabes. These beta males are the ones that keep to the periphery of the colony and are most likely to be the ones pummeled by the dominant male (Blanchard and Blanchard 1990b; Blanchard *et al.* 1988). When confronted by a dominant, these wannabes will engage him in play, but while the play is relatively gentle, the level of roughness can be increased to a level not seen in gamma males (Pellis *et al.* 1993). This increased roughness seems to be a way by which the beta male can test the capacity of the dominant male – if the beta increases the roughness of the play incrementally and the dominant tolerates this, then the relationship can potentially be reversed, but if the dominant retaliates, the subordinate can back down. This use of play fighting can be seen even more strikingly when unfamiliar adults confront one another in a neutral enclosure – the rougher form of play fighting appears to be used to establish their relationship, with escalation to serious fighting only occurring if the playful fighting fails to be a sufficient test (Smith *et al.* 1999). Thus, the fairness in play fighting can be relinquished, and so serve to reinforce one's subordinate status, or it can be bent in one's own favor to test the response of the opponent; this can then lead to the establishment of a new dominance relationship between the pair (Pellis 2002).

Bending this fairness rule in play fighting can therefore be an important tool for social assessment and manipulation, but what is still unclear is whether the use and successfulness in implementing these rules arises from direct mood-induction or from a cognitive evaluation of the situation. Some of the data on rats suggest that there is a cognitive dimension to this rule bending.

Play fighting and the prefrontal cortex

Rats that are denied the opportunity to engage in play fighting as juveniles grow into socially incompetent adults – such males are sexually awkward, are more likely to attract attacks from other males, and even when opportunities are available, fail to adopt strategies that avoid drawing the attention of other males (Byrd and Briner 1999; Einon and Potegal 1991; Moore 1985; Van Dan Berg *et al.* 1999; Von Frijtag *et al.* 2002). Similarly, adult animals, including rats, with damage to the prefrontal cortex, especially the orbital frontal cortex (OFC), exhibit similar social deficiencies (Kolb 1990). A core problem is that without the proper functioning of the OFC, rats fail to modify their behavior in contextually appropriate ways. For example, in colonies, each comprising a dominant male, a subordinate male, and a female, the play behavior of the subordinates with the dominant males and the females was compared before and after ablation of the OFC. As was predicted, when intact subordinate male rats were tested with the females, they did so roughly, but when tested with the dominant, they played gently. Following ablation of the OFC, these subordinate male rats, even though they were just as playful as they were prior to the brain damage, played roughly with both the dominant male and the female – that is, they were no longer able to modulate the intensity of their play with different partners (Pellis *et al.* 2006). Recent evidence connects these juvenile play experiences with adult social skills (Pellis and Pellis 2007). The architecture of the neurons of the prefrontal cortex is modified by social experiences in the juvenile period, but in a somewhat more complex manner than would be anticipated by the above-stated facts.

The experience of play fighting affects the neurons of the medial prefrontal cortex (mPFC), but not those of the OFC. The latter are affected by having experienced a diversity of social partners, whether these partners engaged in play or not (Bell 2008; Bell *et al.* 2007). The absence of play-fighting experience is also known to affect the ability of rats to coordinate their movements with those of a social partner in both sexual (Moore 1985) and nonsexual (Pellis *et al.* 1999) contexts. Similarly, damage to the mPFC of adult rats has been shown to compromise their ability to initiate and coordinate complex sequences of behavior (Hauber *et al.* 1994; Heidbreder and Groenewegen 2003; Kolb and Whishaw 1983), and this includes play fighting (Bell 2008). So, it is likely that either directly through the experience of play (mPFC) or indirectly, by play promoting interactions with multiple partners (OFC), play fighting in the juvenile period facilitates the development of the prefrontal cortex, and in doing so, enhances the development of the skills needed to deal with social situations. This includes the ability to modulate play in socially sensitive ways (Pellis and Pellis 2009). While the precise mechanisms involved are not important for the present argument, what *is* critical to it is that there are cortical controls over the performance of play (Bell 2008; Foroud *et al.* 2004; Kamitakahara *et al.* 2007; Panksepp *et al.* 1994; Pellis *et al.* 1992; 2006).

It is not the case that rats without a cortex do not play – they do. However, while they play just as much, show the age-related waxing and waning of play, and are able to execute all the tactics typical of play, their play appears more rigid – the animals fail to modulate their play in various ways depending on context (Pellis *et al.* 1992). Different aspects of this control appear to involve different parts of the cortex, including the motor cortex (Kamitakahara *et al.* 2007), the OFC (Pellis *et al.* 2006), and the mPFC (Bell 2008). What these studies all demonstrate is that the motivation for play fighting and the means of its production involve subcortical mechanisms. Furthermore, of the dozen or so murid rodents that we have studied, only rats appear to have these added cortical controls over the regulation of play fighting (Pellis and Iwaniuk 2004; Pellis and Pellis 1998a). For other species, such as Syrian golden hamsters, decortication appears to have little effect (e.g., Murphy *et al.* 1981).

There are considerable data both for rats and for some monkeys that play-fighting experience in the juvenile period influences the development of the emotional response system (Pellis and Pellis 2006). As shown above, for some species, such as rats, not only does play-fighting experience when juvenile produce adults that are calmer in the face of calamity, it also produces ones that are better equipped to use the affect modifying properties of play on their social partners in a more effective manner. Given that the latter involves a degree of cognitive evaluation of the partner's status and behavior, it is not surprising that this ability has been shown to depend on cortical circuits (Pellis *et al.* 1993, 2006; Bell 2008). However, the execution of play and its emotional content are dependent on subcortical circuits, and so it is conceivable that there are two steps in the process leading to the evolution of more sophisticated uses of play fighting. Firstly, for all the species of mammals that engage in play fighting, the act of play fighting can modify one's own and one's partner's affective state and this is achieved via subcortical mechanisms (i.e., the emotional system). Such a mechanism may function indirectly, by promoting proximity, which, in turn, may lead to other beneficial consequences. For example, recall that gamma male rats seek out and play gently with the colony's dominant male. This leads to the subordinate being tolerated by the dominant if nearby. This likely occurs because of the positive affective state that the play induces in the dominant via subcortical mechanisms. Functionally, this proximity to the dominant can afford the subordinate the occasional sexual opportunity or food if a new source is discovered. Secondly, cortical mechanisms have evolved that can refine and modify those subcortical systems by evaluating how the context may require the play to be modified (i.e., the cognitive system). Such a mechanism is to be expected in situations where affect-induced proximity may lead to potentially parasitical relationships, which are not necessarily beneficial to both parties. In this way, cortically mediated regulation may become important to attenuate the affect-induction by the partner, but this, in turn, can lead to the evolution of cortical mechanisms that can overcome such induction. A possible example is seen in the relationship of dominant and subordinate adult male rats. Regular play may be insufficient to affect the dominant's mood positively, so the subordinate may need to increase the reward quality of the play by mimicking the juvenile form (Pellis *et al.* 1993).

The more complex the judgments needed so as to modulate the partner's affective state and the more complex the rewards that are to be gained from doing so likely increase the cognitive demands on the animals. This increased cognitive demand may require a more complex regulation of the subcortical affective systems by the cortical cognitive systems. Such a hypothesis is yet to be tested with regard to social play, but there are indications of this dual control system over grooming, especially that seen in primates. There is a growing body of evidence that social grooming does not only reduce aggression and promote tolerance, but also that it may be used for bartering on a wider economic market – that is, grooming can be traded not only for reciprocal grooming, but also for such commodities as aid when attacked, access to food, sex, and infants (Barrett and Henzi 2001). This type of use of grooming would require the animals not only to keep track of who is grooming whom, but also how much, and the occurrence of opportunities so as to be able to cash in such credits for other commodities. Clearly, such exchanges would greatly expand the social networks and information possible as well as the cognitive demand on how and when grooming is to be used (Barrett and Henzi 2006). Simply making the grooming partner feel good is insufficient.

Indeed, while primates as a whole seem to engage in social grooming more frequently than other mammals, there is also a considerable variation between primate species, ranging from as little as 0.1% of the day, to as much as 17% of the day (Lehmann *et al.* 2007). While all primates are likely to be able to influence the affective state of their partners through grooming, our guess is that only some are able to use that affect-induction strategically in a sophisticated enough manner to override the emotional defenses of their partner and gain maximum benefit from the possible commodities to be exchanged. The analogy here is to the play fighting of murid rodents – most play and do so by engaging subcortical mechanisms that likely influence the participants' affective states, but only some have the cognitive wherewithal to use that affect-induction in a strategic manner (Pellis and Iwaniuk 2004; Pellis and Pellis 2009).

Conclusion

Whether social play and social grooming have common origins or not, it seems that both, via touching, can tap into the brain's reward system so as to modify the affective state of their social partners. Such social affect-induction can, in turn, influence the social dynamics possible and so the complexity of the social system. In addition to this emotional system, some lineages of animals have acquired more subtle control over this tool by increasing the role of cognitive control mechanisms that allow a more strategic use of the touch-based affect-induction. If this is so, it would suggest that in evolutionary terms, the affective mechanisms must have appeared before the cognitive mechanisms. The data on play fighting in murid rodents support this scenario. But what is required are more broad ranging, comparative studies of forms of tactile communication and more detailed laboratory studies of the neural mechanisms underlying the *affective* and *cognitive* mechanisms involved. Combining these approaches would lead to a deeper understanding of the role of touch in social behavior and the properties of different forms of touch, such as that experienced through grooming and play.

References

Adams N and Boice R (1983): A longitudinal study of dominance in an outdoor colony of domestic rats. *J Comp Psychol* **97**:24–33.

Adams N and Boice R (1989): Development of dominance in rats in laboratory and seminatural environments. *Behav Processes* **19**:127–42.

Arelis CL (2006): Stress and the power of play. Unpublished MSc thesis. Department of Neuroscience. University of Lethbridge: Lethbridge, AB, Canada.

Barash DP (1973): The social biology of the Olympic marmot. *Anim Behav Monographs* **6**:171–245.

Barash DP (1974): The social behaviour of the hoary marmot (*Marmota marmota*). *Anim Behav* **24**:27–35.

Barnett SA (1975): *The Rat: A Study in Behavior*. Chicago, IL: The University of Chicago Press.

Barrett L and Henzi SP (2001): The utility of grooming in baboon troops. In: Noe R, van Hooff J and Hammerstein P, eds., *Economics in Nature: Social Dilemmas, Mate Choice and Biological Markets*. Cambridge, UK: Cambridge University Press, pp. 119–45.

Barrett L and Henzi SP (2006): Monkeys, markets and minds: biological markets and primate sociality. In: Kappeler PM and van Schaik CP, eds., *Cooperation in Primates and Humans: Mechanisms and Evolution*. Berlin, Germany: Springer, pp. 209–32.

Bekoff M (2001): The evolution of animal play, emotions, and social morality: on science, theology, spirituality, personhood, and love. *Zygon* **36**:615–55.

Bell HC (2008): Playful feedback and the developing brain. Unpublished MSc thesis. Department of Neuroscience. University of Lethbridge: Lethbridge, AB, Canada.

Bell HC, Kolb B and Pellis SM (2007): It's not child's play: brain development is altered by horsing around. Poster at Annual Meeting of the Society for Neuroscience. San Diego, CA.

Blackwell K (1969): Rearing and breeding of Demidoff's galago *Galago demidovii* in captivity. *Int Zoo Yearbook* **9**:24–76.

Blanchard DC and Blanchard RJ (1990a): The colony model of aggression and defense. In: Dewsbury DA, ed., *Contemporary Issues in Comparative Psychology*. Sunderland, MA: Sinauer Associates, pp. 410–30.

Blanchard DC and Blanchard RJ (1990b): Behavioral correlates of chronic dominance–subordinance relationships of male rats in a seminatural situation. *Neurosci Biobehav Rev* **14**:455–62.

Blanchard RJ, Blanchard DC, Takahashi T *et al.* (1977): Attack and defensive behaviour in the albino rat. *Anim Behav* **25**:622–34.

Blanchard RJ, Flannelly KJ and Blanchard DC (1988): Life-span studies of dominance and aggression in established colonies of laboratory rats. *Physiol Behav* **43**:1–7.

Bohm JK and Hendricks B (1997): Effects of interpersonal touch, degree of justification, and sex of subject on compliance with a request. *J Social Psychol* **137**:460–9.

Brueggeman JA (1978): The function of adult play in free-ranging *Macaca mulatta*. In: Smith EO, ed., *Social Play in Primates*. London, UK: Routledge, pp. 169–92.

Byrd KR and Briner WE (1999): Fighting, nonagonistic social behavior, and exploration in isolation-reared rats. *Aggr Behav* **25**:211–23.

Carter CS and Keverne EB (2002): The neurobiology of social affiliation and pair bonding. In: Pfaff DW, ed., *Hormones, Brain, and Behavior*. San Diego, CA: Academic Press, pp. 299–337.

Charles-Dominique P (1977): *Ecology and Behaviour of Nocturnal Primates. Prosimians of Equatorial West Africa*. New York, NY: Columbia University Press.

Charles-Dominique P (1978): Solitary and gregarious prosimians: evolution of social structures in primates. In: Chivers DJ, ed., *Recent Advances in Primatology*, Vol. 3, *Evolution*. London, UK: Academic Press, pp. 139–49.

Charles-Dominique P and Bearder SK (1979): Field studies of Lorisid behavior: methodological aspects. In: Doyle GA and Martin RD, eds., *The Study of Prosimian Behavior*. New York, NY: Academic Press, pp. 567–629.

Crusco A and Wetzel CG (1984): The Midas touch: the effects of interpersonal touch on restaurant tipping. *Personality Social Psychol Bull* **10**:512–17.

Curley JP and Kervene EB (2005): Genes, brains and mammal social bonds. *Trends Ecol Evol* **20**:561–7.

Darwish M, Korányi L, Nyakas C *et al.* (2001): Induced social interaction reduces corticosterone stress response to anxiety in adult and aging rats. *Klin Kísérletes Lab Medicina* **28**:108–11.

De Waal FBM (1995): Sex as an alternative to aggression in the bonobo. In: Abramson P and Pinkerton S, eds., *Sexual Nature, Sexual Culture*. Chicago, IL: The University of Chicago Press, pp. 37–56.

Doyle GA (1974): The behaviour of the lesser bushbaby. In: Martin RD, Doyle GA and Walker AC, eds., *Prosimian Biology*. Pittsburgh, PA: The University of Pittsburgh Press, pp. 213–31.

Doyle GA, Pelletier A and Bekker T (1967): Courtship, mating and parturition in lesser bushbaby (*Galago senegalensis moholi*) under semi-natural conditions. *Folia Primatol* **7**:169–97.

Dugatkin LA and Bekoff M (2003): Play and the evolution of fairness: a game theory model. *Behav Processes* **60**:209–14.

Dunbar RIM (2010): The social role of touch in humans and primates: behavioural function and neurobiological mechanisms. *Neurosci Biobehav Rev* **32**:260–8.

Einon D and Potegal M (1991): Enhanced defense in adult rats deprived of playfighting experience in juveniles. *Aggr Behav* **17**:27–40.

Epps J (1974): Social interactions of *Perodicticus potto* kept in captivity in Kampala, Uganda. In: Martin RD, Doyle GA and Walker AC, eds., *Prosimian Biology*. Pittsburgh, PA: The University of Pittsburgh Press, pp. 233–44.

Erhlich A (1977): Social and individual behaviors in captive greater galago. *Behaviour* **63**:192–214.

Erhlich A and Musicant A (1975): Social and individual behaviors in captive slow lorises. *Behaviour* **60**:195–200.

Ewer RF (1967): *Ethology of Mammals*. New York: Plenum Press.

Field T (2001): *Touch*. Cambridge, UK: Cambridge University Press.

Fisher JD, Rytting M and Heslin R (1976): Hands touching hands: affective and evaluative effects of interpersonal touch. *Sociometry* **39**:416–21.

Fleaggle JG (1988): *Primate Adaptation and Evolution*. New York, NY: Academic Press.

Foroud A and Pellis SM (2002): The development of 'anchoring' in the play fighting of rats: evidence for an adaptive age-reversal in the juvenile phase. *Int J Comp Psychol* **15**:11–20.

Foroud A and Pellis SM (2003): The development of 'roughness' in the play fighting of rats: a Laban Movement Analysis perspective. *Dev Psychobiol* **42**:35–43.

Foroud A, Whishaw IQ and Pellis SM (2004): Experience and cortical control over the pubertal transition to rougher play fighting in rats. *Behav Brain Res* **149**: 69–76.

Geist V (1978): On weapons, combat and ecology. In: Krames L, Pliner P and Alloway T, eds., *Advances in the Study of Communication and Affect*, Vol. 4, *Aggression, Dominance and Individual Spacing*. New York, NY: Plenum Press, pp. 1–30.

Goosen C (1981): On the function of allogrooming in Old World Monkeys. In: Chiarelli AB and Corruccini RS, eds., *Primate Behaviour and Sociobiology*. Berlin, Germany: Springer, pp. 110–20.

Gummert MD and Ho M-RR (2008): The trade balance of grooming and its coordination of reciprocation and tolerance in Indonesian long-tailed macaques (*Macaca fascicularis*). *Primates* **49**:176–85.

Hauber W, Bubser M and Schmidt WJ (1994): 6-hydroxydopamine lesion of the rat prefrontal cortex impairs motor initiation but not motor execution. *Exp Brain Res* **99**:524–8.

Heidbreder CA and Groenewegen HJ (2003): The medial prefrontal cortex in the rat: evidence for a dorso-ventral distinction based upon functional and anatomical characteristics. *Neurosci Biobehav Rev* **27**:555–79.

Hladick CM and Charles-Dominique P (1974): The behaviour and ecology of the sportive lemur (*Lepilemur mustelinus*) in relation to its dietary peculiarities. In: Martin RD, Doyle GA and Walker AC, eds., *Prosimian Biology*. Pittsburgh, PA: The University of Pittsburgh Press, pp. 23–37.

Jones CB (1983): Social organization of captive black howler monkeys (*Aloutta caraya*): social competition and the use of non-damaging behavior. *Primates* **24**:25–39.

Kamitakahara H, Monfils M-H, Forgie ML *et al*. (2007): The modulation of play fighting in rats: role of the motor cortex. *Behav Neurosci* **121**:164–76.

Kolb B (1990): Prefrontal cortex. In: Kolb B and Tees RC, eds., *The Cerebral Cortex of the Rat*. Cambridge, MA: The MIT Press, pp. 437–58.

Kolb B and Whishaw IQ (1983): Dissociation of the contributions of the prefrontal, motor and parietal cortex to the control of movement in the rat. *Can J Psychol* **37**:211–32.

Kurzban R (2001): The social psychophysics of cooperation: nonverbal communication in a public goods game. *J Nonverbal Behav* **25**:241–59.

Lehmann J, Korstjens AH, Dunbar RIM (2007): Group size, grooming and social cohesion in primates. *Anim Behav* **74**:1617–29.

Lim MM and Young L (2006): The neurobiology of social bonds and affiliation. In: Marshall PJ and Fox NA, eds., *The Development of Social Engagement: Neurobiological Perspectives*. Oxford, UK: Oxford University Press, pp. 171–96.

McGlone F, Vallbo AB, Olausson H *et al*. (2007): Discriminative touch and emotional touch. *Can J Exp Psychol* **61**:173–83.

Moore CL (1985): Development of mammalian sexual behavior. In: Gollin ES, ed., *The Comparative Development of Adaptive Skills*. Hillsdale, NJ: Lawrence Erlbaum, pp. 19–56.

Murphy MR, MacLean PD and Hamilton SC (1981): Species-typical behavior of hamsters deprived from birth of the neocortex. *Science* **213**:459–61.

Newell TG (1971): Social encounters in two prosimian species: *Galago crassicaudatus* and *Nycticebus coucang*. *Psychon Soc* **2**:128–30.

Owren MJ and Rendall D (1997): An affect-conditioning model of nonhuman primate vocal signaling. In: Owings DH, Beecher MD and Thompson NS, eds., *Perspectives in Ethology*, Vol. 12, *Communication*. New York, NY: Plenum Press, pp. 299–346.

Pagés E (1978): Home range, behaviour and tactile communication in a nocturnal Malagasy lemur *Mirza coquereli*. In: Chivers D, ed., *Recent Advances in Primatology*, Vol. 3, *Evolution*. New York, NY: Academic Press, pp. 171–7.

Palagi E, Cordoni G and Borgognini Tarli SM (2004): Immediate and delayed benefits of play behavior: new evidence from chimpanzees (*Pan troglodytes*). *Ethology* **110**:949–62.

Palagi E, Paoli T and Borgognini Tarli SM (2006): Short-term benefits of play behavior and conflict prevention in *Pan paniscus*. *Int J Primatol* **27**:1257–70.

Panksepp J (1998): *Affective Neuroscience*. Oxford, UK: Oxford University Press.

Panksepp J, Normansell L, Cox JF *et al.* (1994): Effects of neonatal decortication on the social play of juvenile rats. *Physiol Behav* **56**:429–43.

Paulos RD, Dudzinski KM and Kuczaj SA (2008): The role of touch in select social interactions of Atlantic spotted dolphin (*Stenella frontalis*) and Indo-Pacific bottlenose dolphin (*Tursiops aduncus*). *J Ethol* **26**:153–64.

Pellis SM (1993): Sex and the evolution of play fighting: a review and a model based on the behavior of muroid rodents. *J Play Theory Res* **1**:56–77.

Pellis SM (1997): Targets and tactics: the analysis of moment-to-moment decision making in animal combat. *Aggr Behav* **23**:107–29.

Pellis SM (2002): Keeping in touch: play fighting and social knowledge. In: Bekoff M, Allen C and Burghardt GM, eds., *The Cognitive Animal: Empirical and Theoretical Perspectives on Animal Cognition*. Cambridge, MA: The MIT Press, pp. 421–7.

Pellis SM and Iwaniuk AN (1999): The problem of adult play: a comparative analysis of play and courtship in primates. *Ethology* **105**:783–806.

Pellis SM and Iwaniuk AN (2000): Adult–adult play in primates: comparative analyses of its origin, distribution and evolution. *Ethology* **106**:1083–104.

Pellis SM and Iwaniuk AN (2004): Evolving a playful brain: a levels of control approach. *Int J Comp Psychol* **17**:90–116.

Pellis SM and Pellis VC (1987): Play-fighting differs from serious fighting in both target of attack and tactics of fighting in the laboratory rat *Rattus norvegicus*. *Aggr Behav* **13**:227–42.

Pellis SM and Pellis VC (1988a): Play-fighting in the Syrian golden hamster *Mesocricetus auratus* Waterhouse, and its relationship to serious fighting during post-weaning development. *Dev Psychobiol* **21**:323–37.

Pellis SM and Pellis VC (1988b): Identification of the possible origin of the body target which differentiates play-fighting from serious fighting in Syrian golden hamsters *Mesocricetus auratus*. *Aggr Behav* **14**:437–49.

Pellis SM and Pellis VC (1989): Targets of attack and defense in the play fighting by the Djungarian hamster *Phodopus campbelli*: links to fighting and sex. *Aggr Behav* **15**:217–34.

Pellis SM and Pellis VC (1990): Differential rates of attack, defense and counterattack during the developmental decrease in play fighting by male and female rats. *Dev Psychobiol* **23**:215–31.

Pellis SM and Pellis VC (1991): Role reversal changes during the ontogeny of play fighting in male rats: attack versus defense. *Aggr Behav* **17**:179–89.

Pellis SM and Pellis VC (1992a): Juvenilized play fighting in subordinate male rats. *Aggr Behav* **18**:449–57.

Pellis SM and Pellis VC (1992b): An analysis of the targets and tactics of conspecific attack and predatory attack in northern grasshopper mice *Onychomys leucogaster*. *Aggr Behav* **18**:301–16.

Pellis SM and Pellis VC (1997a): The pre-juvenile onset of play fighting in rats (*Rattus norvegicus*). *Dev Psychobiol* **31**:193–205.

Pellis SM and Pellis VC (1997b): Targets, tactics and the open mouth face during play fighting in three species of primates. *Aggr Behav* **23**:41–57.

Pellis SM and Pellis VC (1998a): The play fighting of rats in comparative perspective: a schema for neurobehavioral analyses. *Neurosci Biobehav Rev* **23**:87–101.

Pellis SM and Pellis VC (1998b): Structure–function interface in the analysis of play. In: Bekoff M and Byers JA, ed., *Animal Play: Evolutionary, Comparative, and Ecological Perspectives*. Cambridge, UK: Cambridge University Press, pp. 115–40.

Pellis SM and Pellis VC (2006): Play and the development of social engagement: a comparative perspective. In: Marshall PJ and Fox NA, eds., *The Development of Social Engagement: Neurobiological Perspectives*. Oxford, UK: Oxford University Press, pp. 247–74.

Pellis SM and Pellis VC (2007): Rough-and-tumble play and the development of the social brain. *Curr Direct Psychol Sci* **16**:95–8.

Pellis SM, Pellis VC (2009): *The Playful Brain: Venturing to the Limits of Neuroscience*. Oxford, UK: Oneworld Press.

Pellis SM, Pellis VC and Dewsbury DA (1989): Different levels of complexity in the playfighting by muroid rodents appear to result from different levels of intensity of attack and defense. *Aggr Behav* **15**:297–310.

Pellis SM, Pellis VC and Whishaw IQ (1992): The role of the cortex in play fighting by rats: developmental and evolutionary implications. *Brain Behav Evol* **39**:270–84.

Pellis SM, Pellis VC and McKenna MM (1993): Some subordinates are more equal than others: play fighting amongst adult subordinate male rats. *Aggr Behav* **19**:385–93.

Pellis SM, Field EF and Whishaw IQ (1999): The development of a sex-differentiated defensive motor-pattern in rats: a possible role for juvenile experience. *Dev Psychobiol* **35**:156–64.

Pellis SM, Pasztor TJ, Pellis VC and Dewsbury DA (2000): The organization of play fighting in the grasshopper mouse (*Onychomys leucogaster*): mixing predatory and sociosexual targets and tactics. *Aggr Behav* **26**:319–34.

Pellis SM, Pellis VC and Foroud A (2005): Play fighting: aggression, affiliation and the development of nuanced social skills. In: Tremblay R, Hartup WW and Archer J, eds., *Developmental Origins of Aggression*. New York, NY: Guilford Press, pp. 47–62.

Pellis SM, Hastings E, Shimizu T *et al.* (2006): The effects of orbital frontal cortex damage on the modulation of defensive responses by rats in playful and non-playful social contexts. *Behav Neurosci* **120**:72–84.

Pellis SM, Pellis VC, Reinhart CJ (2000): The evolution of social play. In: Worthman C, Plotsky P, Schechter D, eds., *Formative Experiences: The Interaction of Caregiving, Culture, and Developmental Psychobiology*. Cambridge, UK: Cambridge University Press.

Roberts P (1971): Social interactions of *Galago crassicaudatus*. *Folia Primatol* **14**:171–81.

Rosenson LM (1973): Group formation in the captive greater bushbaby (*Galago crassicaudatus crassicaudatus*). *Anim Behav* **21**:67–77.

Russell RJ (1975): Body temperature and behavior of captive cheirogaleids. In: Tattersall I and Sussman RW, eds., *Lemur Biology*. New York, NY: Plenum Press, pp. 193–206.

Schino G, Scucchi S, Maestripieri D and Turullazzi PG (1988): Allogrooming as a tension-reduction mechanism: a behavioral approach. *Am J Primatol* **16**:43–60.

Siviy SM and Panksepp J (1987): Sensory modulation of juvenile play in rats. *Dev Psychobiol* **20**:39–55.

Smith LK, Field EF, Forgie ML, *et al.* (1996): Dominance and age-related changes in the play fighting of intact and post-weaning castrated males (*Rattus norvegicus*). *Aggr Behav* **22**:215–26.

Smith LK, Forgie ML and Pellis SM (1998): Mechanisms underlying the absence of the pubertal shift in the playful defense of female rats. *Dev Psychobiol* **33**:147–56.

Smith LK, Fantella S-L and Pellis SM (1999): Playful defensive responses in adult male rats depend upon the status of the unfamiliar opponent. *Aggr Behav* **25**:141–52.

Sparks J (1967): Allogrooming in primates: a review. In: Morris D, ed., *Primate Ethology*. London, UK: Weidenfeld, pp. 148–75.

Sussman RW and Garber PA (2004): Rethinking sociality: cooperation and aggression among primates. In: Sussman RW and Chapman AR, eds., *The Origins of Sociality*. New York, NY: Aldine de Gruyter, pp. 161–90.

Van den Berg CL, Hol T, van Ree JM *et al.* (1999): Play is indispensable for an adequate development of coping with social challenges in the rat. *Dev Psychobiol* **34**:129–38.

Von Frijtag JC, Schot M, van den Bos R *et al.* (2002): Individual housing during the play period results in changed responses to and consequences of a psychosocial stress situation in rats. *Dev Psychobiol* **41**:58–69.

Zak PJ, Kurzban R and Matzner W (2005): Oxytocin is associated with human trustworthiness. *Horm Behav* **48**:522–7.

5

Grooming syntax as a sensitive measure of the effects of subchronic PCP treatment in rats

MARIE-CLAUDE AUDET AND SONIA GOULET

Summary

Grooming is increasingly recognized as a reliable marker of stress-related disturbances in animal models of neuropsychiatric disorders. We previously reported that subchronic exposure to 10 mg/kg of phencyclidine (PCP) for 15 days in rats increased grooming expression under both stressful and appetitive conditions, but impaired grooming syntax only when the behavior was elicited with stressful water sprays directed at the face. For the purpose of this chapter, new indexes from the same rats subjected to the water spray condition were analyzed. Results showed that the PCP group aborted less chains after face washing and spent a lower proportion of time in anterior grooming than control animals. Phencyclidine treatment also increased incorrect chain initiations and enhanced the duration of Phase IV within completed syntactic chains. Finally, PCP-injected rats were less engaged in nongrooming activities, and were more inactive. In a context where grooming was needed rostrally after facial contacts with water sprays, these results indicate that subchronic PCP treatment compromised hygiene efficiency and engendered an unfocused and perseverative grooming, most likely combined with an abnormal stress response. These observations suggest that the two leading approaches in the study of grooming patterning may provide pivotal sets of qualitative observations that help identify hygienic and stress-related irregularities in animal models.

Neurobiology of Grooming Behavior, eds. Allan V. Kalueff, Justin L. LaPorte, and Carisa L. Bergner. Published by Cambridge University Press.

Introduction

Neuropharmacological studies rely in part on animal testing to circumscribe the functional impacts of drugs based on alleged disruptive or therapeutic characteristics. Phencyclidine, a noncompetitive antagonist of glutamatergic N-methyl-D-aspartate (NMDA) receptors, is one of these agents that raises considerable attention due to its addictive, stimulating, and psychotomimetic properties (Bakker and Amini 1961; Lehrmann *et al.* 2006; Luby *et al.* 1959; McGregor *et al.* 2008). Over the past 20 years, a PCP rodent model of schizophrenia has emerged (Javitt and Zukin 1991; Jentsch and Roth 1999; Morris *et al.* 2005; Mouri *et al.* 2007) and appears to share some of its symptomatology with other disorders also of critical scientific relevance such as pathological anxiety (Audet *et al.* 2007a; Turgeon *et al.* 2007) and depression (Noda *et al.* 1995, 1997).

Self-grooming is an innate, naturally occurring behavior in rodents and other species that serves nonhygienic functions in addition to body cleanliness (Spruijt *et al.* 1992). The amount and/or the microstructure of grooming was shown to respond to several environmental and physiological interventions: exposure to novelty and stress (D'Aquila *et al.* 2000; Jolles *et al.* 1979; Kalueff and Tuohimaa 2004, 2005; Komorowska and Pellis 2004; Krebs *et al.* 1996; van Erp *et al.* 1994), basal ganglia lesions (Cromwell and Berridge 1996), hypothalamic electrical stimulation or cytotoxicity (Roeling *et al.* 1991), and multiple neuroactive substances, including 8-OH-DPAT (Hartley and Montgomery 2008), partial and full dopamine D1 agonists (Berridge and Aldridge 2000a, b; Eilam *et al.* 1992; Matell *et al.* 2006; Van Wimersma Greidanus *et al.* 1989; Wachtel *et al.* 1992), amitriptyline, fluoxetine (Enginar *et al.* 2008), desipramine (D'Aquila *et al.* 2000), stressin$_1$-A, urocortin 3 (Zhao *et al.* 2007), scopolamine, 5,7-dihydroxytryptamine, CGS 19755 (Robertson *et al.* 1999), diazepam, and pentylenetetrazole (Kalueff and Tuohimaa 2005).

Surprisingly, only one paper by Audet *et al.* (2006) investigated grooming behavior in the PCP rat model. Locomotion is one of the most commonly used endpoints for rapid automated measurement of behavior under control and drug conditions, including PCP administration. Quantification of locomotor behaviors by means of a video tracking system was performed in some examinations of PCP effects (Martin and van den Buuse 2008; Sams-Dodd 1998a) although most studies involving this agent utilize a box equipped with infrared photocells and a system that compiles activity counts from beam interruptions (Hanania *et al.* 1999; Millan *et al.* 2008; Sams-Dood 1998b; Xu and Domino 1994, 1999). Our past experience with the locomotion paradigm illustrates how ambiguous interpretations resulting from beam counts, without additional information on the animal's state, can at times outweigh the convenience of automated testing. In Audet *et al.* (2007b), we subjected rats treated with 10 mg/kg of PCP once a day for 15 days to locomotor assessment with a compilation of beam interruptions. We showed that drug-exposed rats traveled shorter distances at a lower speed and were more inactive than a control group. However, although these results were quite straightforward, their interpretation was not. Complementary motor and motivational testing had to be carried out before a deficit of behavioral self-generation could be proposed as the probable cause for PCP-induced hypolocomotion. By contrast, the grooming paradigm readily achieves an integrated determination at many levels of behavioral expression and sequencing in rodents. Recently, the analysis of its microstructure showed the reliability of grooming for behavioral appraisal of stress responses (Kalueff and Tuohimaa 2004, 2005). Because its investigation can be adapted to almost any context, grooming is indeed an ideal tool for animal research (see Kalueff *et al.* 2007).

In our previous research (Audet *et al.* 2006), we adopted an exploratory approach to study grooming in Long–Evans rats injected intraperitoneally with 10 mg/kg of PCP once a day for 15 consecutive days and saline-treated controls subjected to one of three inductive conditions (the original report should be consulted for more complete information). Two conditions, repetitive sprays of tap water directed at the face (n = 32 rats) or brief, 10-second presentations of a loud sound of 70 dB (n = 20 rats), were considered stressful; the third one, the consumption of sticky peanut butter that adhered to the forepaws, face, and surrounding fur (n = 22), was appetitive. Rats were not handled during any of the stress or appetitive manipulations. Induction of grooming with water sprays and food was performed in the rats' housing cages and similarly recorded for 15 minutes about 20 hours after the 1st, 8th, and 15th injections. Sound-induced grooming was recorded in a Skinner box for 50 seconds after each auditory stimulation only at injection 15. Afterwards, videotapes were codified frame by frame at one-tenth of the actual speed (i.e., 1/30 s per frame) with Coder2 software (Kappas 1995). In both water and food conditions, percentages of time spent grooming were abnormally high following 8 and 15 injections of PCP, as was the case after 15 injections in the sound condition (Figure 5.1). Although the number of chains initiated was elevated following 8 and 15 expositions to PCP (Audet *et al.* 2006), the relative tendency to instigate chains was equivalent between groups both in the water and food conditions (Figure 5.2). However, induction with water but not food resulted in a lower percentage of complete syntactic chains in the PCP group (Figure 5.3). Audet *et al.* (2006) concluded that subchronic PCP treatment results in enhanced grooming expression that is generalized across contexts, whereas the stereotyped grooming syntax is selectively impaired when PCP exposure is combined with an invasive stressor.

For the purpose of this chapter, we pursued our examination of indexes collected from the same 32 rats exposed to water sprays, the most sensitive

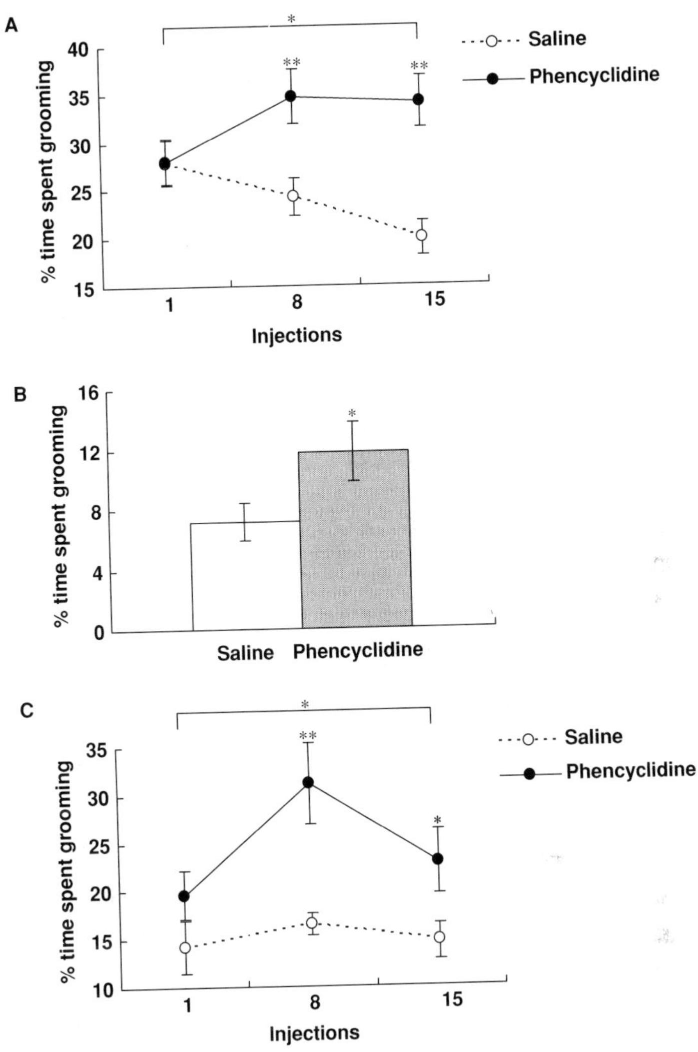

Figure 5.1 Effects of 15 daily injections of 10 mg/kg PCP on mean percentages of time spent grooming under water (A), loud sound (B), and food (C) conditions as a function of the number of injections received. Data represent means ± SEM. Saline: saline-treated group; Phencyclidine: PCP-treated group. * $p < 0.05$; ** $p < 0.01$. From Audet *et al.* (2006).

condition to the effects of PCP. We originally followed the established guidelines from Berridge and collaborators (Aldridge and Berridge 1998; Berridge 1990; Berridge *et al.* 1987; Berridge and Whishaw 1992; Cromwell and Berridge 1996) although more recent and detailed protocols (Kalueff *et al.* 2007) can now be considered. The approach proposed by Berridge *et al.* differentiates sequentially flexible

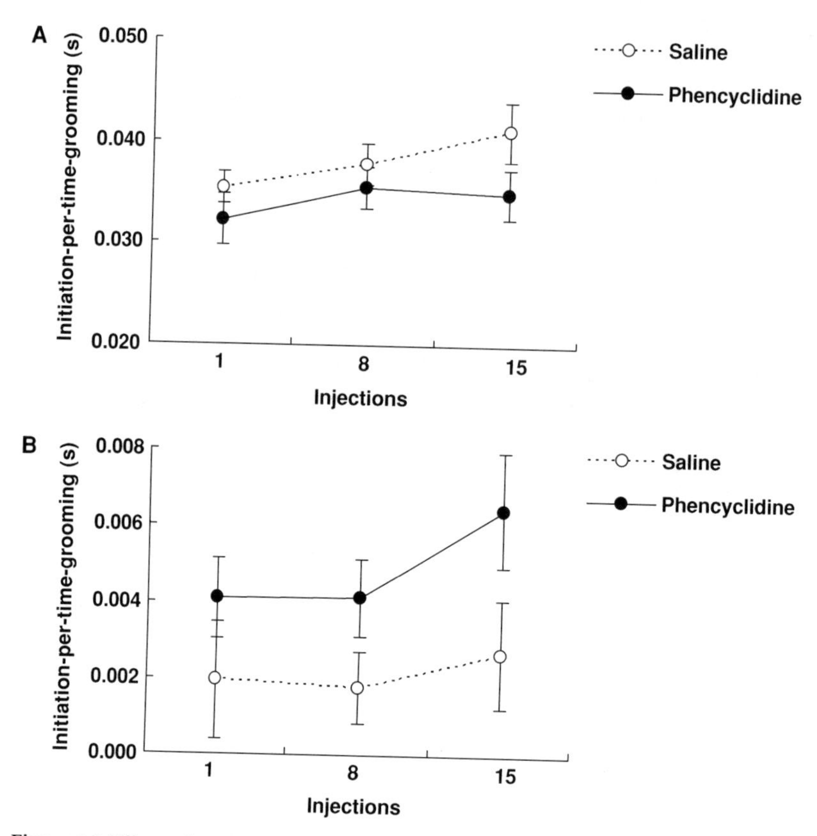

Figure 5.2 Effects of 15 daily injections of 10 mg/kg PCP on the relative probability of initiating syntactic grooming chains under water (A) and food (B) conditions as a function of the number of injections received. Data represent means ± SEM. Saline: saline-treated group; Phencyclidine: PCP-treated group. From Audet *et al.* (2006).

grooming from more rigid, syntactic patterns. The syntactic chain represents the most predictable and stereotyped pattern, connecting up to 25 components into 4 phases that proceed in a cephalo caudal fashion (Berridge 1990; Berridge *et al.* 1987). Specifically, Phase I consists of five to nine rapid elliptical strokes over the tip of the nose, bilaterally, and lasts about one second. Phase II is shorter (0.25 seconds) and involves small asymmetrical strokes over the lower half of the face and up to the eye level. Phase III corresponds to larger bilateral strokes over the ears and lasts two to three seconds. Completed syntactic chains end with Phase IV in which animals perform a postural shift directed to the flanks (Berridge 1990; Berridge *et al.* 1987; but see also Matell *et al.* 2006 and Kalueff *et al.* 2007 for an update on Phase IV). Our definition of completed chains includes those beginning with Phase I that systematically progress through Phases II, III, and IV without

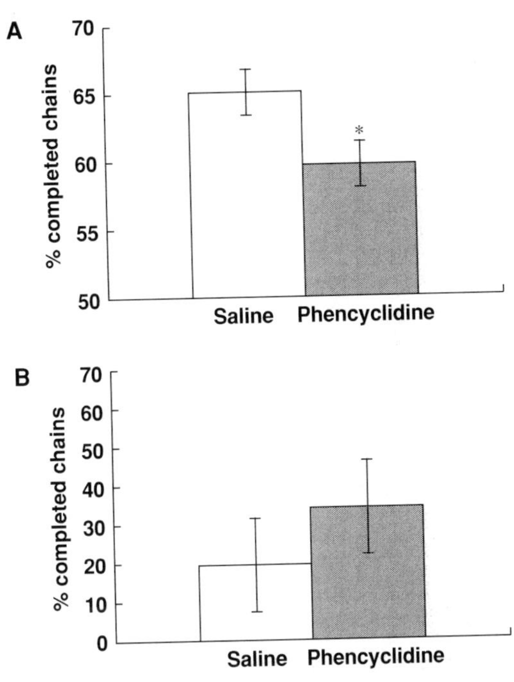

Figure 5.3 Effects of 15 daily injections of 10 mg/kg PCP on mean proportions of completed syntactic grooming chains under water (A) and food (B) conditions. Data represent means ± SEM. Saline: saline-treated group; Phencyclidine: PCP-treated group. * $p < 0.05$. From Audet *et al.* (2006).

interruption. Moreover, Phase IV has to occur within five seconds of the last stroke of Phase I, a criterion proposed by Berridge and Aldridge (2000a).

Usual indexes gathered from grooming records are total duration comprising both nonchain and syntactic grooming, total number of initiated syntactic chains, the relative probability of chain initiation, and the tendency to complete the syntactic pattern after it is initiated (Aldridge and Berridge 1998; Berridge and Aldridge 2000a, b; Berridge and Whishaw 1992; Cromwell and Berridge 1996). In our new analyses, we detailed impairments within syntactic chains that did not progress perfectly through completion and regrouped those into three categories. Chains without Phase III and/or Phase IV were referred to as *aborted chains.* Phase inversions or insertion of extraneous actions (e.g., paw licking, body shaking) constituted *inaccurate chains.* Also examined were *chains lacking Phase I* that consisted of brief sequential episodes of orderly and seemingly well-executed Phases II, III, and IV, but in which Phase I was skipped. According to Berridge (Berridge 1990; Berridge *et al.* 1987; Cromwell and Berridge 1996), Phase I determines which grooming bouts can be considered syntactic chains as it triggers and implements the remaining phases to chain completion. Chains lacking Phase I would then represent flexible transitions from head to body grooming. However, we chose to consider those as syntactic chains because a cephalo caudal and stereotyped progression from Phase II to Phase IV, with all well-executed structural components, was performed. In addition, we analyzed rates of chain abortions, issued from the analysis of chains that did not progress perfectly through completion, and regional distribution of grooming, which are two indices derived from the grooming algorithm of Kalueff and Tuohimaa (2004, 2005). Also examined were phase duration within completed syntactic chains and chains lacking Phase I as well as time invested in nongrooming behaviors, namely exploration (e.g., locomotion, rearing), and inactivity (e.g., stationary position as in lying down on the cage floor or sitting without any active movement).

Data analyses

Ratios of aborted chains, inaccurate chains, and those lacking Phase I as well as the proportion of time spent grooming anterior body parts were analyzed using two-way ANOVAs with group as a between-subject factor (two levels corresponding to saline- and PCP-treated rats) and injection as a within-subject factor (three levels corresponding to the 1st, 8th, and 15th injections). Analyses of simple main effects with Satterthwaithe's correction for the error term and its degrees of freedom (Howell 1997) determined between-group differences at each injection level, when applicable. Student t tests for independent samples with a Bonferonni correction to maintain the alpha at 0.05 served to locate within-group differences at the three-injection levels. Phase duration within completed syntactic chains and chains lacking Phase I at injection 15 was compared using nonparametric Mann–Whitney U tests for independent samples. Finally, time invested in grooming, exploration, or inactivity after 15 injections was subjected to a one-way ANOVA with group as the between-subject factor.

Results

Effects of PCP on syntactic irregularities

Chains that did not progress perfectly through completion were examined for syntactic grooming deficiencies. Figure 5.4 shows proportions of aborted chains (A), inaccurate chains (B), and chains lacking Phase I (C). For both aborted chains and those lacking Phase I, the interaction Group × Injection was significant: $F_{(2,60)} = 4.81$ and 4.64, respectively, $p < 0.05$. The analysis of simple main effects indicated that groups did not differ after the 1st (aborted chains: $F_{(1,90)} < 1$; chains lacking Phase I: $F_{(1,90)} = 1.21$) and the 8th (aborted chains only: $F_{(1,90)} = 2.23$)

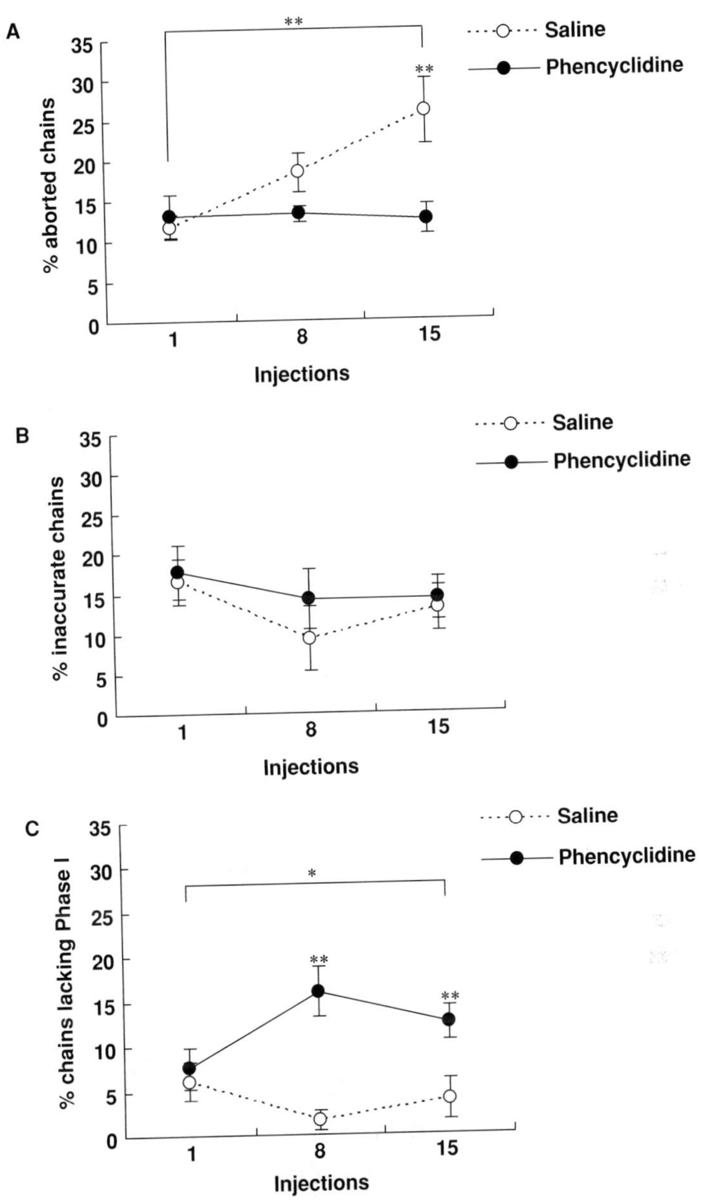

Figure 5.4 Effects of 15 daily injections of 10 mg/kg PCP on mean proportions of aborted chains (A), inaccurate chains (B), and chains lacking Phase I (C) under the water condition as a function of the number of injections received. Data represent means ± SEM. Saline: saline-treated group; Phencyclidine: PCP-treated group. * $p < 0.05$; ** $p < 0.01$.

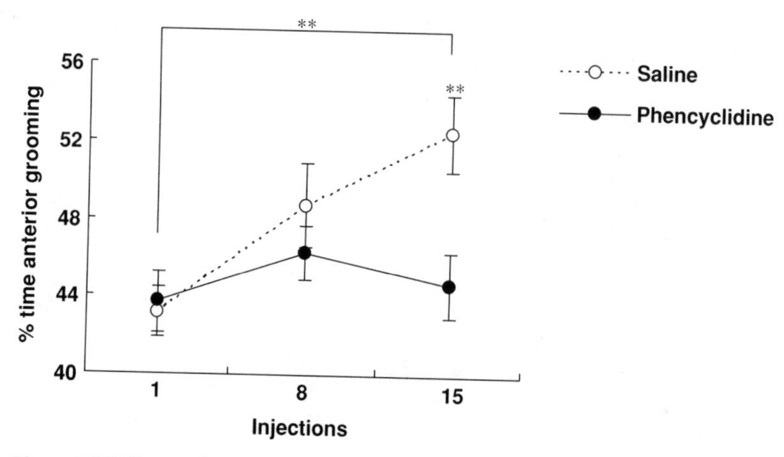

Figure 5.5 Effects of 15 daily injections of 10 mg/kg PCP on mean percentages of time spent grooming anterior body parts under the water condition as a function of the number of injections received. Data represent means ± SEM. Saline: saline-treated group; Phencyclidine: PCP-treated group. ** $p < 0.01$.

injections. Whereas PCP-treated rats suffered less aborted chains than controls after the 15th injection, $F_{(1, 90)} = 15.36$, $p < 0.01$, they produced more of those skipping Phase I after the 8th and the 15th injections, $F_{(1, 90)} = 22.25$ and 7.67, respectively, $p < 0.01$. The proportion of aborted chains was significantly higher after injection 15 than injection 1 in the saline group ($p < 0.01$) but not in the PCP group. Conversely, the proportion of chains lacking Phase I was more elevated at injection 15 compared to injection 1 in PCP-treated rats only ($p < 0.05$).

Effects of PCP on regional grooming distribution

The distribution of grooming between rostral and caudal portions of the body can provide informative clues as to rodents' internal states (Kalueff and Tuohimaa 2005). Figure 5.5 indicates that PCP-treated rats invested a lower proportion of time in the grooming of anterior body parts after 15 injections than their saline counterparts. This difference at injection 15 seems associated with the gradual increase of anterior grooming over recording sessions observed in controls, but not in the PCP group. Indeed, the interaction Group × Injection, $F_{(2, 60)} = 3.73$, $p < 0.05$, was significant. Analyses of simple main effects confirmed that PCP and saline groups did not differ after the 1st and the 8th injections, $F_{(1, 85)} < 1$ and 1.01, but PCP-treated rats spent proportionally less time in anterior grooming than the saline group after 15 injections, $F_{(1, 85)} = 10.44$, $p < 0.01$. Follow-up comparisons within each group revealed that the proportion of rostral grooming significantly increased from injection 1 to 15 in saline- ($p < 0.01$) but not in PCP-treated rats.

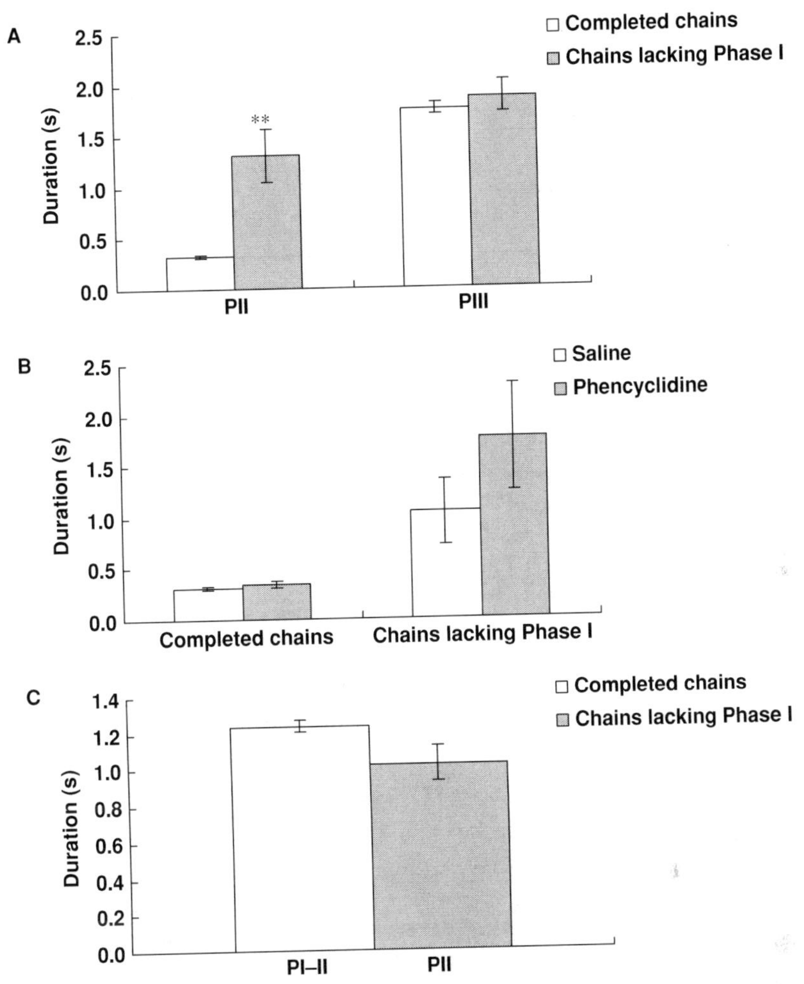

Figure 5.6 Mean duration in seconds (s) of phases within completed syntactic chains and chains lacking Phase I after 15 injections under the water condition. (A) Phases II and III with saline and PCP treatments combined; (B) Phase II in saline and PCP groups separately; (C) Phases I–II and Phase II with saline and PCP treatments combined. Data represent means ± SEM. Saline: saline-treated group; Phencyclidine: PCP-treated group. ** $p < 0.01$.

Phase duration within completed syntactic chains and chains lacking Phase 1

Figure 5.6A shows the duration of Phases II and III within completed chains and those lacking Phase I at injection 15, when PCP and saline treatments were regrouped. Duration of Phase II was longer in completed chains than in chains lacking Phase I ($p < 0.01$). Duration of Phase III, however, was similar between the

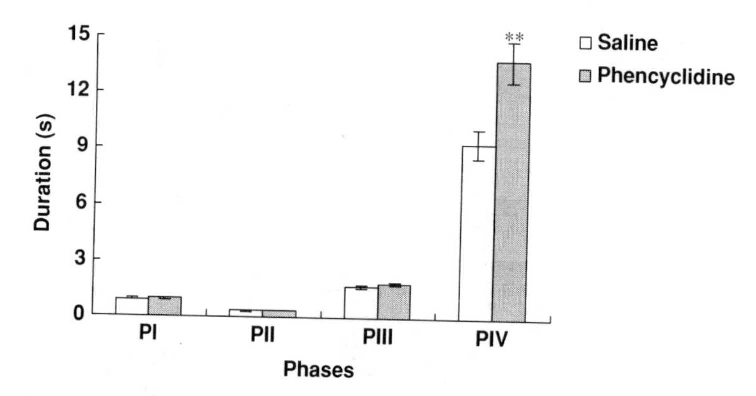

Figure 5.7 Effects of 15 daily injections of 10 mg/kg PCP on duration in seconds (s) of Phases (P) I to IV within syntactically completed chains under the water condition. Data represent means ± SEM. Saline: saline-treated group; Phencyclidine: PCP-treated group. ** $p < 0.01$.

two chains types ($p = 0.21$). To verify if the increased duration of Phase II in chains lacking Phase I was associated with PCP treatment, a comparison between groups was performed. As depicted in Figure 5.6B, Phase II in drug-treated rats did not differ in length from control animals for both completed chains ($p = 0.58$) and chains lacking Phase I ($p = 0.58$). Combined duration of Phases I and II in completed chains was then compared to the duration of Phase II in chains lacking Phase I (Figure 5.6C). Length of Phases I–II within completed chains did not differ from that of Phase II within chains lacking Phase I ($p = 0.11$).

Effects of PCP on phase duration within completed syntactic chains

The duration of individual phases within otherwise "perfectly completed" chains was examined for possible irregularities in the temporal control of grooming actions. Shortening of phase duration within syntactic chains was previously reported in rats subjected to a full dopamine D1 agonist and interpreted as a dysfunctional modulation of phase transitions (Matell *et al.* 2006). As shown in Figure 5.7, lengths of Phases I, II, and III were comparable among PCP and saline groups ($p = 0.43$, 0.58, 0.06, respectively). The duration of Phase IV, on the other hand, was longer after PCP treatment ($p < 0.01$).

Effects of PCP on grooming, exploration, and inactivity

During the 15-min course of a testing session, rats could engage in grooming, but also in other competing activities. Figure 5.8 presents time invested in

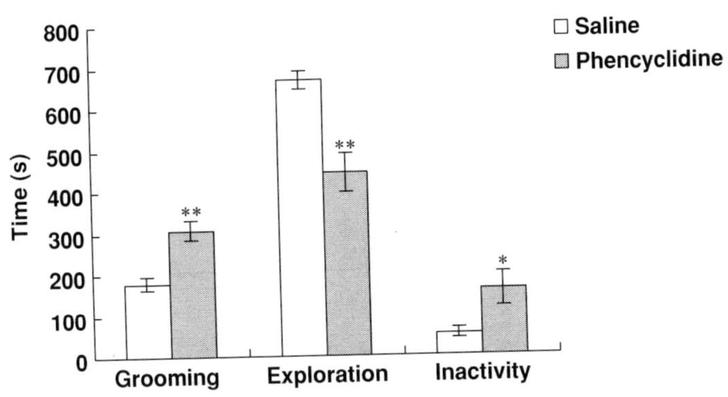

Figure 5.8 Effects of 15 daily injections of 10 mg/kg PCP on mean time in seconds (s) invested in grooming, exploration, and inactivity under the water condition. Data represent means ± SEM. Saline: saline-treated group; Phencyclidine: PCP-treated group. * $p < 0.05$; ** $p < 0.01$.

grooming, in exploration, and in inactivity after 15 injections of PCP or saline vehicle. Relative to the saline group, PCP-treated rats spent more time grooming, $F_{(1, 31)} = 19.18$, $p < 0.01$, but explored their environment less, $F_{(1, 31)} = 18.63$, $p < 0.01$. Moreover, they were more inactive, $F_{(1, 31)} = 6.11$, $p < 0.05$, than saline-injected animals.

Discussion

Our research (Audet *et al.* 2006) was the first to demonstrate a severe disruption of grooming in rats exposed to 10 mg/kg of PCP once a day for 15 days when behavior was stimulated with water sprays to the face and adjacent parts of the body. Eight to 15 injections of PCP significantly increased the time spent grooming post-sprays, as well as after induction with sticky food or a loud noise. However, only after water sprays did the drug also reduce the rates of syntactically completed grooming chains. It was then concluded that the association of repeated PCP administration with an invasive stimulation like water sprays was necessary to impact both the expression of grooming and its stereotyped pattern.

In this chapter, we pursued our analysis of Audet *et al.*'s data in the water spray condition and examined additional indexes to complement our original assertions. In light of the lower proportion of completed syntactic grooming chains in PCP-treated rats relative to controls, the initial step was to specify impairments within the other chains that did not meet the criteria for completion. Inaccuracies in grooming, including phase inversions and insertion of extraneous actions, were as uncommon in both groups. On the other hand, phase omissions within chains were more frequent and revealed important differences between groups after 8 and 15 injections. Phencyclidine treatment engendered higher proportions of syntactic bouts that lacked Phase I but comprised well-ordered, stereotyped Phases II, III, and IV. Chains properly initiated with Phases I and II but missing Phases III and/or IV were those that prevailed in saline-treated rats. The singularity of incorrect chain initiation in PCP-treated rats and of aborted bouts after face washing in controls brought into question the very function served by grooming in these two samples of rats.

In our procedure, water was sprayed several times onto rats' faces with limited spread to their bodies (Audet *et al.* 2006). From a strictly hygienic standpoint, grooming was needed on and around the targeted areas but not so much on the remaining fur. In a new analysis, we showed that at injection 15, the proportion of anterior grooming was higher in the saline group than the PCP. Even more striking was the stability of this measure in drug-exposed rats while it significantly increased across grooming sessions in the saline group in a context where the total time spent grooming in this group actually decreased. Kalueff and Tuohimaa (2005) observed that high-stress rats placed in a brightly lit novel environment displayed a rostral bias in the distribution of grooming. A similar phenomenon was reported in rodents subjected to standard novelty testing (Kalueff and Tuohimaa 2004; Komorowska and Pellis 2004) or to the tricyclic antidepressant amitriptyline that exhibits anxiogenic properties (Enginar *et al.* 2008). Moreover, chain abortions were increased by amitriptyline (Enginar *et al.* 2008) and novelty (Kalueff and Tuohimaa 2004; Komorowska and Pellis 2004) just as observed in light-induced stressed rats in comparison to a low-stress group exposed to a dark box (Kalueff and Tuohimaa 2005). The emergence of an anterior bias and the aforementioned higher rates of aborted chains following Phases I–II in our saline-treated rats thus appear to mirror prior descriptions of grooming in stressed animals. Interestingly, grooming in the PCP group did not sustain a similar rostral increment over sessions. Moreover, this group was less likely than controls to abort grooming chains once initiated.

How can these data be explained? In our grooming procedure, repetitive contact with water sprays was clearly the most anxiogenic factor. One possibility is that proportionally more time was spent grooming the anterior body in control rats because they had been sensitized to water sprays over exposures and, consequently, were more stressed than those treated with PCP. However, this would be surprising considering that the total time spent grooming significantly decreased over sessions in the saline group (Audet *et al.* 2006), clear evidence of habituation. Alternatively, the corollary hypothesis postulating that the lack of rostral bias under PCP would signify low stress does not seem to represent the whole set of data accurately. Firstly, the general overexpression of grooming, as well as the deficit in chain sequencing following PCP (Audet *et al.* 2006), are two typical behavioral markers of stress and anxiety (Kalueff and Tuohimaa 2004, 2005; Komorowska and Pellis 2004). Secondly, compared to controls, the PCP group was more committed to grooming and inactivity than to nongrooming activities such as exploration of the environment, a pattern already described in water-sprayed rats (Hartley and Montgomery 2008) that is strongly associated with physical stress (Pijlman and van Ree 2002). Hence, aberrant hyper-responsiveness to stress in PCP-treated rats stands out as a convincing interpretation of our results.

Another interpretation, which is not mutually exclusive from the preceding one, is that controls but not PCP-injected rats learned to restrict grooming to the most moistened areas of their bodies through repeated exposures to water sprays. The actual need to recondition misted fur on the face and head could, indeed, imply that both the increased rate of aborted chains and the anterior bias in the saline group were adaptive rather than a manifestation of high stress. When water sprays were directed onto rats' backs and abdomens instead of the face region, a *posterior* grooming bias emerged (van Erp *et al.* 1994) in spite of stressful induction. Considering the unique profile of grooming artificially provoked with a stimulation that commands water excess removal and maintenance of certain parts of the body (see van Erp *et al.* 1994 and Gispen and Isaacson 1981 for a valuable comparison between water- and novelty-induced grooming), one can infer from our new data analyses that subchronic PCP administration compromises grooming efficiency and results in unfocused body care compounded, conceivably, with an abnormal stress response.

Syntactic irregularities in rats treated with PCP (Audet *et al.* 2006) potentially arose from their inclination towards sequences lacking Phase I as disclosed in this chapter. In further analyses, we compared phase duration within completed syntactic chains and chains lacking Phase I for each group individually and for the two combined. In this latter case, Phase II lasted longer in chains lacking Phase I than in completed chains but there was no such difference for Phase III. The length of Phase II within the two chain types was not influenced by PCP treatment. We thus suspected that the chain-implementing role usually associated with Phase I was displaced to Phase II: as a matter of fact, when the lasting time of Phase II within chains lacking Phase I was compared with the combined duration of Phases I and II in completed chains, no difference was found. In sum, chains lacking Phase I seem to reflect sequential idiosyncrasies that undergo a significant increase under repeated PCP treatment. This could be due to a problem with syntactic chain initiation per se or with the motoric or coordinated execution of rapid elliptical strokes typical of Phase I that might have been blended with Phase II. Elaborate examination of movement frequency and timing (e.g., Cromwell and Berridge 1996; Matell *et al.* 2006) would be required to characterize grooming microstructure as a whole and to judge if deficiencies at these levels could be responsible for the propensity to skip Phase I after PCP exposure.

Another intriguing outcome is the stronger emphasis on Phase IV in completed syntactic chains following PCP exposure. Our analysis of grooming was adapted from the work of Berridge *et al.* (Aldridge and Berridge 1998; Berridge 1990; Berridge *et al.* 1987; Berridge and Whishaw 1992; Cromwell and Berridge 1996). Phase IV varies more in duration than the other phases due to blending of body licking into subsequent nonchain grooming (see Kalueff *et al.* 2007; Matell *et al.* 2006). In the coding of this phase, we included hind legs, tail, and genital grooming executed after the postural shift to body licking, which was slightly different from Matell *et al.*'s (2006) definition of chain completion. Despite the impossibility of articulating a firm statement, we suspect that the longer duration of Phase IV in the PCP group was linked to a greater focus on these most posterior parts following the postural shift to body licking. This, again, would be reminiscent of what Kalueff and Tuohimaa (2005) recorded in low-stress rats, but we already established that the pattern of results in PCP-treated rats was rather suggestive of increased stress. Another possibility would be the persistence of a functionless behavior triggered by subchronic PCP treatment. Whereas anterior grooming was associated with a clear external incentive, it was not the case for the less misted caudal areas, especially hind legs, tail, and genital parts. Ridley expressed in his 1994 paper (Ridley 1994) that "the ability of higher mammalian species to produce flexible, self-initiated and apparently voluntary behaviour is a considerable achievement and loss of this ability is an important feature of many types of psychopathology" (pp. 221). Enhanced grooming directed to areas mostly spared by water sprays in the PCP group certainly contradicts expectations for optimal adjustment to external stimulation and body state. As a matter of fact, it rather suggests that repeated drug treatment intensified perseverative, nonbeneficial behaviors. Nevertheless, no *super-stereotypy* in the execution of the syntactic chain was observed in our PCP group (see Berridge and Aldridge 2000a, b). The relative probability of initiating a stereotyped syntactic chain (the number of initiated chains divided by the time spent grooming) that contributes to pattern rigidity as a whole (Berridge and Aldridge 2000a, b) was equivalent between treatment conditions (Audet *et al.* 2006). Moreover, although rats under PCP treatment were more likely to complete the stereotyped syntactic chains they initiated, they still displayed an overall syntactic deficit as was addressed earlier (Audet *et al.* 2006).

From an animal modeling perspective, PCP effects on water spray-induced grooming in rats offer an experimental template for the converging examinations of various phenotypes, such as deficient hygiene, but also perseveration and behavioral sensitization. These are all pathological manifestations associated with human schizophrenia (respectively, Weisman *et al.* 1998; Morrens *et al.* 2006;

Collip *et al.* 2008) that arise in other conditions as well, including autism, obsessive–compulsive disorder, and addiction (Ridley 1994). Because grooming facilitates the identification of participating brain regions, neurotransmission circuits, and related genes, it represents an heuristic approach to the testing of anxiolytic, anxiogenic, or psychostimulant properties within pharmacological compounds.

Excessive grooming is not only expressed after exposure to stressors or to anxiogenic situations (Kalueff and Tuohimaa 2004, 2005; Komorowska and Pellis 2004; Krebs *et al.* 1996; van Erp *et al.* 1994), but also after central administration of adrenocorticotropic hormone (ACTH) (Berridge and Aldridge 2000b; Gispen and Isaacson 1981; Jolles *et al.* 1979), or peripheral and central administration of dopamine D1 receptor agonists (Berridge and Aldridge 2000b; Eilam *et al.* 1992; Matell *et al.* 2006; Van Wimersma Greidanus *et al.* 1989; Wachtel *et al.* 1992). Berridge and Aldridge (2000a, b) showed that peripheral and central activation of dopamine D1 receptors dramatically accentuated sequential super-stereotypy. On the other hand, although it markedly increased grooming, central administration of ACTH diminished the predisposition to begin a sequential stereotyped pattern and the likelihood to terminate it (Berridge and Aldridge 2000b). A similar dissociation between the amount of grooming behaviors and control over its stereotyped pattern was observed in our PCP group, thus pointing to plausible ACTH-related disturbances such as the increased plasma levels of ACTH noted with this agent (Pechnick *et al.* 1989, 1990, 2006).

In Audet *et al.* (2006), we cautiously proposed nigrostriatal D1 receptor alterations after repeated PCP treatment. A delineated portion of the anterior dorsolateral neostriatum was found responsible for the serial order of grooming actions (Cromwell and Berridge 1996). Mutant mice lacking the dopamine D1-A receptor also displayed syntactic impairments within grooming (Cromwell *et al.* 1998). Moreover, peripheral administration of a full dopamine D1 agonist decreased Phases I and IV duration within syntactic chains (Matell *et al.* 2006). Therefore, some of our data, including overexpression of grooming, lower proportion of completed syntactic chains, and disrupted phase duration within completed chains, seem consistent with abnormal nigrostriatal dopamine transmission at the level of D1 receptors following PCP treatment. How precisely D1 neurotransmission is modulated by such NMDA blocking agents is unclear and necessitates further investigation. Also relevant to the current outcomes is the modulation of purposeless behavioral manifestations by interventions targeting D2 receptors (Garner and Mason 2002). Since PCP-treated rats displayed both exaggerated grooming (expressed towards nonrelevant body parts especially) and poor syntax, one can speculate that the two agonistic cortico–striatal circuit loops were disturbed by the drug.

As first demonstrated by Audet *et al.* (2006) and confirmed by original data presented in this chapter, the grooming paradigm brings a remarkable contribution to the study in rats of PCP's effects. Selection of indexes based on two leading approaches to the analysis of grooming patterning gave rise to two pivotal sets of informative data. On the one hand, Berridge *et al.*'s pioneering work on grooming syntax served to identify stress-related abnormalities, such as the overexpression of grooming and the lower proportion of completed syntactic chains, in our PCP group. On the other hand, Kalueff *et al.*'s recent grooming analysis algorithm was highly helpful in addressing the hygienic and perseverative irregularities related to drug treatment. Also of particular importance to grooming assessment is the method used to trigger that behavior. Induction with water sprays represents a probative method to tax both the hygienic and the stress-related impacts of pharmacological agents, but also of cerebral or genetic insults. Grooming surpasses more widely used tests such as locomotion due to the availability of qualitative indexes that are concomitant with behavioral expression, thus promoting inferential interpretations of anomalies. Different parameters may significantly affect the presence and strength of our reported effects, including the dose and type of the pharmacological agent, the extent of fur moistening, the co-occurrence of stressful stimuli, etc. It thus seems important to manipulate such variables with an experimental strategy and to take those into account during data interpretation in future research.

References

Aldridge JW and Berridge KC (1998): Coding of serial order by neostriatal neurons: a "natural action" approach to movement sequence. *J Neurosci* **18**:2777–87.

Audet MC, Goulet S and Doré FY (2006): Repeated subchronic exposure to phencyclidine elicits excessive atypical grooming in rats. *Behav Brain Res* **167**:103–10.

Audet MC, Goulet S and Doré FY (2007a): Enhanced anxiety follows withdrawal from subchronic exposure to phencyclidine in rats. *Behav Brain Res* **176**:358–61.

Audet MC, Goulet S and Doré FY (2007b): Transient hypolocomotion in rats repeatedly exposed to phencyclidine: an appraisal of motor function and motivation. *Prog Neuropsychopharmacol Biol Psychiatry* **31**:142–50.

Bakker CB and Amini FB (1961): Observations on the psychotomimetic effects of Sernyl. *Compr Psychiatry* **2**:269–80.

Berridge K (1990): Comparative fine structure of action: rules of form and sequence in the grooming patterns of six rodent species. *Behaviour* **113**:21–56.

Berridge KC and Aldridge JW (2000a): Super-stereotypy I: enhancement of a complex movement sequence by systemic dopamine D1 agonists. *Synapse* **37**:194–204.

Berridge KC and Aldridge JW (2000b): Super-stereotypy II: enhancement of a complex movement sequence by intraventricular dopamine D1 agonists. *Synapse* **37**:205–15.

Berridge KC and Whishaw IQ (1992): Cortex, striatum and cerebellum: control of serial order in a grooming sequence. *Exp Brain Res* **90**:275–90.

Berridge KC, Fentress JC and Parr H (1987): Natural syntax rules control action sequence of rats. *Behav Brain Res* **23**:59–68.

Collip D, Myin-Germeys I and Van Os J (2008): Does the concept of "sensitization" provide a plausible mechanism for the putative link between the environment and schizophrenia? *Schizophr Bull* **34**:220–5.

Cromwell HC and Berridge KC (1996): Implementation of action sequences by a neostriatal site: a lesion mapping study of grooming syntax. *J Neurosci* **16**: 3444–58.

Cromwell HC, Berridge KC, Drago J and Levine MS (1998): Action sequencing is impaired in D1A-deficient mutant mice. *Eur J Neurosci* **10**:2426–32.

D'Aquila PS, Peana AT, Carboni V and Serra G (2000): Exploratory behaviour and grooming after repeated restraint and chronic mild stress: effect of desipramine. *Eur J Pharmacol* **399**:43–7.

Eilam D, Talangbayan H, Canaran G and Szechtman H (1992): Dopaminergic control of locomotion, mouthing, snout contact, and grooming: opposing roles of D1 and D2 receptors. *Psychopharmacology (Berl)* **106**:447–54.

Enginar N, Hatipoglu I and Firtina M (2008): Evaluation of the acute effects of amitriptyline and fluoxetine on anxiety using grooming analysis algorithm in rats. *Pharmacol Biochem Behav* **89**:450–5.

Garner JP and Mason GJ (2002): Evidence for a relationship between cage stereotypies and behavioural disinhibition in laboratory rodents. *Behav Brain Res* **136**:83–92.

Gispen WH and Isaacson RL (1981): ACTH-induced excessive grooming in the rat. *Pharmacol Ther* **12**:209–46.

Hanania T, Hillman GR and Johnson KM (1999): Augmentation of locomotor activity by chronic phencyclidine is associated with an increase in striatal NMDA receptor function and an upregulation of the NR1 receptor subunit. *Synapse* **31**:229–39.

Hartley JE and Montgomery AM (2008): 8-OH-DPAT inhibits both prandial and waterspray-induced grooming. *J Psychopharmacol* **22**:746–52.

Howell DC (1997): *Statistical Methods for Psychology*, 4th edn. Belmont, CA: Duxbury Press.

Javitt DC and Zukin SR (1991): Recent advances in the phencyclidine model of schizophrenia. *Am J Psychiatry* **148**:1301–8.

Jentsch JD and Roth RH (1999): The neuropsychopharmacology of phencyclidine: from NMDA receptor hypofunction to the dopamine hypothesis of schizophrenia. *Neuropsychopharmacology* **20**:201–25.

Jolles J, Rompa-Barendregt J and Gispen WH (1979): ACTH-induced excessive grooming in the rat: the influence of environmental and motivational factors. *Horm Behav* **12**:60–72.

Kalueff AV and Tuohimaa P (2004): Grooming analysis algorithm for neurobehavioural stress research. *Brain Res Brain Res Protoc* **13**:151–8.

Kalueff AV and Tuohimaa P (2005): The grooming analysis algorithm discriminates between different levels of anxiety in rats: potential utility for neurobehavioural stress research. *J Neurosci Methods* **143**:169–77.

Kalueff AV, Aldridge JW, LaPorte JL, Murphy DL and Tuohimaa P (2007): Analyzing grooming microstructure in neurobehavioral experiments. *Nat Protoc* **2**:2538–44.

Kappas A (1995): *Coder2*. Québec (Canada): École de psychologie, Université Laval.

Komorowska J and Pellis SM (2004): Regulatory mechanisms underlying novelty-induced grooming in the laboratory rat. *Behav Processes* **67**:287–93.

Krebs H, Macht M, Meyers P, Weijers HG and Janke W (1996): Effects of stressful noise on eating and non-eating behavior in rats. *Appetite* **26**:192–202.

Lehrmann E, Colantuoni C, Deep-Soboslay A *et al.* (2006): Transcriptional changes common to human cocaine, cannabis and phencyclidine abuse. *PLoS ONE* **1**:e114.

Luby ED, Cohen BD, Rosenbaum G, Gottlieb JS and Kelley R (1959): Study of a new schizophrenomimetic drug; sernyl. *AMA Arch Neurol Psychiatry* **81**:363–9.

Martin S and van den Buuse M (2008). Phencyclidine-induced locomotor hyperactivity is enhanced in mice after stereotaxic brain serotonin depletion. *Behav Brain Res* **191**:289–93.

Matell MS, Berridge KC and Aldridge JW (2006): Dopamine D1 activation shortens the duration of phases in stereotyped grooming sequences. *Behav Processes* **71**:241–9.

McGregor IS, Callaghan PD and Hunt GE (2008): From ultrasocial to antisocial: a role for oxytocin in the acute reinforcing effects and long-term adverse consequences of drug use? *Br J Pharmacol* **154**:358–68.

Millan MJ, Loiseau F, Dekeyne A *et al.* (2008): S33138 (*N*-[4-[2-[(3*aS*,9*bR*)-8-cyano-1,3*a*,4,9*b*-tetrahydro[1] benzopyrano[3,4-*c*]pyrrol-2(3*H*)-yl)-ethyl]phenyl-acetamide), a preferential dopamine D3 versus D2 receptor antagonist and potential antipsychotic agent: III. Actions in models of therapeutic activity and induction of side effects. *J Pharmacol Exp Ther* **324**:1212–26.

Morrens M, Hulstijn W, Lewi PJ, De Hert M and Sabbe BG (2006): Stereotypy in schizophrenia. *Schizophr Res* **84**:397–404.

Morris BJ, Cochran SM and Pratt JA (2005): PCP: from pharmacology to modelling schizophrenia. *Curr Opin Pharmacol* **5**:101–6.

Mouri A, Noda Y, Enomoto T and Nabeshima T (2007): Phencyclidine animal models of schizophrenia: approaches from abnormality of glutamatergic neurotransmission and neurodevelopment. *Neurochem Int* **51**:173–84.

Noda Y, Yamada K, Furukawa H and Nabeshima T (1995): Enhancement of immobility in a forced swimming test by subacute or repeated treatment with phencyclidine: a new model of schizophrenia. *Br J Pharmacol* **116**:2531–7.

Noda Y, Mamiya T, Furukawa H and Nabeshima T (1997): Effects of antidepressants on phencyclidine-induced enhancement of immobility in a forced swimming test in mice. *Eur J Pharmacol* **324**:135–40.

Pechnick RN, George R and Poland RE (1989): Naloxone does not antagonize PCP-induced stimulation of the pituitary-adrenal axis in the rat. *Life Sci* **44**:143–7.

Pechnick RN, Chun BM, George R, Hanada K and Poland RE (1990): Determination of the loci of action of phencyclidine on the CNS–pituitary–adrenal axis. *J Pharmacol Exp Ther* **254**:344–9.

Pechnick RN, Bresee CJ and Poland RE (2006): The role of antagonism of NMDA receptor-mediated neurotransmission and inhibition of the dopamine reuptake in the neuroendocrine effects of phencyclidine. *Life Sci* **78**:2006–11.

Pijlman FT and van Ree JM (2002): Physical but not emotional stress induces a delay in behavioural coping responses in rats. *Behav Brain Res* **136**:365–73.

Ridley RM (1994): The psychology of perseverative and stereotyped behaviour. *Prog Neurobiol* **44**:221–31.

Robertson BJ, Boon F, Cain DP and Vanderwolf CH (1999): Behavioral effects of anti-muscarinic, anti-serotonergic, and anti-NMDA treatments: hippocampal and neocortical slow wave electrophysiology predict the effects on grooming in the rat. *Brain Res* **838**:234–40.

Roeling TA, van Erp AM, Meelis W, Kruk MR and Veening JG (1991): Behavioural effects of NMDA injected into the hypothalamic paraventricular nucleus of the rat. *Brain Res* **550**:220–4.

Sams-Dodd F (1998a): Effects of continuous D-amphetamine and phencyclidine administration on social behaviour, stereotyped behaviour, and locomotor activity in rats. *Neuropsychopharmacology* **19**:18–25.

Sams-Dood F (1998b): A test of the predictive validity of animal models of schizophrenia based on phencyclidine and D-amphetamine. *Neuropsychopharmacology* **18**:293–304.

Spruijt BM, van Hooff JA and Gispen WH (1992): Ethology and neurobiology of grooming behavior. *Physiol Rev* **72**:825–52.

Turgeon SM, Lin T and Subramanian M (2007). Subchronic phencyclidine exposure potentiates the behavioral and c-Fos response to stressful stimuli in rats. *Pharmacol Biochem Behav* **88**:73–81.

van Erp AM, Kruk MR, Meelis W and Willekens-Bramer DC (1994): Effect of environmental stressors on time course, variability and form of self-grooming in the rat: handling, social contact, defeat, novelty, restraint and fur moistening. *Behav Brain Res* **65**:47–55.

Van Wimersma Greidanus TB, Maigret C, Torn M *et al.* (1989): Dopamine D-1 and D-2 receptor agonists and antagonists and neuropeptide-induced excessive grooming. *Eur J Pharmacol* **173**:227–31.

Wachtel SR, Brooderson RJ and White FJ (1992): Parametric and pharmacological analyses of the enhanced grooming response elicited by the D1 dopamine receptor agonist SKF 38393 in the rat. *Psychopharmacology (Berl)* **109**:41–8.

Weisman AG, Nuechterlein KH, Goldstein MJ and Snyder KS (1998): Expressed emotion, attributions, and schizophrenia symptom dimensions. *J Abnorm Psychol* **107**:355–9.

Xu X and Domino EF (1994): Phencyclidine-induced behavioral sensitization. *Pharmacol Biochem Behav* **47**:603–8.

Xu X and Domino EF (1999): A further study on asymmetric cross-sensitization between MK-801 and phencyclidine-induced ambulatory activity. *Pharmacol Biochem Behav* **63**:413–16.

Zhao Y, Valdez GR, Fekete EM *et al.* (2007): Subtype-selective corticotropin-releasing factor receptor agonists exert contrasting, but not opposite, effects on anxiety-related behavior in rats. *J Pharmacol Exp Ther* **323**:846–54.

6

Modulatory effects of estrogens on grooming and related behaviors

RACHEL A. HILL AND WAH CHIN BOON

Summary

Both the estrogen-synthesizing enzyme and estrogen receptors are present in the brain. The distribution of estrogen-sensitive cells in the brain corresponds to regions that control sexual differentiation, masculine and feminine sexual behaviors, and aggressive behaviors as well as grooming. All these are indicative that estrogens play diverse roles in several neural circuits. From observing animal models and psychiatric patients, it is evident that estrogens have modulatory effects on grooming and related behaviors.

Introduction

Traditionally viewed as female reproductive hormones, estrogens have in the past decade been shown to have widespread biological actions in both males and females. Estrogens are C-18 steroids derived from cholesterol and occur naturally in the human body in three different forms: 17β-estradiol (E_2), estrone (E_1), and estriol (E_3) (Young *et al.* 1964). The last step in estrogen biosynthesis is the aromatization of androgens, including testosterone and androstenedione, to estradiol and estrone respectively. In addition to the reproductive organs, estrogens are expressed in many tissues including: breast, fat, muscle, bone, and brain (Carreau *et al.* 1999; Jenkins *et al.* 1993; Sasano and Ozaki 1997; Sasano *et al.* 1997, 1998; Simpson *et al.* 1997a). Investigations on the mechanism of action of estrogens have escalated greatly over the past decade due to the discovery of three types of

Neurobiology of Grooming Behavior, eds. Allan V. Kalueff, Justin L. LaPorte, and Carisa L. Bergner. Published by Cambridge University Press.

estrogen receptors: estrogen receptor α (ERα; Green *et al.* 1986) estrogen receptor β (ERβ; Kuiper *et al.* 1996), and membrane ER (ERX; Watson *et al.* 1999). However, the last is probably ERα in most cases.

The enzyme aromatase, which converts androgens into estrogens is expressed abundantly throughout the brain in several discrete regions (Harada and Yamada 1992; Hutchison *et al.* 1997; Lephart *et al.* 2001). In addition, the ERs are differentially expressed throughout the brain, with some sex differences in expression in specific nuclei of the hypothalamus (Mitra *et al.* 2003). The presence of aromatase and ERs throughout the brain suggested a possible neuroprotective role for estrogens, and indeed a flourishing amount of literature has reported widespread neuroprotective and neuromodulative estrogenic effects. These neuroprotective actions of estrogens have been implicated in a therapeutic setting for neurodegenerative diseases such as Alzheimer's disease (Tang *et al.* 1996; Xu *et al.* 1998) and Parkinson's disease (Cyr *et al.* 2002), and in brain injuries as a result of stroke or ischemic attacks (Wise *et al.* 2001b). Estrogens have also been shown to play a role in memory and cognition tasks (Henderson *et al.* 1996; Wise *et al.* 2001b), as well as sexual and aggressive behaviors (reviewed by Mong and Pfaff 2003).

Several effects of estrogens have been suggested to be responsible for these neuroprotective actions, including modulation of synaptogenesis, protection from free radicals, anti-inflammatory effects, regulation of calcium channels, and protection via increasing cerebral blood flow (Amantea *et al.* 2005). A vast amount of literature has also been reported on the anti-apoptotic properties of estrogens (Amantea *et al.* 2005; Garcia-Segura *et al.* 1999; Manthey and Behl 2006). The majority of these effects are mediated through the classic nuclear ER activation (Zhao *et al.* 2004), and in addition, rapid actions appear to be regulated by the recently identified ERX (Amantea *et al.* 2005). Estrogen receptor-independent effects appear to be due to the antioxidant free-radical scavenging properties of the estrogen steroid structure (Wise *et al.* 2001b).

While several studies have reported that estrogens play a pivotal role in neuroprotection in females as demonstrated in studies of postmenopausal compared to premenopausal women suffering from brain injuries sustained by stroke (Wise *et al.* 2001a), estrogens may also play a role in the male brain. For example, the higher prevalence of Parkinson's disease in the male may be attributed to lower estrogen levels in the brain (Cyr *et al.* 2002). In addition, male sexual behavior (Robertson *et al.* 2001) and aggression (Toda *et al.* 2001) have been shown to be under the control of local steroid hormone levels. More recently, some intriguing data have shown that male estrogen-deficient (i.e., the aromatase knockout, ArKO) mice exhibit compulsive-like behaviors such as grooming and excessive running-wheel activity (Hill *et al.* 2007). These studies demonstrate a role for estrogens in the modulation of compulsive-like behaviors. This chapter will review the effects of estrogens on neurobiology and behaviors.

Aromatase and estrogen receptor distribution in the brain

Aromatase expression in the brain

Aromatase is abundantly expressed throughout the mouse brain, with the highest level of aromatase activity reported in the hypothalamic region of the diencephalon (Harada and Yamada 1992; Hutchison *et al.* 1997). The diencephalon region contains the thalamus, metathalamus, habenula, and hypothalamus. The development of aromatase expression within the diencephalon is genetically controlled from approximately embryonic day 12 onwards (Abe-Dohmae *et al.* 1997; Harada and Yamada 1992). From this point aromatase mRNA expression levels increase rapidly during fetal and neonatal development and peak at around three to four days postnatal, then levels gradually decrease to lower adult levels (Harada and Yamada 1992). Similar aromatase expression patterns are found in the hippocampus of mice, whereby aromatase is increased during the first two postnatal weeks and then decreases gradually to lower adult levels (Ivanova and Beyer 2000). In both the hippocampus and the diencephalon regions aromatase mRNA expression is higher in males than females (Ivanova and Beyer 2000; Yamada *et al.* 1993). This is particularly evident in the diencephalon where males express 150% higher aromatase levels than females (Yamada *et al.* 1993). More specifically, the hypothalamic region, which resides within the diencephalon, shows a much higher aromatase expression level in males when compared with female embryonic mice (Hutchison *et al.* 1997). Aromatase activity and mRNA are also expressed in the embryonic striatum and levels tend to increase postnatally in this region as well. However, no sex difference in aromatase expression was found within the striatum (Kuppers and Beyer 1998). In the cortex, however, aromatase expression has been reported to be either very low (Hutchison *et al.* 1997) or undetectable (Abe-Dohmae *et al.* 1997). Aromatase is also not detectable in the cerebellum or medulla (Harada and Yamada 1992).

In general, aromatase is normally only found to be expressed in neurons of the brain; however, it is interesting to note that upon induced brain injury in the mouse, by either kainic acid injections or by a penetrating injury, aromatase becomes expressed in reactive glia in the brain regions affected by the injury (Garcia-Segura *et al.* 1999). Not surprisingly, testosterone treatment increases aromatase activity within the neurons of the hypothalamus, and this is an androgen-receptor (AR) mediated event (Beyer and Hutchison 1997; Hutchison *et al.* 1997). Furthermore, tonic estrogen treatment (in the form of a pellet) was found to upregulate aromatase expression in the hippocampus, whereas 17β-estradiol injections actually decreased aromatase expression in the same region (Iivonen *et al.* 2006). This study emphasizes the importance of selecting the correct method of steroid treatment.

In the nonhuman primate rhesus monkey, aromatase is also expressed at high levels within particular regions of the hypothalamus and in addition the amygdala. These regions include the medial preoptic area–anterior hypothalamus (MPO-AH), ventromedial hypothalamus (VMH), bed nucleus of the stria terminalis (BnST), cortical amygdala, and medial amygdala (Roselli and Resko 2001). Similar to the rodent, testosterone acts through the AR to mediate aromatase expression within the nonhuman primate brain (Roselli and Resko 2001). Of particular interest is a very strong labeling of aromatase and AR mRNA containing cells within the central component of the MPO in a region analogous to the sexually dimorphic nucleus of the medial preoptic area (SDN-MPO) found in the rat (Roselli and Resko 1997; Selmanoff *et al.* 1976).

In the human, aromatase is expressed in the temporal and frontal neocortex, the hippocampus and the subcortical white matter of the temporal lobe, and the hypothalamus (Stoffel-Wagner 2001). Aromatase gene expression in the human brain is regulated by a brain specific exon 1 promoter (Honda *et al.* 1994). However, a more recent report has shown different patterns of utilization of exon 1 in different areas of the human brain tissue, for example, hypothalamic or limbic structures utilize exon 1f and/or 1d, while other areas including the pons and frontal lobe utilize exon 1b (Sasano *et al.* 1998). While levels of aromatase mRNA expression have been reported to be higher in males within the MPO of rats (Lephart *et al.* 2001), sheep (Roselli *et al.* 2004), and quail (Balthazart *et al.* 2003), no sex differences have been reported in aromatase expression throughout the human brain (Stoffel-Wagner 2001).

Estrogen receptor α and β expression in the brain

In the mouse, ERα and ERβ are abundantly expressed throughout the brain within the cell nuclei of select regions including the olfactory bulb, cerebral cortex, septum, preoptic area, BnST, amygdala, paraventricular nucleus (PVN), thalamus, ventral tegmental area, substantia nigra, dorsal raphe, locus coeruleus, and cerebellum. Extracellular immunoreactive ER staining has also been detected in the olfactory bulb, CA3 stratum lucidum and CA1 stratum radiatum of the hippocampus, and the cerebellum (Mitra *et al.* 2003). Interestingly, the expression patterning of ERs within the brain is similar to aromatase, whereby levels occur prenatally and increase until birth (Couse *et al.* 1997).

While both receptors are generally expressed in a similar distribution in the mouse, there are a few differences in expression. Nuclear ERα immunoreactivity was found to be the predominant subtype in the hippocampus, and most of the hypothalamus, whereas very little ERα may be detected in the cerebral cortex and cerebellum (Mitra *et al.* 2003). High levels of ERα mRNA were detected in the pituitary while no ERβ mRNA expression was noted. This same study reported that levels of ERβ expression were approximately half that of ERα mRNA expression in the hypothalamus (Couse *et al.* 1997). However, in the PVN (a region within the hypothalamus), ERβ appears to be the more predominant receptor subtype in terms of protein levels (Zhang *et al.* 2004). These discrepancies in the literature may be due to differences in protein and mRNA expression as while Couse and colleagues measured mRNA, Zhang was measuring protein expression. In addition, Couse and colleagues (1997) extracted RNA from the whole hypothalamus and measured ER expression by RNase protection assay techniques. While this is a highly accurate technique, the hypothalamus contains several different nuclei that serve many different functions, thus the distribution of ERs in this region may change considerably. Indeed, mRNA expression of ERα and ERβ, as determined by *in situ* hybridization, in the rat has proven many differences in ER subtype distribution within specific hypothalamic regions (Shughrue *et al.* 1997). For example, while the medial preoptic area displays similar levels of ERα to ERβ, the paraventricular nucleus shows no ERα expression but high ERβ mRNA expression, and the arcuate nucleus of the hypothalamus contains a high expression of ERα but relatively low expression of ERβ (Shughrue *et al.* 1997). This extensive study of mRNA expression of the two ER subtypes in the rat brain confirms the dominant expression of ERβ in the olfactory bulb and the mesencephalon and cortex as well as the eye, while ERα is the more predominant receptor subtype in the subfornical organ, periventricular nucleus, the arcuate nucleus, and the ventromedial nucleus (VMN) of the hypothalamus, the dentate nucleus, and dorsomedial tegmental nucleus (Shughrue *et al.* 1997). Such a thorough study of ER expression in the rat central nervous system (CNS) should certainly be replicated in the mouse CNS.

Interestingly, sex differences in ERβ expression have been reported in the mouse MPO-AH, with males showing higher levels than females – similar to the aromatase expression patterns previously described in the rat (Lephart *et al.* 2001). In addition, a higher level of both ERα and ERβ have been reported in adult females compared to male mice (Sharma and Thakur 2006). Sex differences in ERβ mRNA and protein expression levels have been reported in the anteroventral periventricular nucleus (AVPV) of the rat preoptic area, in this case with females expressing much larger numbers of ERβ positive cells than males (Orikasa *et al.* 2002). In humans, more nuclear ERα was found in the diagonal band of Broca and caudally in the medial mammillary nucleus, suprachiasmatic nucleus, and VMN, while more intense nuclear ERα was found in the SDN-MPO, PVN, and lateral hypothalamus of males than females (Kruijver *et al.* 2002).

The majority of animal studies have analyzed the physiological effects of estrogens by methods of ovariectomy followed by placebo or 17β-estradiol replacement. A major problem with such studies is that local levels of brain aromatase may still convert circulating androgens to estrogens. Hence their estrogen-deficient model may not be completely void of estrogens. The generation of the ArKO mouse and the estrogen receptor knockout (ERKO) mice has enabled researchers to characterize the effects of a total lack of estrogens within the brain. Brain and behavioral phenotypes arising from these knockout animals are discussed below.

Animal models

Brain and behavior phenotypes of the aromatase knockout (ArKO) mouse model

In order to study the physiological effects of estrogens, an ArKO mouse was generated (Fisher *et al.* 1998). The final step in estrogen biosynthesis is catalyzed by the enzyme cytochrome P450 aromatase, the product of the *Cyp19* gene. Therefore disruption to the *Cyp19* gene would lead to an aromatase deficiency and consequently a total lack of estrogens. Aromatase knockout mice were generated by the insertion of a neomycin resistant cassette to exon IX of the *Cyp19* gene (Fisher *et al.* 1998). Exon IX was selected for targeted disruption because, firstly, this exon is highly conserved among all aromatase cDNA reported to date (Simpson *et al.* 1997b). Secondly, almost all naturally occurring mutations in the human *CYP19* gene, which cause aromatase deficiency in both men and women, have been located in exons IX and X (Carani *et al.* 1997; Conte *et al.* 1994; Ito *et al.* 1993; Morishima *et al.* 1995; reviewed by Jones *et al.* 2007). Finally, computer alignment of the aromatase amino acid sequence indicated that this sequence within exon IX contained several important structural features of the enzyme, including the K and K′ helices, and a number of β-sheet regions (Fisher *et al.* 1998; Graham-Lorence *et al.* 1995).

Early studies of the brain of ArKO mice reported an upregulation of ERα expression in the MPO of the hypothalamus of ArKO mice when compared to wild type (WT) littermates (Agarwal *et al.* 2000). Aromatase knockout mice also presented with a significant decrease in arginine vasopressin (AVP) immunoreactive structures in the BnST and the medial amygdala, and in the lateral septum and PVN of the hypothalamus, while no changes in AVP immunoreactivity were found in the MPO (Plumari *et al.* 2002), indicating the involvement of estrogens in the regulation of the AVP system in the mouse forebrain. In regards to female ArKO mice, reduced levels of lordosis have been correlated with impairments in olfactory investigations of odors (Bakker *et al.* 2002, 2003). In addition, the female ArKO mice show greater ischemic damage after middle cerebral artery occlusion, when compared to WT controls, demonstrating a neuroprotective role for estrogens (McCullough *et al.* 2003). Aromatase knockout females also show decreased active behaviors such as struggling and swimming and increased passive behavior such as floating during a forced swim test situation. Such behaviors are indicative of depressive-like symptoms, and these behaviors were accompanied by a concomitant decrease in serotonergic activity within the hippocampus of ArKO females compared to WT controls (Dalla *et al.* 2004). Interestingly, male ArKO mice show normal levels of activity, exploration, anxiety, and depressive-like symptomology (Dalla *et al.* 2005). Disruptions within the hippocampus have also been described by Boon *et al.* (2005), who reported significantly higher transcript levels of the N-methyl-D-aspartate (NMDA) receptor subunit NR2B in female ArKO compared to WT controls.

Both male and female ArKO mice performed significantly worse than WT controls in tests for short-term memory by use of the Y-maze test, thus revealing an important role for estrogens in the memory process (Martin *et al.* 2003). Male ArKO mice only show an age-related reduction in prepulse inhibition (PPI), and significantly greater amphetamine-induced hyperactivity, thus indicating a neuroprotective effect of estrogens, particularly in the aging brain, in mechanisms involved in PPI and locomotor activity regulation (van den Buuse *et al.* 2003). This led authors to postulate a role for estrogens in the regulation of dopaminergic activity in the male (van den Buuse *et al.* 2003).

Indeed, estradiol but not testosterone had been shown to significantly reduce the behavioral changes (oral stereotypies, grooming and sitting behavior) induced by both dopamine antagonist haloperidol (catalepsy) and dopamine agonist apomorphine, and this effect was more pronounced in neonatally treated animals (Häfner *et al.* 1991). These results suggested a downregulation of dopamine neurotransmission by estradiol, which is supported by the results of ^{3}H-sulpiride binding determinations in brain homogenates from the same animals: estradiol caused a 2.8-fold reduction of dopamine receptor affinity to sulpiride, a selective antagonist of postsynaptic dopamine D2 receptors.

Recent studies have demonstrated compulsive-like behaviors in the mouse upon estrogen deficiency. Adult male ArKO mice display a decrease in normal ambulatory activity when compared to WT, while running-wheel activity was significantly increased in male ArKO mice compared to WT (Hill *et al.* 2007). This specific behavior of excessive running-wheel activity was returned to WT levels upon estrogen replacement, although this was not statistically significant when compared to KO levels (Hill *et al.* 2007). Previous studies have consistently reported that running-wheel activity increases with increasing levels of estrogens (Fahrbach *et al.* 1985, King 1979, Morgan and Pfaff 2001; Roy and Wade 1975). However, most of these studies were carried out on females only. Hill and colleagues (2007) found a decrease in running-wheel activity in female ArKO mice, compared to WT, although this was not significant. This concurs with previous reports that estrogens increase running-wheel activity (Fahrbach *et al.* 1985; King 1979; Morgan and Pfaff 2001; Roy and Wade 1975). In contrast, male ArKO mice showed increased running-wheel activity when compared to WT animals.

Excessive grooming behavior has been referred to as an obsessive–compulsive related behavior in both animals and humans (Ferris *et al.* 2001; Greer and Capecchi 2002; Nordstrom and Burton 2002; Rapoport 1991). Analysis of grooming behaviors revealed extreme grooming activities in the male ArKO mice that are at a significantly higher level than those of their WT littermates (Hill *et al.* 2007). This excessive grooming behavior of the male ArKO mice was sufficiently restored to WT levels following three weeks of 17β-estradiol replacement (Hill *et al.* 2007), thus indicating that in adult males, induced grooming activity via a water-mist spray may be regulated by levels of estrogens. No differences in grooming behavior were observed in female ArKO compared to WT, indicating that estrogen actions on this obsessive–compulsive disorder (OCD) related behavior are specific to males only (Hill *et al.* 2007). Clinical trials using flutamide (an androgen receptor antagonist) treatment for OCD patients proved no changes in measures of obsessions and compulsions over an eight-week trial period (Altemus *et al.* 1999), suggesting that any effects of steroids on OCD symptoms are more likely to be mediated through estrogen receptors (Altemus *et al.* 1999).

The enzyme catechol-O-methyltransferase (COMT) is involved in the inactivation of catecholaminergic neurotransmitters such as dopamine (Axelrod and Tomchick 1958). Hill *et al.* (2007) found that male ArKO mice show a decrease in COMT expression in the hypothalamic region. Decreases in COMT enzyme activity may lead to increases in dopamine levels, which are paralleled by increases in grooming and running-wheel behaviors. Development of an excessive grooming phenotype has also been reported in the dopamine transporter knockdown mouse model (Berridge *et al.* 2005), which is hyperdopaminergic. It should be noted that low COMT activity may be a risk factor for OCD in male patients (Karayiorgou *et al.* 1997; Pooley *et al.* 2007); Furthermore, it has been established that the hypothalamus is involved in grooming (for a review see Kruk *et al.* 1998) and running-wheel activities (Rhodes *et al.* 2003). Therefore all the above data support the correlation between the low hypothalamic COMT level and excessive grooming/wheel-running activities we observed in six-month-old male ArKO mice.

Current literature strongly implicates the frontal cortex in the regulation of OCD behaviors (for a review see Graybiel and Rauch 2000), especially from evidence gathered through neuroimaging (Friedlander and Desrocher 2006). However, Western blot data did not show any differences between the COMT protein levels in the frontal cortex of either genotype (Hill *et al.* 2007). Given that in the COMT knockout (COMT$^{-/-}$) mouse, dopamine accumulation in nerve endings has been found to be more evident in the striatum and hypothalamus than in the cortex (Huotari *et al.* 2002), it would appear that the effects of decreased COMT on dopamine levels are more evident in the hypothalamic region, as Hill *et al.* (2007) have confirmed. Behavioral phenotypes arising from these increased dopamine levels in the hypothalamus may be due to direct effects on specific areas such as the medial preoptic area previously shown to regulate both grooming (Lumley *et al.* 2001) and running-wheel behaviors (Fahrbach *et al.* 1985; King 1979).

No obsessive–compulsive behavior has been reported in the COMT$^{-/-}$ mouse although sexual dimorphisms have been reported. The female COMT$^{-/-}$ has been reported to display impaired emotional reactivity, while male heterozygous COMT$^{-/+}$ mice tended to be significantly more aggressive (Gogos *et al.* 1998). Attenuation of the startle response by PPI is a measure of sensorimotor gating, and it has been reported that OCD patients display PPI deficit (Kumari *et al.* 2001). However, PPI experiments on the COMT$^{-/-}$ mice revealed no significant effect of genotype in either male or female animals (Gogos *et al.* 1998). In contrast to the COMT$^{-/-}$ mice, male ArKO exhibit PPI deficits in an age-dependent manner (van den Buuse *et al.* 2003). The lack of obsessive–compulsive related behaviors in the COMT$^{-/-}$ mice could be explained by the age of the animals that were studied – since the OCD phenotype of male ArKO did not become apparent until six months of age. Both grooming and running-wheel studies were performed at five and twelve weeks in the ArKO and WT of both sexes but no significant differences between the two genotypes were observed (Hill *et al.* 2007). Therefore, by the same token, no OCD behavior would be apparent in the reported behavioral studies (Gogos *et al.* 1998) at 11 to 16 weeks in COMT$^{-/-}$ mice. We expect that any excessive or compulsive behaviors would not be noticeable until six months of age in these COMT$^{-/-}$ animals. Hill *et al.* (2007) observed that only six-month-old male ArKO mice show OCD-related behaviors paralleled by decreased COMT levels in the hypothalamus.

As estrogens are required during neuronal development, we cannot rule out the possibility that the OCD-related phenotype presented in the male ArKO mice is a result of estrogen deficiency during development. Nonetheless, just three weeks of 17β-estradiol replacement did restore the COMT levels in the ArKO hypothalamus with concomitant amelioration of the compulsive behaviors (Hill *et al.* 2007).

Thus, decreased estrogen levels correlate with a specific decrease in hypothalamic COMT expression and development of compulsive behaviors in male mice. This relationship, if proven to be true in male OCD patients, could open up a potentially new avenue for clinical intervention.

Brain and behavior phenotypes of estrogen receptor knockout (ErKO) mouse models

The first ERα knockout (ERαKO) mouse was generated by targeted insertional disruption of exon 2 of the ERα gene, which disrupts the reading frame (Lubahn *et al.* 1993). The ERβ knockout (ERβKO) mouse was generated in 1998 by insertion of a neomycin resistance gene into exon 3 of the ERβ coding gene (Krege *et al.* 1998). In order to study the differential effects of the two ER subtypes, Dupont and colleagues (2000) independently generated their own ERαKO, ERβKO, and double ERαβKO models. Although the ERβKO mouse was generated in a similar fashion (Krege *et al.* 1998), the ERαKO mouse was generated by targeted disruption of exon 3 with a Cre-lox system (Dupont *et al.* 2000). Double ERαβKO heterozygous mice ($ER\alpha^{+/-}$ / $ER\beta^{+/-}$) were obtained from breeding $ER\alpha^{+/-}$ females with $ER\beta^{-/-}$ males. These mice were then bred to obtain homozygous ERαβKO mice (Dupont *et al.* 2000)

Several studies have utilized the ERKO models in order to study the potential ER-mediated effects of estrogens as a neuroprotective agent. Studies in general have been based upon previous findings of estrogen-regulated neuroprotection. One such basic understanding is that estrogens have been shown to modulate serotonin levels in the dorsal raphe nucleus (Alves *et al.* 2000; Gundlah *et al.* 2005). However, differential effects of ER-mediated estrogen regulation may be seen depending on the specific brain region. For instance, 17β-estradiol treatment effectively increased tryptophan hydroxylase-1 (TPH-1) (serotonin synthesizing enzyme) mRNA expression in the midbrain of WT and ERαKO mice, but not ERβKO, indicating ERβ is responsible for mediating estrogen-regulated TPH expression in the mouse dorsal raphe nucleus (Gundlah *et al.* 2005). This study correlates with previous reports that show that estrogens and progesterone modulation of serotonergic function is regulated via ERβ, as no differences in progesterone receptor, ERα, and TPH immunoreactive colocalization expression were seen in WT compared to ERαKO mice (Alves *et al.* 2000). However, the hippocampus cells immunoreactive for progesterone receptors after estrogen induction were only found in WT animals, but not in ERαKO animals, demonstrating a role for ERα in estrogen-induced progesterone and possible TPH expression in the hippocampus (Alves *et al.* 2000). These studies point out the different effects of estrogens depending on ER subtype expression within the particular brain region of interest.

Estrogens have also been shown to increase the numbers of phospho-cAMP response element binding protein (CREB) immunoreactive cells in the mouse brain, again in region-specific manners, and this is also thought to be due to differential ER subtype expression (Abraham *et al.* 2004). By comparing data from ERαKO and ERβKO animals, it could be deduced whether this estrogen-mediated effect is mediated through ERα or ERβ. In brain areas such as the medial septum, the effect seemed to be ERβ mediated, as the estrogen response is absent in ERβKO mice but normal in ERαKO mice. In contrast, in the ventrolateral aspect of the hypothalamic ventromedial nucleus, the response is mediated through ERα as no phospho-CREB increase was found in the ERαKO but it was observed in the ERβKO. In the medial preoptic nucleus, which expresses both ER subtypes, the estrogen-induced phosphorylation of CREB was normal in both ERαKO and ERβKO mice (Abraham *et al.* 2004) but absent in the double ERαβKO mice, suggesting a compensatory mechanism may be in place here when one ER is dysfunctional (Abraham *et al.* 2004). Suggestions of compensatory mechanisms by ERs have also been inferred in regards to the hypothalamic–pituitary–adrenal and gonadal axes of ERβKO mice. Several neurotransmitters in this region coexpress ERβ; however, in the ERβKO mouse, the region is relatively unaffected, indicating compensation by ERα (Bodo and Rissman 2006). This study reiterates the varying effects of ER-mediated functions depending on the specific brain region.

The ERαKO mouse has also been an efficient model to determine the predominant ERα involvement in estrogen regulation of satiation signaling, feeding effects (Geary *et al.* 2001), and in estrogen regulation of gonadotropin signaling pathways and luteinizing hormone (LH) secretion from the pituitary (Couse *et al.* 2003; Lindzey *et al.* 1998). The ERα subtype was also shown to be the predominant receptor in estrogen-mediated neuroprotection from ischemic and stroke-induced insults, by comparing ERαKO to WT mice (Ardelt *et al.* 2005; Dubal *et al.* 2001). Significant morphological abnormalities have been found in the brains of ERβKO mice. There is a regional neuronal hypocellularity, with severe neuronal deficit in the somatosensory cortex, particularly in layers II, III, IV, and V, and this is accompanied by an intense proliferation of astroglial cells in the limbic system but not in the cortex (Wang *et al.* 2001). This phenotype is exacerbated with age (Wang *et al.* 2001). Put together, these studies highlight the usefulness of all three ERKO models (ERαKO, ERβKO, and ERαβKO) in determining the pathways involved in estrogen-mediated neuroprotection.

In fact, some studies have found that the use of ERαKO and ERβKO models has actually led to the discovery that neither ER subtype may be involved in estrogen-regulated neuroprotective mechanisms, but rather the proposed third membrane-bound estrogen receptor is responsible. Indeed, the membrane-bound, or ERX, receptor is believed to be responsible for estrogen-regulated activation of the mitogen-activated protein kinase (MAPK) cascade (Singh *et al.* 2000; Toran-Allerand 2000). Furthermore, the ERX receptor, which is believed to be situated within the caveolar-like microdomains of postnatal but not adult WT and ERKO neocortex and uterine plasma membranes, is re-expressed in the adult brain following ischemic stroke injury (Toran-Allerand *et al.* 2002). Such studies represent great prospects in rapid-acting neuroprotective actions of estrogens in the therapeutic setting.

Sexual behavior studies on ERKO models revealed that male ERαKO mice showed reduced levels of intromission and ejaculation, while mounting behaviors were unaffected (Ogawa *et al.* 1997); in contrast male ERβKO mice displayed all components of sexual behavior, with no apparent effect of genotype (Ogawa *et al.* 1999). However, the double ERαβKO does not show any components of sexual behavior, including mounting behavior (Ogawa *et al.* 2000). The sexual behavior phenotype of the ERαβKO is similar to the male ArKO sexual behavior phenotype. Female ERαKO mice are infertile and do not appear to engage in reproductive behaviors for two reasons: one is that the stud male believes that the ERαKO females are in fact intruder males and proceed to attack the females rather than mount; and secondly, even after treatment with strong cutaneous stimuli, which usually promote chlordosis behavior, the ERαKO female does not display any sexual behavioral characteristics such as immobilization or vertebral dorsiflexion (see review by Ogawa *et al.* 1996). This was further confirmed by the work of Rissman and colleagues who found by use of the ERαKO model that ERα is indeed required for sexual receptivity but not essential for female attractivity (Rissman *et al.* 1997). However, male ERαKO mice do not exhibit the same social preferences for female mice as do their WT littermates (Rissman *et al.* 1999). Other behavioral phenotypes of the male ERαKO include a loss of aggressive behavior and typical female rather than typical male emotional responses in open-field tests (for a review see Ogawa *et al.* 1996). Male ERβKO mice, on the other hand, display a loss of visuospatial learning, and, further, a role for ERβ has been suggested in the defeminization of sexual behavior (for a review see Bodo and Rissman 2006).

Previous studies on both male and female ERαKO and ERβKO mice revealed that while there were no differences in running-wheel activity in either untreated ERαKO or ERβKO compared to WT, estradiol benzoate treatment increased running-wheel activity in both ten- to twelve-week-old male and female WT and ERαKO mice, but not in ERβKO, suggesting that ERα is the predominant receptor involved in estrogen-stimulated running-wheel activity (Ogawa *et al.* 2003). However, caution should be taken when interpreting the results of knockout mice; the absence of one receptor may result in the development of compensatory mechanisms or altered hormonal profiles.

By using estrogen receptor selective agonists Lund *et al.* (2005) have demonstrated that the ERβ-selective agonist diarylpropionitrile (DPN) significantly decreased the time spent grooming in both male and female gonadectomized rats during open-field tests, whereas the ERα-selective agonist propyl-pyrazole-triol (PPT) increased these behaviors. Thus, desmonstrating the dichotomous neurobiological response to estrogens may be mediated by the existence of two distinct ERs.

Mechanisms of modulation

Estrogens exert their neuroprotective actions via several different mechanisms, which may be ER-dependent, ER-independent, genomic, or nongenomic. While genomic activities are those in which the direct targeting of estrogen response genes occurs through the ER, the nongenomic activities are those where ER activation results in cross-talk between ERs and other signaling pathways such as MAPK, phosphatidyl-inositol-3-kinase (PI-3K), and glucogen synthase kinase 3β (GSK-3β) (Manthey and Behl 2006). The mechanisms of action of estrogen may be determined by the dose of 17β-estradiol administered, as while pharmacological levels of estradiol protect the brain via ER-independent mechanisms, physiological levels of estradiol protect the brain through ER-dependent mechanisms (for reviews see Dhandapani and Brann 2002; Wise *et al.* 2001a; Marin *et al.* 2005).

Estrogen receptor-dependent mechanisms

In a study (Manthey and Behl 2006) designed to address genomic ER-dependent mechanisms, a human neuroblastoma cell line was stably transfected with either ERα or ERβ and a large-scale DNA-chip analysis was performed to identify genes that were up- or downregulated by ERα or ERβ. It was found that ERα transfectants displayed a strong upregulation of insulin-like growth factor binding protein-5 (IGFBP-5) that carries out anti-apoptotic functions. Other genes found to be up- or downregulated in ERα transfectants included genes that play a role in synaptic plasticity and detoxification, such as calmodulin and γ-synuclein (Manthey and Behl 2006). Signal transduction pathway genes such as IGFBP-5, protein kinase C (PKC) α and MAP kinase kinase 3 (MKK3) were all upregulated in ERα transfectants, while PKC-βI and PKC-βII were both downregulated (Manthey and Behl 2006). Expression of genes involved in stress responses such as heat shock proteins, and inflammatory responses such as janus kinase (JAK) were also ER mediated; while JAK-1 is regulated by ERα, JAK-2 is regulated by ERβ (Manthey and Behl 2006). In addition, genes involved in Alzheimer's disease were found to be downregulated in both ER subtype transfectants including amyloid precursor protein (APP), amyloid precursor-like protein 1 (APLP1), Presenilin 1, and Fe65, while a disintegrin/metalloprotease (ADAM), which is involved in the reduction of amyloidogenic APP processing, was upregulated in both ER transfectants (Manthey and Behl 2006). This study clearly demonstrated that ER-dependent genomic mechanisms are employed to mediate the effects of estrogens on neuroprotection in Alzheimer's disease, stress responses, and also pathways involved in promoting neuronal survival. Neurotrophins such as brain-derived neurotrophic factor (BDNF), erkA, and GAP-43 may all be regulated by levels of estradiol, and BDNF expression has been implicated in increased striatal dopaminergic neurons in Parkinson's models, and increased dopamine D3 receptor expression in schizophrenia models (Dluzen 2000; Rao and Kolsch 2003).

Estrogens are also involved in the regulation of several messenger signaling molecules including cAMP, MAPK, or PI-3K/Akt pathway factors following middle cerebral artery occlusion (MCAO) mouse and rat models. However, it is still unclear as to whether these pathways are activated via ER-dependent or ER-independent mechanisms (for a review see Wise *et al.* 2001a). Estrogens also regulate phosphorylation of the transcription factor CREB via ER-activated mechanisms. Estrogens have also been shown to increase the numbers of phospho-CREB immunoreactive cells in the mouse brain within 15 minutes of 17β-estradiol treatment (Abraham *et al.* 2004) desmonstrating a nongenomic action of ER. The CREB protein plays a role in schizophrenia as a messenger involved in intracellular signal transduction processes, which are regulated by dopaminergic and serotonergic receptors (Rao and Kolsch 2003).

Another ER-mediated neuroprotective effect of estrogens is via regulation of growth factors, particularly insulin-like growth factor-1 (IGF-1). Interaction of IGF-1 and estradiol signaling pathways has been shown to play a role in hypothalamic neuronal differentiation (Duenas *et al.* 1996). Insulin-like growth factor-1 is coupled to the PI-3K and MAPK transduction pathway and exerts neuroprotective effects through neuronal differentiation and synaptic plasticity, and dentritic growth, and has been shown to be involved in the estrogen-regulated neuroprotective effects in animal models of Parkinson's disease and in brain injury induced by kainic acid in ovariectomized rats. These beneficial effects of estrogens through IGF-1 activation have been shown to be blocked by ER antagonists and therefore offer nongenomic ER-dependent neuroprotection (for a review see Amantea *et al.* 2005).

Estrogens may also have direct actions on glutamate receptors such as NMDA and α-amino-3-hydroxy-5-methyl-4-isoxazolepropionic acid (AMPA) receptors, which are dysfunctional in schizophrenic patients (Rao and Kolsch 2003), on dopaminergic receptors in striatal regions affected by Parkinson's disease, and in the modulation and clearance of amyloid β peptide production reminiscent of Alzheimer's disease (Barron *et al.* 2006). Furthermore, estrogens may potentially modulate thermoregulatory-related reponses to neurotoxicity (Dluzen 2000).

Estrogens and human pathological behaviors

Sex hormones exert powerful influences on both normal and pathological behaviors. The possible involvement of estrogens in modulating pathological behaviors is supported by reports that revealed that in female OCD patients, their symptoms tend to worsen during menstruation, after menopause, or within a month of childbirth (Lochner *et al.* 2004). In addition, gender may be involved in the development of OCD, as male patients have been shown to display symptoms with more severe outcomes at a younger age compared with females (Castle *et al.* 1995; Lochner *et al.* 2004). Similarly, Tourette's syndrome is more prevalent in boys than girls (Kadesjö and Gillberg 2000) and is characterized by complex tics consisting of relatively normal movements made out of context, without meaning, and repetitive to the point of self-injury. They share similarities with the compulsive repetitive behaviors and obsessive thoughts of OCD. Schizophrenia is another mental disorder that presents gender differences: schizophrenic women have been consistently found to have a later age of onset and a less severe clinical course of illness as compared to schizophrenic men (for a review see Leung and Chue 2000). The findings summarized in this review suggest that estrogens may act as a protective modulator in compulsive behavior and schizophrenia by enhancing the vulnerability threshold for psychosis through the downward regulation of dopamine neurotransmission. Such mechanisms could explain, at least in part, the later onset and the more favorable course of schizophrenia or obsessive–compulsive behavior in female patients.

Conclusion

Understanding how sex hormones modulate behavior is important in psychiatry, for a better understanding of our society and, potentially, for the veterinary industry. There is ample evidence of the actions of estrogens in disorders of the dopamine system including OCD and Tourette's syndrome and there is emerging evidence that estrogens are protective in neurodegeneration. Animal models, such as the ArKO mouse and ERKO mouse, are invaluable tools in understanding the precise mode of action of estrogens on neurons or neuronal circuitry.

Acknowledgment

This work is supported by the Australian NHMRC project grant 494813.

References

Abe-Dohmae S, Takagi Y and Harada N (1997): Autonomous expression of aromatase during development of mouse brain is modulated by neurotransmitters. *J Steroid Biochem Mol Biol* **61**:299–306.

Abraham IM, Todman MG, Korach KS and Herbison AE (2004): Critical in vivo roles for classical estrogen receptors in rapid estrogen actions on intracellular signaling in mouse brain. *Endocrinology* **145**:3055–61.

Agarwal VR, Sinton CM, Liang C *et al.* (2000): Upregulation of estrogen receptors in the forebrain of aromatase knockout (ArKO) mice. *Mol Cell Endocrinol* **162**: 9–16.

Altemus M, Greenberg BD, Keuler D, Jacobson KR and Murphy DL (1999): Open trial of flutamide for treatment of obsessive–compulsive disorder. *J Clin Psychiatry* **60**:442–5.

Alves SE, McEwen BS, Hayashi S *et al.* (2000): Estrogen-regulated progestin receptors are found in the midbrain raphe but not hippocampus of estrogen receptor alpha (ER alpha) gene-disrupted mice. *J Comp Neurol* **427**:185–95.

Amantea D, Russo R, Bagetta G and Corasaniti MT (2005): From clinical evidence to molecular mechanisms underlying neuroprotection afforded by estrogens. *Pharmacol Res* **52**:119–32.

Ardelt AA, McCullough LD, Korach KS *et al.* (2005): Estradiol regulates angiopoietin-1 mRNA expression through estrogen receptor-alpha in a rodent experimental stroke model. *Stroke* **36**:337–41.

Axelrod J and Tomchick R (1958): Enzymatic O-methylation of epinephrine and other catechols. *J Biol Chem* **233**:702–5.

Bakker J, Honda S, Harada N and Balthazart J (2002): Sexual partner preference requires a functional aromatase (cyp19) gene in male mice. *Horm Behav* **42**: 158–71.

Bakker J, Honda S, Harada N and Balthazart J (2003): The aromatase knockout (ArKO) mouse provides new evidence that estrogens are required for the development of the female brain. *Ann N Y Acad Sci* **1007**:251–62.

Balthazart J, Baillien M, Charlier TD, Cornil CA and Ball GF (2003): Multiple mechanisms control brain aromatase activity at the genomic and non-genomic level. *J Steroid Biochem Mol Biol* **86**:367–79.

Barron AM, Fuller SJ, Verdile G and Martins RN (2006): Reproductive hormones modulate oxidative stress in Alzheimer's disease. *Antioxid Redox Signal* **8**: 2047–59.

Berridge KC, Aldridge JW, Houchard KR and Zhuang X (2005): Sequential super-stereotypy of an instinctive fixed action pattern in hyper-dopaminergic mutant mice: a model of obsessive compulsive disorder and Tourette's. *BMC Biol* **14**:3–4.

Beyer C and Hutchison JB (1997): Androgens stimulate the morphological maturation of embryonic hypothalamic aromatase-immunoreactive neurons in the mouse. *Brain Res Dev Brain Res* **98**:74–81.

Bodo C and Rissman EF (2006): New roles for estrogen receptor beta in behavior and neuroendocrinology. *Front Neuroendocrinol* **27**:217–32.

Boon WC, Diepstraten J, van der Burg J *et al.* (2005): Hippocampal NMDA receptor subunit expression and watermaze learning in estrogen deficient female mice. *Brain Res Mol Brain Res* **140**:27–32.

Carani C, Qin K, Simoni M *et al.* (1997): Effect of testosterone and estradiol in a man with aromatase deficiency. *N Engl J Med* **337**:91–5.

Carreau S, Genissel C, Bilinska B and Levallet J (1999): Sources of oestrogen in the testis and reproductive tract of the male. *Int J Androl* **22**:211–23.

Castle DJ, Deale A and Marks IM (1995): Gender differences in obsessive compulsive disorder. *Aust N Z J Psychiatry* **29**:114–17.

Conte FA, Grumbach MM, Ito Y, Fisher CR and Simpson ER (1994): A syndrome of female pseudohermaphrodism, hypergonadotropic hypogonadism, and multicystic ovaries associated with missense mutations in the gene encoding aromatase (P450arom). *J Clin Endocrinol Metab* **78**:1287–92.

Couse JF, Lindzey J, Grandien K, Gustafsson JA and Korach KS (1997): Tissue distribution and quantitative analysis of estrogen receptor-alpha (ERalpha) and estrogen receptor-beta (ERbeta) messenger ribonucleic acid in the wild-type and ERalpha-knockout mouse. *Endocrinology* **138**:4613–21.

Couse JF, Yates MM, Walker VR and Korach KS (2003): Characterization of the hypothalamic-pituitary-gonadal axis in estrogen receptor (ER) null mice reveals hypergonadism and endocrine sex reversal in females lacking ERalpha but not ERbeta. *Mol Endocrinol* **17**:1039–53.

Cyr M, Calon F, Morissette M and Di Paolo T (2002): Estrogenic modulation of brain activity: implications for schizophrenia and Parkinson's disease. *J Psychiatry Neurosci* **27**:12–27.

Dalla C, Antoniou K, Papadopoulou-Daifoti Z, Balthazart J and Bakker J (2004): Oestrogen-deficient female aromatase knockout (ArKO) mice exhibit depressive-like symptomatology. *Eur J Neurosci* **20**:217–28.

Dalla C, Antoniou K, Papadopoulou-Daifoti Z, Balthazart J and Bakker J (2005): Male aromatase-knockout mice exhibit normal levels of activity, anxiety and "depressive-like" symptomatology. *Behav Brain Res* **163**:186–93.

Dhandapani KM and Brann DW (2002): Protective effects of estrogen and selective estrogen receptor modulators in the brain. *Biol Reprod* **67**:1379–85.

Dluzen DE (2000): Neuroprotective effects of estrogen upon the nigrostriatal dopaminergic system. *J Neurocytol* **29**:387–99.

Dubal DB, Zhu H, Yu J *et al.* (2001): Estrogen receptor alpha, not beta, is a critical link in estradiol-mediated protection against brain injury. *Proc Natl Acad Sci USA* **98**:1952–7.

Duenas M, Torres-Aleman I, Naftolin F and Garcia-Segura LM (1996): Interaction of insulin-like growth factor-I and estradiol signaling pathways on hypothalamic neuronal differentiation. *Neuroscience* **74**:531–9.

Dupont S, Krust A, Gansmuller A *et al.* (2000): Effect of single and compound knockouts of estrogen receptors alpha (ERalpha) and beta (ERbeta) on mouse reproductive phenotypes. *Development* **127**:4277–91.

Fahrbach SE, Meisel RL and Pfaff DW (1985): Preoptic implants of estradiol increase wheel running but not the open field activity of female rats. *Physiol Behav* **35**:985–92.

Ferris CF, Rasmussen MF, Messenger T and Koppel G (2001): Vasopressin-dependent flank marking in golden hamsters is suppressed by drugs used in the treatment of obsessive–compulsive disorder. *BMC Neurosci* **2**:10.

Fisher CR, Graves KH, Parlow AF and Simpson ER (1998): Characterization of mice deficient in aromatase (ArKO) because of targeted disruption of the cyp19 gene. *Proc Natl Acad Sci USA* **95**:6965–70.

Friedlander L and Desrocher M (2006): Neuroimaging studies of obsessive–compulsive disorder in adults and children. *Clin Psychol Rev* **26**:32–49.

Garcia-Segura LM, Wozniak A, Azcoitia I *et al.* (1999): Aromatase expression by astrocytes after brain injury: implications for local estrogen formation in brain repair. *Neuroscience* **89**:567–78.

Geary N, Asarian L, Korach KS, Pfaff DW and Ogawa S (2001): Deficits in E2-dependent control of feeding, weight gain, and cholecystokinin satiation in ER-alpha null mice. *Endocrinology* **142**:4751–7.

Gogos JA, Morgan M, Luine V *et al.* (1998): Catechol-O-methyltransferase-deficient mice exhibit sexually dimorphic changes in catecholamine levels and behavior. *Proc Natl Acad Sci USA* **95**:9991–6.

Graham-Lorence S, Amarneh B, White RE, Peterson JA and Simpson ER (1995): A three-dimensional model of aromatase cytochrome P450. *Protein Sci* **4**:1065–80.

Graybiel AM and Rauch SL (2000): Toward a neurobiology of obsessive–compulsive disorder. *Neuron* **28**:343–47.

Green S, Walter P, Greene G *et al.* (1986): Cloning of the human oestrogen receptor cDNA. *J Steroid Biochem* **24**:77–83.

Greer JM and Capecchi MR (2002): Hoxb8 is required for normal grooming behavior in mice. *Neuron* **33**:23–34.

Gundlah C, Alves SE, Clark JA *et al.* (2005): Estrogen receptor-beta regulates tryptophan hydroxylase-1 expression in the murine midbrain raphe. *Biol Psychiatry* **57**: 938–42.

Harada N and Yamada K. (1992): Ontogeny of aromatase messenger ribonucleic acid in mouse brain: fluorometrical quantitation by polymerase chain reaction. *Endocrinology* **131**:2306–12.

Häfner H, Behrens S, De Vry J and Gattaz WF (1991): An animal model for the effects of estradiol on dopamine-mediated behavior: implications for sex differences in schizophrenia. *Psychiatry Res* **38**:125–34.

Henderson VW, Watt L and Buckwalter JG (1996): Cognitive skills associated with estrogen replacement in women with Alzheimer's disease. *Psychoneuroendocrinology* **21**:421–30.

Hill RA, McInnes KJ, Gong EC *et al.* (2007): Estrogen deficient male mice develop compulsive behavior. *Biol Psychiatry* **61**:359–66.

Honda S, Harada N and Takagi Y (1994): Novel exon 1 of the aromatase gene specific for aromatase transcripts in human brain. *Biochem Biophys Res Commun* **198**: 1153–60.

Huotari M, Santha M, Lucas LR *et al.* (2002): Effect of dopamine uptake inhibition on brain catecholamine levels and locomotion in catechol-O-methyltransferase-disrupted mice. *J Pharmacol Exp Ther* **303**:1309–16.

Hutchison JB, Beyer C, Hutchison RE and Wozniak A (1997): Sex differences in the regulation of embryonic brain aromatase. *J Steroid Biochem Mol Biol* **61**:315–22.

Iivonen S, Heikkinen T, Puolivali J *et al.* (2006): Effects of estradiol on spatial learning, hippocampal cytochrome P450 19, and estrogen alpha and beta mRNA levels in ovariectomized female mice. *Neuroscience* **137**:1143–52.

Ito Y, Fisher CR, Conte FA, Grumbach MM and Simpson ER (1993): Molecular basis of aromatase deficiency in an adult female with sexual infantilism and polycystic ovaries. *Proc Natl Acad Sci USA* **90**:11673–7.

Ivanova T and Beyer C (2000): Ontogenetic expression and sex differences of aromatase and estrogen receptor-alpha/beta mRNA in the mouse hippocampus. *Cell Tissue Res* **300**:231–7.

Jenkins C, Michael D, Mahendroo M and Simpson E. (1993): Exon-specific northern analysis and rapid amplification of cDNA ends (RACE) reveal that the proximal promoter II (PII) is responsible for aromatase cytochrome P450 (CYP19) expression in human ovary. *Mol Cell Endocrinol* **97**:R1–6.

Jones ME, Boon WC, McInnes K *et al.* (2007): Recognizing rare disorders: aromatase deficiency. *Nat Clin Pract Endocrinol Metab* **3**:414–21.

Kadesjö B and Gillberg C (2000): Tourette's disorder: epidemiology and comorbidity in primary school children. *J Am Acad Child Adolesc Psychiatry* **39**:548–55.

Karayiorgou M, Altemus M, Galke BL *et al.* (1997): Genotype determining low catechol-O-methyltransferase activity as a risk factor for obsessive–compulsive disorder. *Proc Natl Acad Sci USA* **94**:4572–5.

King JM (1979): Effects of lesions of the amygdala, preoptic area, and hypothalamus on estradiol-induced activity in the female rat. *J Comp Physiol Psychol* **93**:360–7.

Krege JH, Hodgin JB, Couse JF *et al.* (1998): Generation and reproductive phenotypes of mice lacking estrogen receptor beta. *Proc Natl Acad Sci USA* **95**:15677–82.

Kruijver FP, Balesar R, Espila AM, Unmehopa UA and Swaab DF (2002): Estrogen receptor-alpha distribution in the human hypothalamus in relation to sex and endocrine status. *J Comp Neurol* **454**:115–39.

Kruk MR, Westphal KG, Van Erp AM *et al.* (1998): The hypothalamus: cross-roads of endocrine and behavioural regulation in grooming and aggression. *Neurosci Biobehav Rev* **23**:163–77.

Kuiper GG, Enmark E, Pelto-Huikko M, Nilsson S and Gustafsson JA (1996): Cloning of a novel receptor expressed in rat prostate and ovary. *Proc Natl Acad Sci USA* **93**:5925–30.

Kumari V, Kaviani H, Raven PW, Gray JA and Checkley SA (2001): Enhanced startle reactions to acoustic stimuli in patients with obsessive–compulsive disorder. *Am J Psychiatry* **158**:134–6.

Kuppers E and Beyer C (1998): Expression of aromatase in the embryonic and postnatal mouse striatum. *Brain Res Mol Brain Res* **63**:184–8.

Lephart ED, Lund TD and Horvath TL (2001): Brain androgen and progesterone metabolizing enzymes: biosynthesis, distribution and function. *Brain Res Brain Res Rev* **37**:25–37.

Leung A and Chue P (2000) Sex differences in schizophrenia, a review of the literature. *Acta Psychiatr Scand Suppl* **401**:3–38.

Lindzey J, Wetsel WC, Couse JF *et al.* (1998): Effects of castration and chronic steroid treatments on hypothalamic gonadotropin-releasing hormone content and pituitary gonadotropins in male wild-type and estrogen receptor-alpha knockout mice. *Endocrinology* **139**:4092–101.

Lochner C, Hemmings SM, Kinnear CJ *et al.* (2004): Gender in obsessive–compulsive disorder: clinical and genetic findings. *Eur Neuropsychopharmacol* **14**:105–13.

Lubahn DB, Moyer JS, Golding TS *et al.* (1993): Alteration of reproductive function but not prenatal sexual development after insertional disruption of the mouse estrogen receptor gene. *Proc Natl Acad Sci USA* **90**:11162–6.

Lumley LA, Robison CL, Chen WK, Mark B and Meyerhoff JL (2001): Vasopressin into the preoptic area increases grooming behavior in mice. *Physiol Behav* **73**:451–5.

Lund TD, Rovis T, Chung WC and Handa RJ (2005): Novel actions of estrogen receptor on anxiety-related behaviors. *Endocrinology* **146**:797–807.

Manthey D and Behl C (2006): From structural biochemistry to expression profiling: neuroprotective activities of estrogen. *Neuroscience* **138**:845–50.

Marin R, Guerra B, Alonso R, Ramirez CM and Diaz M (2005): Estrogen activates classical and alternative mechanisms to orchestrate neuroprotection. *Curr Neurovasc Res* **2**:287–301.

Martin S, Jones M, Simpson E and van den Buuse M (2003): Impaired spatial reference memory in aromatase-deficient (ArKO) mice. *Neuroreport* **14**:1979–82.

McCullough LD, Blizzard K, Simpson ER, Oz OK and Hurn PD (2003): Aromatase cytochrome P450 and extragonadal estrogen play a role in ischemic neuroprotection. *J Neurosci* **23**:8701–5.

Mitra SW, Hoskin E, Yudkovitz J *et al.* (2003): Immunolocalization of estrogen receptor beta in the mouse brain: comparison with estrogen receptor alpha. *Endocrinology* **144**:2055–67.

Mong JA and Pfaff DW (2003) Hormonal and genetic influences underlying arousal as it drives sex and aggression in animal and human brains. *Neurobiol Aging* **24** Suppl 1:S83–8; discussion S91–2.

Morgan MA and Pfaff DW (2001): Effects of estrogen on activity and fear-related behaviors in mice. *Horm Behav* **40**:472–82.

Morishima A, Grumbach MM, Simpson ER, Fisher C and Qin K (1995): Aromatase deficiency in male and female siblings caused by a novel mutation and the physiological role of estrogens. *J Clin Endocrinol Metab* **80**:3689–98.

Nordstrom EJ and Burton FH (2002): A transgenic model of comorbid Tourette's syndrome and obsessive–compulsive disorder circuitry. *Mol Psychiatry* **7**: 617–25.

Ogawa S, Gordan JD, Taylor J *et al.* (1996): Reproductive functions illustrating direct and indirect effects of genes on behavior. *Horm Behav* **30**:487–94.

Ogawa S, Lubahn DB, Korach KS and Pfaff DW (1997): Behavioral effects of estrogen receptor gene disruption in male mice. *Proc Natl Acad Sci USA* **94**:1476–81.

Ogawa S, Chan J, Chester AE *et al.* (1999): Survival of reproductive behaviors in estrogen receptor beta gene-deficient (betaERKO) male and female mice. *Proc Natl Acad Sci USA* **96**:12887–92.

Ogawa S, Chester AE, Hewitt SC *et al.* (2000): Abolition of male sexual behaviors in mice lacking estrogen receptors alpha and beta (alpha beta ERKO). *Proc Natl Acad Sci USA* **97**:14737–41.

Ogawa S, Chan J, Gustafsson JA, Korach KS and Pfaff DW (2003): Estrogen increases locomotor activity in mice through estrogen receptor alpha: specificity for the type of activity. *Endocrinology* **144**:230–9.

Orikasa C, Kondo Y, Hayashi S, McEwen BS and Sakuma Y (2002): Sexually dimorphic expression of estrogen receptor beta in the anteroventral periventricular nucleus of the rat preoptic area: implication in luteinizing hormone surge. *Proc Natl Acad Sci USA* **99**:3306–11.

Plumari L, Viglietti-Panzica C, Allieri F *et al.* (2002): Changes in the arginine–vasopressin immunoreactive systems in male mice lacking a functional aromatase gene. *J Neuroendocrinol* **14**:971–8.

Pooley EC, Fineberg N and Harrison PJ (2007): The met(158) allele of catechol-O-methyltransferase (COMT) is associated with obsessive–compulsive disorder in men: case-control study and meta-analysis. *Mol Psychiatry* **12**:556–61.

Rao ML and Kolsch H (2003): Effects of estrogen on brain development and neuroprotection – implications for negative symptoms in schizophrenia. *Psychoneuroendocrinology* **28** Suppl 2:83–96.

Rapoport JL (1991): Recent advances in obsessive–compulsive disorder. *Neuropsychopharmacology* **5**:1–10.

Rhodes JS, Garland T Jr and Gammie SC (2003): Patterns of brain activity associated with variation in voluntary wheel-running behavior. *Behav Neurosci* **117**:1243–56.

Rissman EF, Early AH, Taylor JA, Korach KS and Lubahn DB (1997): Estrogen receptors are essential for female sexual receptivity. *Endocrinology* **138**:507–10.

Rissman EF, Wersinger SR, Fugger HN and Foster TC (1999): Sex with knockout models: behavioral studies of estrogen receptor alpha. *Brain Res* **835**:80–90.

Robertson KM, Simpson ER, Lacham-Kaplan O and Jones ME (2001): Characterization of the fertility of male aromatase knockout mice. *J Androl* **22**:825–30.

Roselli CE and Resko JA (1997): Sex differences in androgen-regulated expression of cytochrome P450 aromatase in the rat brain. *J Steroid Biochem Mol Biol* **61**: 365–74.

Roselli CE and Resko JA (2001): Cytochrome P450 aromatase (CYP19) in the non-human primate brain: distribution, regulation, and functional significance. *J Steroid Biochem Mol Biol* **79**:247–53.

Roselli CE, Larkin K, Schrunk JM and Stormshak F (2004): Sexual partner preference, hypothalamic morphology and aromatase in rams. *Physiol Behav* **83**:233–45.

Roy EJ and Wade GN (1975): Role of estrogens in androgen-induced spontaneous activity in male rats. *J Comp Physiol Psychol* **89**:573–9.

Sasano H and Ozaki M (1997): Aromatase expression and its localization in human breast cancer. *J Steroid Biochem Mol Biol* **61**:293–8.

Sasano H, Uzuki M, Sawai T *et al.* (1997): Aromatase in human bone tissue. *J Bone Miner Res* **12**:1416–23.

Sasano H, Takashashi K, Satoh F, Nagura H and Harada N (1998): Aromatase in the human central nervous system. *Clin Endocrinol (Oxf)* **48**:325–9.

Selmanoff MK, Pramik-Holdaway MJ and Weiner RI (1976): Concentrations of dopamine and norepinephrine in discrete hypothalamic nuclei during the rat estrous cycle. *Endocrinology* **99**:326–9.

Sharma PK and Thakur MK (2006): Expression of estrogen receptor (ER) alpha and beta in mouse cerebral cortex: effect of age, sex and gonadal steroids. *Neurobiol Aging* **27**:880–7.

Shughrue PJ, Lane MV and Merchenthaler I (1997): Comparative distribution of estrogen receptor-alpha and -beta mRNA in the rat central nervous system. *J Comp Neurol* **388**:507–25.

Simpson ER, Michael MD, Agarwal VR *et al.* (1997a): Cytochromes P450 11: expression of the CYP19 (aromatase) gene: an unusual case of alternative promoter usage. *FASEB J* **11**:29–36.

Simpson ER, Zhao Y, Agarwal VR *et al.* (1997b): Aromatase expression in health and disease. *Recent Prog Horm Res* **52**:185–213; discussion 213–14.

Singh M, Setalo G Jr, Guan X, Frail DE and Toran-Allerand CD (2000): Estrogen-induced activation of the mitogen-activated protein kinase cascade in the cerebral cortex of estrogen receptor-alpha knock-out mice. *J Neurosci* **20**:1694–700.

Stoffel-Wagner B (2001): Neurosteroid metabolism in the human brain. *Eur J Endocrinol* **145**:669–79.

Tang MX, Jacobs D, Stern Y *et al.* (1996): Effect of oestrogen during menopause on risk and age at onset of Alzheimer's disease. *Lancet* **348**:429–32.

Toda K, Saibara T, Okada T, Onishi S and Shizuta Y (2001): A loss of aggressive behaviour and its reinstatement by oestrogen in mice lacking the aromatase gene (Cyp19). *J Endocrinol* **168**:217–20.

Toran-Allerand CD (2000): Novel sites and mechanisms of oestrogen action in the brain. *Novartis Found Symp* **230**:56–69; discussion 69–73.

Toran-Allerand CD, Guan X, MacLusky NJ *et al.* (2002): ER-X: a novel, plasma membrane-associated, putative estrogen receptor that is regulated during development and after ischemic brain injury. *J Neurosci* **22**:8391–401.

van den Buuse M, Simpson ER and Jones ME (2003): Prepulse inhibition of acoustic startle in aromatase knock-out mice: effects of age and gender. *Genes Brain Behav* **2**:93–102.

Wang L, Andersson S, Warner M and Gustafsson JA (2001): Morphological abnormalities in the brains of estrogen receptor beta knockout mice. *Proc Natl Acad Sci USA* **98**:2792–6.

Watson CS, Norfleet AM, Pappas TC and Gametchu B (1999): Rapid actions of estrogens in GH3/B6 pituitary tumor cells via a plasma membrane version of estrogen receptor-alpha. *Steroids* **64**:5–13.

Wise PM, Dubal DB, Wilson ME, Rau SW and Bottner M (2001a): Minireview: neuroprotective effects of estrogen-new insights into mechanisms of action. *Endocrinology* **142**:969–73.

Wise PM, Dubal DB, Wilson ME *et al.* (2001b): Estradiol is a protective factor in the adult and aging brain: understanding of mechanisms derived from in vivo and in vitro studies. *Brain Res Brain Res Rev* **37**:313–19.

Xu H, Gouras GK, Greenfield JP *et al.* (1998): Estrogen reduces neuronal generation of Alzheimer beta-amyloid peptides. *Nat Med* **4**:447–51.

Yamada K, Harada N, Tamaru M and Takagi Y (1993): Effects of changes in gonadal hormones on the amount of aromatase messenger RNA in mouse brain diencephalon. *Biochem Biophys Res Commun* **195**:462–8.

Young WC, Goy RW and Phoenix CH (1964): Hormones and sexual behavior. *Science* **143**:212–18.

Zhang JQ, Su BY and Cai WQ (2004): Immunolocalization of estrogen receptor beta in the hypothalamic paraventricular nucleus of female mice during pregnancy, lactation and postnatal development. *Brain Res* **997**:89–96.

Zhao L, Wu TW and Brinton RD (2004): Estrogen receptor subtypes alpha and beta contribute to neuroprotection and increased Bcl-2 expression in primary hippocampal neurons. *Brain Res* **10**:22–34.

7

Lack of barbering behavior in the phospholipase Cβ1 mutant mouse: a model animal for schizophrenia

HEE-SUP SHIN, DAESOO KIM, AND HAE-YOUNG KOH

Summary

Abnormal phospholipid metabolism has been implicated in the pathogenesis of schizophrenia, and phospholipase C (PLC) β1 was shown to be reduced in specific brain areas of patients with schizophrenia. However, the causal relationship of the PLCβ1 gene with the behavioral symptoms of schizophrenia remains unclear. Recent studies with the knockout (KO) mice for the PLCβ1 gene have revealed an array of interesting phenotypes, which along with other previous information makes the PLCβ1-KO mouse a good candidate for an animal model for schizophrenia. This also suggests that the PLCβ1-linked signaling pathways may be involved in the neural system whose function is disrupted in the pathogenesis of schizophrenia. In this chapter we will introduce various studies relevant to this issue, highlighting the social withdrawal phenotypes of the mutant, such as the lack of barbering behaviors.

Introduction

An animal model for a disease is expected to display endophenotypes, which are quantifiable phenotypes relevant to symptoms of the disease to be modeled (Braff and Freedman 2002; Gould and Gottesman 2006; van den Buuse *et al.* 2005). The endophenotypes currently pursued in schizophrenia models are: locomotive hyperactivity, sensorimotor gating deficit, deficits in social interaction, and cognitive deficits (e.g., learning and memory). Genetically modified mice

Neurobiology of Grooming Behavior, eds. Allan V. Kalueff, Justin L. LaPorte, and Carisa L. Bergner. Published by Cambridge University Press.

targeted on candidate susceptibility genes have so far been generated as animal models for schizophrenia. These mouse models display at least one or two of the endophenotypes listed above (Gainetdinov *et al.* 2001; Gerber *et al.* 2001; Kellendonk *et al.* 2006; Lijam *et al.* 1997; Miyakawa *et al.* 2003; Robertson *et al.* 2006; Yee *et al.* 2005).

Phospholipase Cβ1 is a G-protein-coupled receptor (GPCR)-associated enzyme and hydrolyzes phosphatidylinositol 4,5-bisphosphate to produce second messengers, diacylglycerol and inositol 1,4,5-trisphosphate (IP_3). The second messengers in turn activate protein kinase C (PKC) or calcium-dependent cellular components, respectively, thereby inducing a whole array of cellular responses. Three different groups of neurotransmitter receptors – muscarinic, metabotropic glutamate, and serotonin receptors – are known to be coupled to PLCβ1 for their signal transduction in neuronal cells (Chuang *et al.* 2001; Kim *et al.* 1997; Rhee and Bae 1997). Phospholipase Cβ1 is expressed in select areas of the brain such as the cerebral cortex, hippocampus, amygdala, lateral septum, and olfactory bulb (Watanabe *et al.* 1998). Therefore, it is implicated for participation in diverse brain functions, and possibly for involvement in psychiatric disorders. Besides, possible genetic association of PLCβ1 with schizophrenia has been reported in linkage studies (20p12; Arinami *et al.* 2005; Peruzzi *et al.* 2002), and its abnormal expression patterns were observed in the brains of patients with schizophrenia (Lin *et al.* 1999; Shirakawa *et al.* 2001). These results implicate a derangement of the PLCβ1-dependent phospholipid signaling in schizophrenia.

In as much as PLCβ1 is an enzyme involved in phospholipid metabolism, it is interesting to note that phospholipids are required not only for the structure of neural membranes but also for the signal transduction processes that link receptor occupancy and actual neuronal response (Oude Weernink *et al.* 2006). The importance of membrane phospholipid composition in normal functioning of the brain (e.g., neurotransmission) was proposed following the observation that phospholipid composition affects the activities of ion channels, transporters, and receptors (Bourre *et al.* 1991; Spector and York 1985). Furthermore, there is evidence for abnormalities in the phospholipid metabolism in schizophrenic patients (for a review see du Bois *et al.* 2005).

All this information raised a possibility that a defect in PLCβ1 signaling may result in the development of schizophrenia-like conditions. The availability of KO mice for this gene (Kim *et al.* 1997) has allowed testing of this hypothesis. Indeed, we found that PLCβ1$^{-/-}$ mice demonstrate various endophenotypes regarded relevant to schizophrenia in humans (Koh *et al.* 2008; McOmish *et al.* 2008b). Uniquely among reported mouse models for schizophrenia, PLCβ1$^{-/-}$ mice demonstrate endophenotypes in all the four categories: locomotive hyperactivity, sensorimotor gating deficit, deficits in social interaction, and cognitive deficits. Furthermore, previous studies showed a derangement in the cortical development in PLCβ1$^{-/-}$ mice (Hannan *et al.* 2001). The PLCβ1$^{-/-}$ mouse, therefore, may serve as an ideal animal model for studying the brain functions disrupted in schizophrenia and other related disorders, and for developing drugs to treat the disease.

A major part of the description in this chapter is based on one of our recent reports (Koh *et al.* 2008), to which readers are referred for further details.

Locomotive hyperactivity

Among the positive symptoms of human schizophrenia, locomotive hyperactivity may be the only endophenotype that can be easily measured in the mouse model of the disease. Casual inspection of home-cage behaviors suggested that PLCβ1-KO mice are more active than their wild type (WT) littermates. To confirm their hyperactivity, locomotor activity in an open-field arena (a square-floored rectangular box made of white acrylic, 40 × 40 × 50 cm) was monitored. To start the test each mouse was gently placed at the center of the open-field kit under diffused lighting. The distance of spontaneous movement during a one-hour period was monitored at five-minute intervals via digital video recording. The video data were analyzed by software that tracks the horizontal movement of the weight center of the object using the contrast between the object and the background. The PLCβ1$^{-/-}$ mice showed increased locomotor activity for most of the monitoring period relative to WT mice. The total distance traveled by PLCβ1$^{-/-}$ mice in one hour was significantly higher than that of WT mice. The PLCβ1$^{-/-}$ mice showed a tendency of early increase in locomotor activity and an evident habituation. After habituation, at 55- and 60-minute bins, the locomotor activities were not significantly different between WT and PLCβ1$^{-/-}$ mice. These results show that the PLCβ1$^{-/-}$ mice show locomotor hyperactivity in a novel environment.

In both humans and rodents, amphetamine-induced increase in the locomotor activity is known to involve increased dopamine release in the ventral striatum (Creese and Iversen 1975; Drevets *et al.* 2001). Amphetamine-induced neurochemical effect in the striatum was found to be greater in patients with schizophrenia than in controls (Laruelle *et al.* 1996). Especially, this enhancement was correlated with psychosis (Laruelle *et al.* 1999). These findings suggested that an enhanced striatal dopamine release is the main neural correlate of psychosis, which became the basis for using locomotor hyperactivity as a model for positive symptoms of schizophrenia. In the open-field test, PLCβ1$^{-/-}$ mice exhibited increased locomotor activity, even without treatment with psychostimulants. Many other putative mice models have been known to exhibit locomotor hyperactivity at baseline and/or in response to novelty.

Sensorimotor gating deficit and its reversal by an antipsychotic drug

Sensorimotor gating is essentially a protection mechanism against sensory information overload (Geyer *et al.* 2001). Measures of failure in such inhibition include prepulse inhibition (PPI) of the acoustic startle response (ASR), where a weaker and shorter prepulse stimulus suppresses the response to a subsequent startle stimulus (Turetsky *et al.* 2007). Human patients with schizophrenia are impaired in PPI, and it has been reported that there is a substantial covariation between the severity of positive symptoms and the degree of the PPI deficit (Weike *et al.* 2000).

The PPI test is one of the most widely used neurological tests in animals with suspected neurological defects. Mice of around 11 to 15 weeks old, were tested for PPI by using an acoustic startle chamber (Coulbourn Instruments, Allentown, PA, USA). All the animals used were behaviorally naïve in that no other kinds of tests had been performed on them previously, and only a single session of the PPI test was performed on each animal. The startle reflex was triggered by a pulse stimulus in the form of a 40-ms, 120-dB burst of white noise (startle stimulus, SS). Inhibition of the SS-elicited startle response was achieved by using a 20-ms prepulse stimulus (PP) of various intensities (74, 82, and 90 dB of white noise) that preceded the SS by 100 ms. The test was composed of a series of seven blocks, each of which was a "semi-random" mixture of eight different trial types (no stimulus, SS only, three PP only, three PP plus SS), separated by 10- to15-s intertrial intervals. The percentage prepulse inhibition (% PPI) was calculated as [1 – (response to PP–SS coupling/response to SS only)] × 100. In the pulse-alone trials, the ASR of the PLCβ1$^{-/-}$ mice was somewhat lower than that of the WT, but the difference was not statistically significant. However, a significant attenuation of PPI was observed in the PLCβ1$^{-/-}$ mice compared to the WT at all prepulse intensities.

Haloperidol is a dopaminergic D2 receptor antagonist and is used to treat schizophrenia. Furthermore, it has been observed that the PPI in antipsychotics-medicated patients does not significantly differ from that of the healthy control group (Weike *et al.* 2000). Therefore, we tried to see if the PPI impairment of PLCβ1$^{-/-}$ mice can be ameliorated by treatment with haloperidol. Haloperidol (0.2 mg/kg body weight) was intraperitoneally administered 45 minutes before the test. There was no significant effect of haloperidol on the ASR measured in SS pulse-alone trials. However, the haloperidol treatment reversed the decreased PPI in the PLCβ1$^{-/-}$ mice to a level similar to that of the WT mice treated with either vehicle or haloperidol. The responsiveness to haloperidol of the impaired PPI phenotype of PLCβ1$^{-/-}$ mice may be analogous to that in human schizophrenic patients. Like locomotor hyperactivity, PPI deficit has been observed in most of the putative mice models of schizophrenia, although its responsiveness to antipsychotics has been rarely reported (Russig *et al.* 2004; Yee *et al.* 2005).

In addition to haloperidol, clozapine has also been used to reverse the impaired PPI of PLCβ1$^{-/-}$ mice (McOmish *et al.* 2008a). In this study, however, haloperidol was ineffective in reversing the PPI impairment, presumably due to inadequate drug doses and the different protocols used.

Social withdrawal phenotypes of PLCβ1$^{-/-}$ mice

From anecdotal observations of the mice housed as uniform genotypes, it appeared that PLCβ1$^{-/-}$ mice interacted less frequently among themselves than WT mice, suggesting a possible abnormality in social interaction in PLCβ1$^{-/-}$ mice. Therefore, we have tried to examine the social behavior of the mutant mouse.

Lack of barbering (whisker trimming) behavior

Casual inspection of physical appearance revealed a striking difference between PLCβ1$^{-/-}$ and WT mice. Most WT male and female mice were completely devoid of long whiskers (Figure 7.1A left), whereas all the PLCβ1$^{-/-}$ mice had full sets of long whiskers (Figure 7.1A right). At the age of weaning, all PLCβ1$^{-/-}$ and WT mice had full sets of long whiskers, but as they got older WT mice lost long whiskers, probably resulting from mutual whisker trimming. These preliminary observations suggested that PLCβ1$^{-/-}$ mice lack whisker trimming behavior, which may indicate reduced social interaction.

To quantify this behavior, WT and PLCβ1$^{-/-}$ mice were housed in pairs as either uniform or mixed genotypes and were scored for the presence of whiskers longer than 0.5 cm at three months of coupling. Five pairs of WT;WT, five pairs of WT;PLCβ1$^{-/-}$, and four pairs of PLCβ1$^{-/-}$;PLCβ1$^{-/-}$ were tested. At three months of coupling, the relative amount of long whiskers (longer than 0.5 cm) on individual mice was obtained by cutting all the whiskers and measuring the lengths [(number of long whiskers/number of all whiskers) × 100], and the results were compared between groups. In the uniform genotype housing, the relative amount of long whiskers was significantly greater in PLCβ1$^{-/-}$ mice (92.4 ± 2.76%) compared to that in WT mice (10.3 ± 3.1%) ($t = 18.63$, $df = 16$, $p < 0.0001$) (Figure 7.1B, +/+;+/+, −/−;−/−). While WT mice had significantly more long whiskers in the mixed (76.8 ± 8.6 %) than in the uniform genotype housing ($t = 7.1$, $df = 13$, $p < 0.0001$) (Figure 7.1B, +/+;+/+, +/+;−/−), PLCβ1$^{-/-}$ mice had significantly more long whiskers in the uniform than in the mixed genotype housing (6.3 ± 1.4 %) ($t = 20.28$, $df = 11$, $p < 0.0001$) (Figure 7.1B, +/+;−/−, −/−;−/−). These results demonstrated that WT mice trimmed the whiskers of their cage-mates,

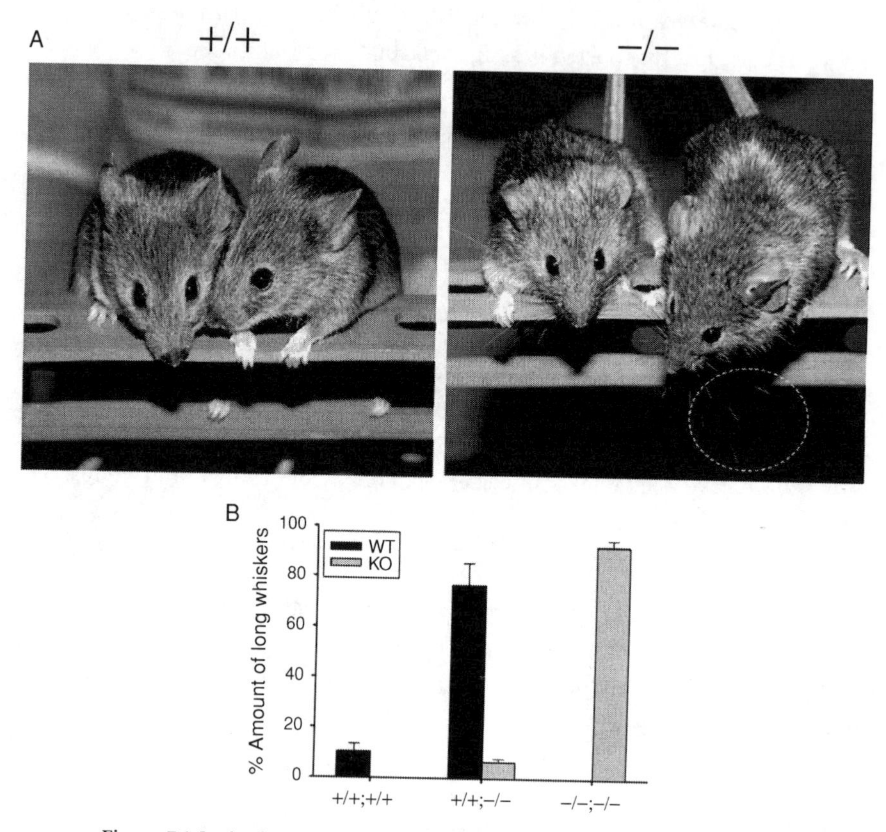

Figure 7.1 Lack of mutual barbering behavior in PLCβ1$^{-/-}$ mice. Wild type and PLCβ1$^{-/-}$ mice housed in pairs of either uniform or mixed genotypes were scored for the presence of whiskers longer than 0.5 cm at three months of coupling. (A) In the uniform genotype housing, the WT mice were completely devoid of long whiskers (left, +/+). However, the PLCβ1$^{-/-}$ mice had full sets of long whiskers (right, −/−). (B) In the uniform genotype housing (+/+;+/+, −/−;−/−), the relative amount of long whiskers was significantly higher in the PLCβ1$^{-/-}$ mice (KO) compared to the WT. While WT mice had significantly more long whiskers in the mixed (+/+;−/−) than in the uniform genotype housing (+/+;+/+), PLCβ1$^{-/-}$ mice had significantly more long whiskers in the uniform (−/−;−/−) than in the mixed genotype housing (+/+;−/−). All bars with error bars are mean ± SEM. From Koh *et al.* 2008.

but PLCβ1$^{-/-}$ mice rarely did so. This interpretation was confirmed by 2 × 2 × 2 (genotype × gender × pairing) ANOVA for percentage control of long whiskers, which showed a significant interaction between genotype and pairing ($F_{1,20} = 184$, $p < 0.001$). No interaction was observed between gender and either genotype or pairing (F = 0.171 and 0.011, respectively, $p > 0.05$).

Socially recessive trait determined by the tube test for social dominance

Social dominance was tested for 24 WT and 24 PLCβ1$^{-/-}$ mice as described (Messeri *et al.* 1975). The whole social dominance tube kit was made of clear acrylic. Two waiting chambers, 10 × 10 × 10 cm, were connected through a tube, 30 × 3 × 3 cm, between them with a sliding door at the entrance to each chamber. A WT and a PLCβ1$^{-/-}$ mouse of the same gender were put in each of the waiting chambers, and then were released to enter the tube by opening the doors. A subject was considered a winner when it remained in the tube while its opponent completely retreated from the tube. The winner was given a score "1," and the loser, "0." Each mouse went through a single ten-minute session. In most cases, a single contact between the two mice happened during a session. In cases where multiple contacts happened during a session, a subject was declared winner if it won in more than one out of three contacts. Since the scores of the WT and PLCβ1$^{-/-}$ subjects in a pair were not independent, a χ^2 one-sample analysis was used to determine whether the number of wins by PLCβ1$^{-/-}$ mice was significantly different from an outcome expected by chance. Each mouse was tested once with a mouse of the opposite genotype and the same gender. Only 2 of 24 PLCβ1$^{-/-}$ mice tested (8.33%) won over the opponent, which is significantly less than expected by chance ($\chi^2 = 16.7, p < 0.001$; Figure 7.2). This result demonstrates that PLCβ1$^{-/-}$ mice are socially recessive to WT mice when matched against each other.

Lack of nesting behavior

Casual inspection of mice cages housing uniform genotypes (four or five mice per cage) revealed that WT mice always build fluffy nests with the wooden chips provided, at a corner of the cage floor, whereas PLCβ1$^{-/-}$ mice do not build anything so that the cage floor always looks even and flat. Based on this preliminary observation, nesting behavior was quantified using commercial cotton nesting material. Each mouse was placed alone in a cage with 3 × 3 cm pieces of cotton nesting material evenly spread on top of the ordinary wooden chips. An hour later, photographs were taken of the floor of each cage to inspect whether there was a nest made of the cotton material. Within an hour after being placed in a cage with the cotton pieces, each of the nine WT mice tested built a nest, but none of the nine PLCβ1$^{-/-}$ mice tested did (Figure 7.3).

Social withdrawal is one of the negative symptoms of human schizophrenia. Whisker trimming, also called barbering behavior, is observed in both male and female mice from many of the commonly used inbred mouse strains. This behavior seems to reflect a cooperative social activity, and also to be associated with social dominance (Strozik and Festing 1981). The PLCβ1$^{-/-}$ mice that lack barbering

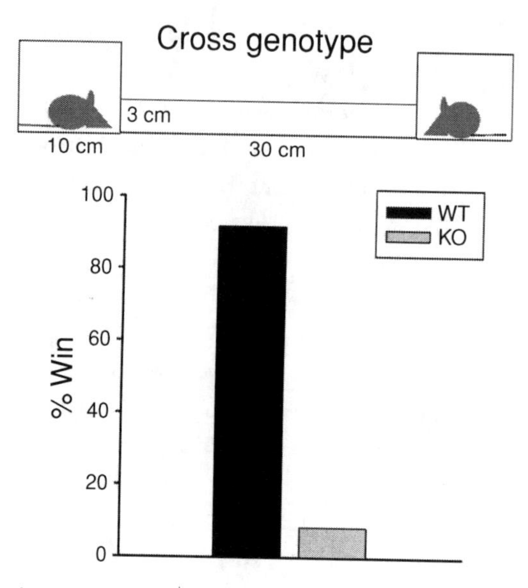

Figure 7.2 PLCβ1$^{-/-}$ mice are socially recessive. Social dominance was tested using the acrylic tube kit (upper panel). Each mouse was tested once with a mouse of the opposite genotype and the same gender. Only 2 of 24 PLCβ1$^{-/-}$ mice tested (KO, 8.33%) won, showing that wild type mice (WT) are more dominant than PLCβ1$^{-/-}$ mice. From Koh *et al.* 2008.

behavior, also exhibited a socially recessive trait in the social dominance tube test, suggesting that the lack of barbering in PLCβ1$^{-/-}$ mice is related to a low level of social interaction. Nest building is an activity shared by all members of the home cage and it provides an area for group sleep/huddling (Schneider and Chenoweth 1970). Thus, it is regarded as a behavioral measure of social interaction (Crawley 2004; Long *et al.* 2004). The lack of barbering and nesting behavior, and the socially recessive trait, can be endophenotypes relevant to the negative symptoms found in many schizophrenia patients. A similar set of phenotypes relevant to social withdrawal were also observed in mice lacking disheveled homolog 1 (Dvl1-KO) (Lijam *et al.* 1997).

Cognitive deficits of the PLCβ1$^{-/-}$ mice

Working memory deficit: delayed nonmatch to sample (DNMTS) T-maze test

Working memory deficit is often the major contributor to the morbidity of schizophrenia. We have evaluated PLCβ1$^{-/-}$ mice for the working memory capacity using the DNMTS T-maze test according to the procedure as previously described (Dias and Aggleton 2000; Kellendonk *et al.* 2006) with minor modifications. In this test, in order to obtain food, a mouse has to remember the arm in the T-maze that

Figure 7.3 Lack of nest building behavior in PLCβ1$^{-/-}$ mice. Nesting behavior was tested using commercial cotton nesting material acutely provided in the cage. Within an hour after being placed in a cage with the cotton pieces, each of all the wild type mice (+/+, n = 9) tested built a nest, but none of the PLCβ1$^{-/-}$ mice (–/–, n = 9) tested did. From Koh *et al.* 2008.

it has just visited and choose to go to the other arm of the maze. A correct choice is scored when the mouse visits the alternate arm in the choice run. The criterion for completing the task was reached when 11 correct choices out of 12 consecutive trials (92%) were made. Each mouse was tested for up to 28 days, during which the sequence of baiting sides (right or left) was randomized. All of the eight WT mice tested reached the criterion (11 correct choices out of 12 consecutive trials) with 6.38 ± 0.34 days of delay. In contrast, none of the eight PLCβ1$^{-/-}$ mice tested ever reached the criterion within 28 days. However, PLCβ1$^{-/-}$ mice performed normally in the simple right–left discrimination T-maze test where the baiting target was fixed at one side, either left or right, for all trials throughout the whole experimental period.

The delayed alternation tasks such as DNMTS T-maze have been widely used to assess working memory function in rats and mice, and impaired working memory is considered a core cognitive deficit in schizophrenia (Elvevag and Goldberg 2000). The inability of PLCβ1$^{-/-}$ mice to reach the criterion within a given period of time, with no significant impairment in the simple right–left discrimination task, suggests that they may have a working memory deficit. A similar deficit was observed in striatal dopamine receptor D2 knockup mice (Kellendonk *et al.* 2006).

Deficits in hippocampus-dependent long-term memory

The PLCβ1$^{-/-}$ mice also showed an impaired performance in the Morris water maze, along with a lack of cholinergic, Type-II, theta rhythms on the hippocampal field recordings (Shin *et al.* 2005). A similar behavioral deficit was observed in human schizophrenic patients in tests with a virtual Morris water maze (Hanlon *et al.* 2006). Impaired spatial learning has also been observed in dopamine transporter knockout mice (DAT-KO) in a study using the eight-arm radial maze (Gainetdinov *et al.* 1999). Recently, the hippocampus-dependent fear conditioning was also shown to be impaired in PLCβ1$^{-/-}$ mice (McOmish *et al.* 2008a).

Discussion

Schizophrenia is characterized by positive symptoms (delusion, hallucination, disorganized speech, and psychomotor hyperactivity), negative symptoms (flat affect, avolition, poverty of speech and language, and social withdrawal), and cognitive impairments (deficits in attention, planning and abstract thinking, and short- and long-term memory deficits) (Andreasen 1995; Lewis and Lieberman 2000). Due to the heterogeneity in potential pathogenetic mechanisms suggested by heavy genetic linkage studies that identified multiple susceptibility loci and alleles (Gogos and Gerber 2006), there can not exist a single ideal mouse model of schizophrenia that can represent all the aspects of this disease. Instead, behavioral paradigms with relevance to schizophrenia are sought on the basis of multiple overlapping criteria. These are indirect behavioral measures that resemble the features of the disorder, and abnormalities in a behavioral pattern found in already existing models, as well as directly measurable abnormalities that are almost exactly the same in mice and human patients (Arguello and Gogos 2006; Powell and Miyakawa 2006).

The PLCβ1$^{-/-}$ mice exhibited some of the endophenotypes related to all the three categories of symptoms of schizophrenia – positive, negative, and cognitive symptoms. The pattern of severe behavioral abnormalities of these KO mice is strikingly similar to those of existing putative mice models of schizophrenia. This finding is another encouraging example supporting the idea that various kinds of genetic mutations could result in a similar behavioral symptom in mice, which may potentially correspond to psychiatric disorders in humans. The PLCβ1$^{-/-}$ mouse is still among the very few cases of animal models of schizophrenia proposed to date, e.g., NMDA receptor subunit NR1 knockdown (Cheli *et al.* 2006; Mohn *et al.* 1999), stable tubule only peptide (STOP) knockout (Andrieux *et al.* 2002; Brun *et al.* 2005; Fradley *et al.* 2005), and conditional calcineurin (CN) knockout mice (Miyakawa *et al.* 2003; Zeng *et al.* 2001).

The results suggest that PLCβ1-linked signaling pathways are relevant to the physiology of neural functions disrupted in schizophrenia, and thus add significance to the previous works implicating PLCβ1 and related phospholipid metabolism in the pathogenesis of schizophrenia (Arinami *et al.* 2005; du Bois *et al.* 2005; Peruzzi *et al.* 2002; Lin *et al.* 1999; Shirakawa *et al.* 2001). In addition, some of the endophenotypes of PLCβ1$^{-/-}$ mice, the locomotor hyperactivity and sensorimotor gating deficits, are subject to beneficial modulation by environmental enrichment (McOmish *et al.* 2008b): an observation with important implications in the management of schizophrenic patients. The PLCβ1$^{-/-}$ mice may be used for future experiments to discover basic pathogenetic mechanisms underlying schizophrenia, which will eventually help develop novel therapeutics.

References

Andreasen NC (1995): Symptoms, signs, and diagnosis of schizophrenia. *Lancet* **346**:477–81.

Andrieux A, Salin PA, Vernet M *et al.* (2002): The suppression of brain cold-stable microtubules in mice induces synaptic defects associated with neuroleptic-sensitive behavioral disorders. *Genes Dev* **16**:2350–64.

Arguello PA and Gogos JA (2006): Modeling madness in mice: one piece at a time. *Neuron* **52**:179–96.

Arinami T, Ohtsuki T, Ishiguro H *et al.* (2005): Genomewide high-density SNP linkage analysis of 236 Japanese families supports the existence of schizophrenia susceptibility loci on chromosomes 1p, 14q, and 20p. *Am J Hum Genet* **77**: 937–44.

Bourre JM, Dumont O, Piciotti M *et al.* (1991): Essentiality of omega 3 fatty acids for brain structure and function. *World Rev Nutr Diet* **66**:103–17.

Braff DL and Freedman R (2002): *Endophenotypes in Studies of the Genetics of Schizophrenia.* Philadelphia: Lippincott Williams & Wilkins.

Brun P, Begou M, Andrieux A *et al.* (2005): Dopaminergic transmission in STOP null mice. *J Neurochem* **94**:63–73.

Cheli V, Adrover M, Blanco C *et al.* (2006): Knocking-down the NMDAR1 subunit in a limited amount of neurons in the rat hippocampus impairs learning. *J Neurochem* **97** Suppl 1:68–73.

Chuang SC, Bianchi R, Kim D, Shin HS and Wong RK (2001): Group I metabotropic glutamate receptors elicit epileptiform discharges in the hippocampus through PLCbeta1 signaling. *J Neurosci* **21**:6387–94.

Crawley JN (2004): Designing mouse behavioral tasks relevant to autistic-like behaviors. *Ment Retard Dev Disabil Res Rev* **10**:248–58.

Creese I and Iversen SD (1975): The pharmacological and anatomical substrates of the amphetamine response in the rat. *Brain Res* **83**:419–36.

Dias R and Aggleton JP (2000): Effects of selective excitotoxic prefrontal lesions on acquisition of nonmatching- and matching-to-place in the T-maze in the rat: differential involvement of the prelimbic-infralimbic and anterior cingulate cortices in providing behavioural flexibility. *Eur J Neurosci* **12**:4457–66.

Drevets WC, Gautier C, Price JC *et al.* (2001): Amphetamine-induced dopamine release in human ventral striatum correlates with euphoria. *Biol Psychiatry* **49**: 81–96.

du Bois TM, Deng C and Huang XF (2005): Membrane phospholipid composition, alterations in neurotransmitter systems and schizophrenia. *Prog Neuropsychopharmacol Biol Psychiatry* **29**:878–88.

Elvevag B and Goldberg TE (2000): Cognitive impairment in schizophrenia is the core of the disorder. *Crit Rev Neurobiol* **14**:1–21.

Fradley RL, O'Meara GF, Newman RJ *et al.* (2005): STOP knockout and NMDA NR1 hypomorphic mice exhibit deficits in sensorimotor gating. *Behav Brain Res* **163**:257–64.

Gainetdinov RR, Wetsel WC, Jones SR *et al.* (1999): Role of serotonin in the paradoxical calming effect of psychostimulants on hyperactivity. *Science* **283**:397–401.

Gainetdinov RR, Mohn AR and Caron MG (2001): Genetic animal models: focus on schizophrenia. *Trends Neurosci* **24**:527–33.

Gerber DJ, Sotnikova TD, Gainetdinov RR *et al.* (2001): Hyperactivity, elevated dopaminergic transmission, and response to amphetamine in M1 muscarinic acetylcholine receptor-deficient mice. *Proc Natl Acad Sci USA* **98**:15312–17.

Geyer MA, Krebs-Thomson K, Braff DL and Swerdlow NR (2001): Pharmacological studies of prepulse inhibition models of sensorimotor gating deficits in schizophrenia: a decade in review. *Psychopharmacology (Berl)* **156**:117–54.

Gogos JA and Gerber DJ (2006): Schizophrenia susceptibility genes: emergence of positional candidates and future directions. *Trends Pharmacol Sci* **27**: 226–33.

Gould TD and Gottesman, II (2006): Psychiatric endophenotypes and the development of valid animal models. *Genes Brain Behav* **5**:113–19.

Hanlon FM, Weisend MP, Hamilton DA *et al.* (2006): Impairment on the hippocampal-dependent virtual Morris water task in schizophrenia. *Schizophr Res* **87**:67–80.

Hannan AJ, Blakemore C, Katsnelson A *et al.* (2001): PLC-beta1, activated via mGluRs, mediates activity-dependent differentiation in cerebral cortex. *Nat Neurosci* **4**:282–8.

Kellendonk C, Simpson EH, Polan HJ *et al.* (2006): Transient and selective overexpression of dopamine D2 receptors in the striatum causes persistent abnormalities in prefrontal cortex functioning. *Neuron* **49**:603–15.

Kim D, Jun KS, Lee SB *et al.* (1997): Phospholipase C isozymes selectively couple to specific neurotransmitter receptors. *Nature* **389**:290–3.

Koh HY, Kim D, Lee J, Lee S and Shin HS (2008): Deficits in social behavior and sensorimotor gating in mice lacking phospholipase Cβ1. *Genes Brain Behav* **7**: 120–8.

Laruelle M, Abi-Dargham A, van Dyck CH *et al.* (1996): Single photon emission computerized tomography imaging of amphetamine-induced dopamine release in drug-free schizophrenic subjects. *Proc Natl Acad Sci USA* **93**:9235–40.

Laruelle M, Abi-Dargham A, Gil R, Kegeles L and Innis R (1999): Increased dopamine transmission in schizophrenia: relationship to illness phases. *Biol Psychiatry* **46**:56–72.

Lewis DA and Lieberman JA (2000): Catching up on schizophrenia: natural history and neurobiology. *Neuron* **28**:325–34.

Lijam N, Paylor R, McDonald MP *et al.* (1997): Social interaction and sensorimotor gating abnormalities in mice lacking Dvl1. *Cell* **90**:895–905.

Lin XH, Kitamura N, Hashimoto T, Shirakawa O and Maeda K (1999): Opposite changes in phosphoinositide-specific phospholipase C immunoreactivity in the left prefrontal and superior temporal cortex of patients with chronic schizophrenia. *Biol Psychiatry* **46**:1665–71.

Long JM, LaPorte P, Paylor R and Wynshaw-Boris A (2004): Expanded characterization of the social interaction abnormalities in mice lacking Dvl1. *Genes Brain Behav* **3**:51–62.

McOmish CE, Burrows EL, Howard M and Hannan AJ (2008a): PLC-beta1 knockout mice as a model of disrupted cortical development and plasticity: behavioral endophenotypes and dysregulation of RGS4 gene expression. *Hippocampus* **18**:824–34.

McOmish CE, Burrows E, Howard M *et al.* (2008b): Phospholipase C-beta1 knockout mice exhibit endophenotypes modeling schizophrenia which are rescued by environmental enrichment and clozapine administration. *Mol Psychiatry* **13**:661–72.

Messeri P, Eleftheriou BE and Oliverio A (1975): Dominance behavior: a phylogenetic analysis in the mouse. *Physiol Behav* **14**:53–8.

Miyakawa T, Leiter LM, Gerber DJ *et al.* (2003): Conditional calcineurin knockout mice exhibit multiple abnormal behaviors related to schizophrenia. *Proc Natl Acad Sci USA* **100**:8987–92.

Mohn AR, Gainetdinov RR, Caron MG and Koller BH (1999): Mice with reduced NMDA receptor expression display behaviors related to schizophrenia. *Cell* **98**:427–36.

Oude Weernink PA, Han L, Jakobs KH and Schmidt M (2006): Dynamic phospholipid signaling by G protein-coupled receptors. *Biochim Biophys Acta* **1768**:888–900.

Peruzzi D, Aluigi M, Manzoli L *et al.* (2002): Molecular characterization of the human PLC beta1 gene. *Biochim Biophys Acta* **1584**:46–54.

Powell CM and Miyakawa T (2006): Schizophrenia-relevant behavioral testing in rodent models: a uniquely human disorder? *Biol Psychiatry* **59**:1198–207.

Rhee SG and Bae YS (1997): Regulation of phosphoinositide-specific phospholipase C isozymes. *J Biol Chem* **272**:15045–8.

Robertson GS, Hori SE and Powell KJ (2006): Schizophrenia: an integrative approach to modelling a complex disorder. *J Psychiatry Neurosci* **31**:157–67.

Russig H, Spooren W, Durkin S, Feldon J and Yee BK (2004): Apomorphine-induced disruption of prepulse inhibition that can be normalised by systemic haloperidol is insensitive to clozapine pretreatment. *Psychopharmacology (Berl)* **175**:143–7.

Schneider CW and Chenoweth MB (1970): Effects of hallucinogenic and other drugs on the nest-building behaviour of mice. *Nature* **225**:1262–3.

Shin J, Kim D, Bianchi R, Wong RK and Shin HS (2005): Genetic dissection of theta rhythm heterogeneity in mice. *Proc Natl Acad Sci USA* **102**:18165–70.

Shirakawa O, Kitamura N, Lin XH, Hashimoto T and Maeda K (2001): Abnormal neurochemical asymmetry in the temporal lobe of schizophrenia. *Prog Neuropsychopharmacol Biol Psychiatry* **25**:867–77.

Spector AA and Yorek MA (1985): Membrane lipid composition and cellular function. *J Lipid Res* **26**:1015–35.

Strozik E and Festing MF (1981): Whisker trimming in mice. *Lab Anim* **15**:309–12.

Turetsky BI, Calkins ME, Light GA *et al.* (2007): Neurophysiological endophenotypes of schizophrenia: the viability of selected candidate measures. *Schizophr Bull* **33**:69–94.

van den Buuse M, Garner B, Gogos A and Kusljic S (2005): Importance of animal models in schizophrenia research. *Aust N Z J Psychiatry* **39**:550–7.

Watanabe M, Nakamura M, Sato K *et al.* (1998): Patterns of expression for the mRNA corresponding to the four isoforms of phospholipase C beta in mouse brain. *Eur J Neurosci* **10**:2016–25.

Weike AI, Bauer U and Hamm AO (2000): Effective neuroleptic medication removes prepulse inhibition deficits in schizophrenia patients. *Biol Psychiatry* **47**:61–70.

Yee BK, Keist R, von Boehmer L *et al.* (2005): A schizophrenia-related sensorimotor deficit links alpha 3-containing GABAA receptors to a dopamine hyperfunction. *Proc Natl Acad Sci USA* **102**:17154–9.

Zeng H, Chattarji S, Barbarosie M *et al.* (2001): Forebrain-specific calcineurin knockout selectively impairs bidirectional synaptic plasticity and working/episodic-like memory. *Cell* **107**:617–29.

8

Grooming after cerebellar, basal ganglia, and neocortical lesions

ROBERT LALONDE AND C. STRAZIELLE

Summary

Electrical stimulation of the midline cerebellum and striatum elicits grooming in rats. Lesioning methods with either surgery or genetic mutations indicate that these brain regions contribute to grooming behaviors. *Grid2*^{Lc} mutant mice with selective cerebellar atrophy and *Girk2*^{Wv} mutants with combined cerebellar and substantia nigra atrophy display different effects on grooming. While *Grid2*^{Lc} mutants were affected in grooming completion but not serial ordering, the reverse was true in *Girk2*^{Wv} mutants. Our results implicate cerebello–neocortical pathways in the completion of grooming chains, and a striato–pallido–neocortical pathway in the serial ordering of grooming chains.

Introduction and methodological considerations

The role of the cerebellum and basal ganglia on grooming is of some importance considering that grooming implies movement. It is therefore expected that part of the neural circuitry underlying grooming involves some aspect of motor function. In view of the importance of the cerebellum and basal ganglia in balance and posture (Lalonde and Strazielle 2007a), there is a special challenge in interpreting lesion effects of these brain regions on grooming. This is achievable by measuring serial ordering of grooming sequences. It is well established that rodents groom in a cephalocaudal order, anterior before posterior body parts (Richmond and Sachs 1980; Sachs 1988). Different types of grooming components

Neurobiology of Grooming Behavior, eds. Allan V. Kalueff, Justin L. LaPorte, and Carisa L. Bergner. Published by Cambridge University Press.

may also be measured, such as face washing; licking of forelimbs, abdomen, back, and hindlimbs; as well as body shaking and scratching (Vanderwolf *et al.* 1978). Lesions may selectively affect some grooming components in a fashion inexplainable by motor deficits. A grooming sequence may end when the animal starts to ambulate (Vanderwolf *et al.* 1978) or after a specified amount of time has elapsed (Kalueff and Tuohimaa 2004), so that the number and the duration of elements per bout are estimated. By using the first method, the possibility that these measures decrease as a result of motor dysfunction is minimized. Kalueff and Tuohimaa (2004) demonstrated that the microstructure of grooming elements is altered in mice exposed to unfamiliar environments and intruders, likely due to the role of emotion. Thus, it is worth bearing in mind that altered grooming may be caused either by motor or by emotion-related factors.

The role of the cerebellum and basal ganglia on grooming was first established with the electrical stimulation technique. Here we discuss the effects of cerebellar and basal ganglia damage on grooming tendencies and compare the data with what is known on neocortical lesions.

Cerebellum

Electrical stimulation

Electrical stimulation of the midline (vermis and fastigial nucleus) but not of the lateral cerebellum has been shown to evoke grooming in rats and cats (Berntson and Paulicci 1979; Berntson *et al.* 1973; Berntson and Torello 1982; Lisander and Martner 1975). Grooming was also observed after stimulation of the locus coeruleus (Micco 1974), a cerebellar afferent region (Snider 1975). Grooming elicited by midline cerebellar stimulation in cats did not consist of motor automatisms, in that it appeared indistinguishable from natural grooming and, like natural grooming, varied according to environmental stimuli (Berntson *et al.* 1988; Berntson and Paulicci 1979). For example, fur licking was observed with a clean coat surface and biting when it contained foreign material (Berntson *et al.* 1988). When a furry object was presented to the cat, the predominant response was self-grooming, probably as a result of emotion-related displacement activity, although allogrooming was also seen. One grooming component missing after cerebellar midline or locus coeruleus stimulation was face washing, indicating that this component comprises a separate neural circuitry from body licking.

Surgical lesions

Although fastigial stimulation elicits grooming, lesions of either the fastigial or the lateral nucleus did not affect the number of grooming bouts in rats (Berntson and Schumacher 1980). Thus, the fastigial nucleus participates in grooming without forming an essential part of its neural pathway. In view of the importance of grooming to an animal's health, other brain regions likely compensate for this damage. Because only a total score was presented, it remains to be determined whether the fastigial nucleus is involved in specific grooming components. Since face washing was not evoked after cerebellar or brainstem stimulation (Berntson *et al.* 1988), this measure cannot be expected to be affected after lesions in these areas, unless motor coordination deficits intervene. Berridge and Whishaw (1992) examined the effects of cerebellar cortical ablation on grooming chains composed of different face washing components and body licking in rats, and found no long-term deficit.

Mutant mice with cerebellar neuropathology

The autosomal semi-dominant *Lurcher* mutation causes a gain in malfunction of *Grid2* encoding the GluR2 ionotropic glutamate receptor (Zuo *et al.* 1997) functionally related with the amino-methyl-isoxazoleproprionate (AMPA) receptor (Landsend *et al.* 1997). In the normal brain, GluR2 mRNA is predominantly expressed on Purkinje cells (Takayama *et al.* 1996), the main cell type undergoing degeneration in *Grid2*^{Lc} mutants. The massive degeneration of granule cells is secondary to the lack of trophic influence exerted by the Purkinje cells (Vogel *et al.* 1991). Other secondary consequences of Purkinje cell atrophy include loss of inferior olive (Heckroth and Eisenman 1991) and deep cerebellar nuclei (Heckroth 1994). The neuropathology leads to cerebellar ataxia and deficient motor coordination, without any decrease of motor activity (Lalonde and Strazielle 2007a, b).

Grooming in Grid2^{Lc} *mutants*

We compared grooming behaviors in *Grid2*^{Lc} mouse mutants to wild type (WT) controls (Strazielle and Lalonde 1998) following the method of Vanderwolf *et al.* (1978), in which the entire body of the mouse was submerged for a few seconds in a basin of water (28°C) and then placed in a dry empty cage. The number and duration of grooming episodes were measured for five minutes, including face washing; licking of the forelimbs, abdomen (including genitals), back, and hindlimbs; as well as body shaking and scratching (the latter two behaviors were measured inside and outside sequences). Each grooming bout ended only when the mice started to walk with all four limbs in motion from any part of the cage, not after a fall, which was frequently observed in *Grid2*^{Lc} but not in WT mice. A cephalocaudal sequence score was calculated by assigning a value of "1" to a bout beginning with face washing or forelimb licking and a value of "0" to licking more posterior body parts. When a bout began with body shaking or scratching, the next behavior in the chain was considered, since these components fall outside the cephalocaudal sequence.

Despite ataxia, *Grid2*^{Lc} mutants were able to execute all seven grooming components. In line with the normal cephalocaudal order of grooming sequences (Richmond and Sachs 1980; Sachs 1988), WT controls began a bout with either face washing or forelimb licking 86% of the time and *Grid2*^{Lc} mice, 87% of the time. Thus, the cephalocaudal order was preserved in the mutants. However, total grooming components and total grooming bouts decreased in *Grid2*^{Lc} mutants relative to WT, as well as grooming components per bout of five or more components. All grooming components were reduced except body shaking. The total time spent grooming also decreased in mutants, including every component except face washing and body shaking.

Unaltered time spent face washing in mutants is in line with the finding that this component is not elicited after cerebellar or brainstem stimulation (Berntson *et al.* 1988). Normal values for body shaking are probably attributable to the fact that standing on hindlegs is not required, thereby lessening its vulnerability to a loss in postural control. Frequent losses in balance were undoubtedly a factor in reducing the number and duration of most grooming components in *Grid2*^{Lc} mutants. However, the decrease in grooming components per bout cannot be explained by a loss in motor control, because no bout ever ended with a fall. Instead, these results point to the conclusion that combined cerebellar and brainstem damage affects the completion of grooming sequences without altering the normal cephalocaudal sequence. Our data differ from the absence of any effect in the completion of grooming sequences found in rats with cerebellar cortical ablation (Berridge and Whishaw 1992). Aside from the use of different species and lesioning methods, this discrepancy may be due to different methods of analysis, as the latter paradigm places a heavy emphasis on face washing, which once again is not elicited after cerebellar or brainstem stimulation (Berntson *et al.* 1988).

Basal ganglia electrical stimulation

In addition to cerebellar/brainstem stimulation, it has been shown that repeated low-frequency stimulation of the caudate nucleus causes grooming in cats (Shishliannikova 1980). This behavior was suppressed after administration of haloperidol, an antagonist of dopamine receptors.

Neurotoxic lesions and dopamine receptor stimulation

Unlike cerebellar ablation (Berridge and Whishaw 1992), kainic acid-induced lesions of the anterior and posterior striatum reduced grooming-chain completion (Berridge and Fentress 1987). A similar defect was found after bilateral lesions of the substantia nigra caused by 6-hydroxydopamine (6-OHDA), underlining the importance of dopamine on this behavior (Berridge 1989). One possible reason for these effects is that unlike cerebellar or locus coeruleus stimulation (Berntson *et al.* 1988), face washing can be induced with a D1 though not a D2 receptor agonist (Kropf and Kuschinsky 1993). The impact of dopamine is further emphasized by findings that intrastriatal infusion of D1 and D2 receptor agonists cause different grooming behaviors after bilateral 6-OHDA lesions (Hartgraves and Randall 1986). Although unilateral 6-OHDA lesions of the substantia nigra had no effect on the number of grooming bouts (Fornaguera and Schwarting 1999), it remains to be determined whether the same negative result is found with respect to components per bout.

Aside from a brief mention on the absence of any effect on general grooming ability after lesions of the pedunculopontine tegmental nucleus (Winn 1998), we know of no analysis of other basal ganglia structures. To pursue this inquiry, we and others evaluated the *Weaver* mutation.

Mutant mice with combined midbrain and cerebellar lesions

The autosomal semi-dominant *Weaver* mutation comprises a base pair substitution of *Girk2*, encoding an inward rectifying K^+ channel highly expressed in the cerebellum and ventral midbrain (Patil *et al.* 1995). In the normal adult mouse brain, Girk2 mRNA and protein were detectable in cerebellar granule cells, Purkinje cells, and deep nuclei, as well as the substantia nigra pars compacta, the ventral tegmentum, and the olfactory tubercle (Schein *et al.* 1998). The main cell depletion occurs in *Girk2*^{Wv} granule cells (Hirano and Dembitzer 1973). It is milder in ectopically positioned Purkinje cells (Blatt and Eisenman 1985). Unlike *Grid2*^{Lc} mice (Heckroth and Eisenman 1991), inferior olive neuronal number was maintained in *Girk2*^{Wv} mice (Blatt and Eisenman 1985), presumably because of the more limited Purkinje cell depletion. The atrophy of granule and medium to large (Purkinje and Golgi) cells is more severe in the midline cerebellar cortex (Herrup and Trenkner 1987) and deep cerebellar nuclei (Maricich *et al.* 1997) than in lateral regions. In addition, neurons in the substantia nigra and the ventral tegmentum degenerate (Triarhou *et al.* 1988), reducing dopamine concentrations in the dorsal and ventral striatum, respectively (Roffler-Tarlov and Graybiel 1986). The neuropathology causes cerebellar ataxia and deficient motor coordination, but unlike *Grid2*^{Lc}, *Girk2*^{Wv} mice are hypoactive, either because dysfunctional Purkinje cells lead to a worse outcome than their absence or because of the additional basal ganglia atrophy (Lalonde and Strazielle 2007b).

Grooming in Girk2^{Wv} *mutants*

Coscia and Fentress (1993) and Bolivar *et al.* (1996) evaluated grooming in *Girk2*^{Wv} mutants and WT (*Wv*/+ or +/+) controls during developmental

Table 8.1 *Grooming behaviors (mean, S.D.) of* Girk2^{Wv} *mutants and wild type mice*

Measures	Wild type	*Girk2*^{Wv} mutants
Total components	70.5 (27.2)	57.5 (35.9)
Bouts	11.8 (3.2)	8.9 (4.4)
Components per bout	6.3 (2.7)	6.7 (4.3)
Face washing	22.2 (8.7)	**7.8 (9.3)***
Forelimb licking	9.8 (4.8)	9.8 (5.8)
Abdomen licking	6.7 (2.6)	11.5 (6.7)
Hindlimb licking	1.1 (2.8)	2.8 (2.3)
Back licking	16.8 (6.0)	17.8 (10.4)
Body shaking inside a bout	11.0 (4.4)	6.0 (5.2)
Body shaking outside a bout	15.5 (7.5)	**4.5 (3.2)***
Scratching inside a bout	2.9 (3.6)	1.8 (2.0)
Scratching outside a bout	0.7 (1.1)	1.3 (1.3)

* $p < 0.01$ (unpaired t-test).

stages, from postnatal week 2 to 3, when postural deficits are already apparent. Unlike our *Grid2*^{Lc} study (Strazielle and Lalonde 1998), their method of analysis included the possibility of grooming bouts ending with a fall. The results showed that the number of grooming bouts was higher in *Girk2*^{Wv} mutants than in controls, at least partly because of the higher susceptibility of the mutants to falling (Bolivar *et al.* 1996; Coscia and Fentress 1993). In addition, grooming components per bout decreased in *Girk2*^{Wv} mutants (Bolivar *et al.* 1996). The latter result was not ascribed to postural defects, because this difference occurred irrespective of whether mutants fell or not (Coscia and Fentress 1993). The lower number of grooming components per bout is similar to findings in rats with striatal (Berridge and Fentress 1987) or substantia nigra (Berridge 1989) lesions.

We assessed six-month-old *Girk2*^{Wv} mutants (obtained from Jackson Laboratory, Bar Harbor, ME, USA) using the same paradigm as the *Grid2*^{Lc} study (Strazielle and Lalonde 1998). Ten homozygous *Girk2*^{Wv} mutants were compared with ten nonataxic controls, either WT or heterozygotes of the B6CBA-Aw-J strain, equally distributed for gender (50%). Despite ataxia, *Girk2*^{Wv} mutants were able to execute all seven grooming components. As shown in Table 8.1, *Girk2*^{Wv} mutants had normal values for total grooming components, grooming bouts, and grooming components per bout. The lack of an intergroup difference with respect to grooming components per bout is in contrast with the Coscia and Fentress (1993) report in the same mutant but with a different method, in which three face-washing measures and one body-licking measure were used. In our study, a wider range of body grooming components was compiled. As seen with *Grid2*^{Lc} (Strazielle and Lalonde 1998), we found that *Girk2*^{Wv} mice had fewer face-washing episodes. Thus, we believe that the discrepancy between the Coscia and Fentress (1993) study and ours is due to their heavy emphasis on face washing. Unlike *Grid2*^{Lc}, body shaking was also reduced in *Girk2*^{Wv} mice, perhaps due to their dopaminergic deficiency, as this anomaly is only observed in the latter (Strazielle *et al.* 1998).

The main intergroup difference occurred on serial ordering of grooming chains. While controls began to groom with face washing or forelimb licking 68.9% (SD = 21.1) of the time (somewhat lower than in the previous study with a similar B6CBA strain, perhaps due to the inclusion of heterozygotes), *Girk2*^{Wv} homozygotes began to groom in this way only 32.5% (SD = 26.3) of the time ($p < 0.01$). As seen in Figure 8.1, when each component was assigned a rank order for all bouts combined (including repetitions), *Girk2*^{Wv} mutants had fewer face-washing events specifically on the second behavior in a series and forelimb licking on the fourth.

Thus, a double dissociation occurred in regard to the two cerebellar mutants. While *Grid2*^{Lc} mutants were affected in grooming completion but not serial ordering, the reverse was true in *Girk2*^{Wv} mutants. The lack of any effect on grooming completion in *Girk2*^{Wv} mice may be due to their milder Purkinje cell atrophy, while disturbance in the cephalocaudal sequence may be ascribed to basal ganglia damage.

Lesions of the cerebral cortex

As found in *Grid2*^{Lc} mutants (Strazielle and Lalonde 1998), grooming episodes per bout decreased relative to sham-operated controls in adult (Kolb and Whishaw 1981; Vanderwolf *et al.* 1978; Whishaw *et al.* 1981) and neonatal (Kolb and Whishaw 1981) decorticate rats, as well as in adult deneocorticate rats (Vanderwolf *et al.* 1978), all these experiments using the Vanderwolf *et al.* (1978) method. These results implicate a cerebello–neocortical pathway in the completion of grooming chains. In addition, neocortical lesions in adult rats increased the number of fragmented grooming chains (Cannon *et al.* 1992) determined with the same method used in evaluating striatal and substantia nigra lesions (Berridge 1989; Berridge and Fentress 1987). These results implicate a striato–pallido–neocortical pathway in chain completion of face-washing components. However, it must be noted that decreases in grooming components per bout (Whishaw *et al.* 1981) and increased fragmented grooming chains (Cannon *et al.* 1992) in adult decorticates were found only after water immersion and not in dry rats. Thus, as mentioned by Kalueff and Tuohimaa (2004), the microstructure of grooming appears to differ depending on stress levels.

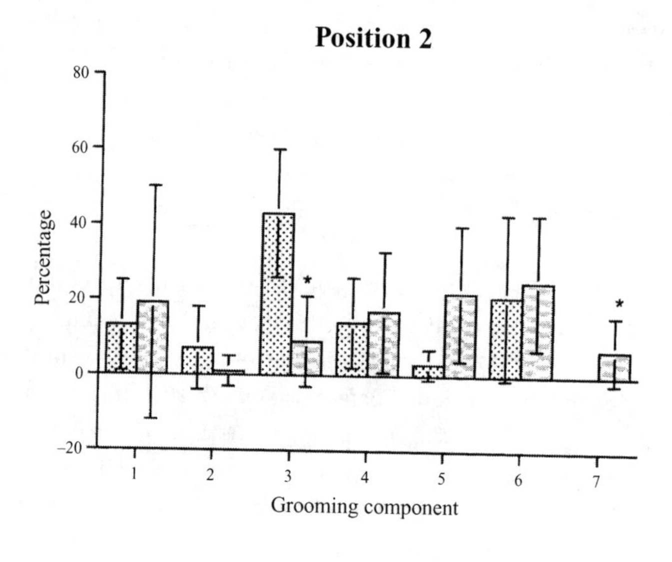

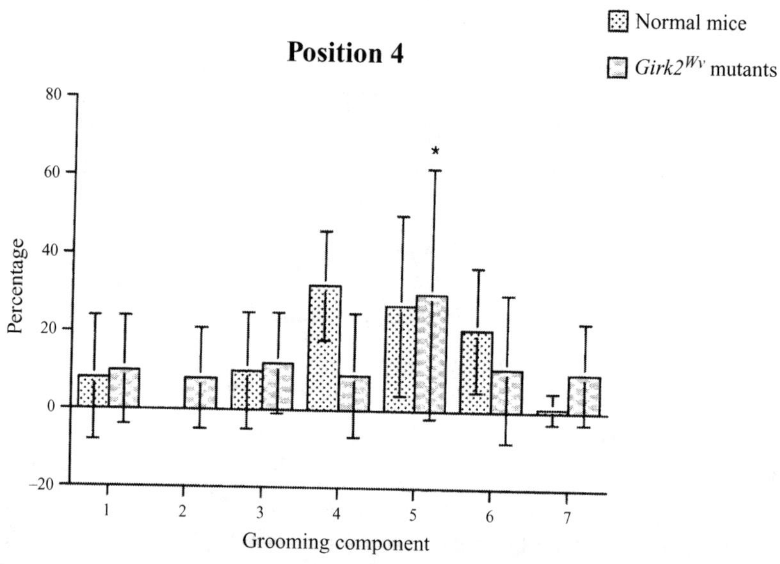

Figure 8.1 Grooming components in serial positions 2 and 4 (based on sequential order of all grooming bouts): 1 = body shaking; 2 = scratching; 3 = face washing; 4 = forelimb licking; 5 = abdomen licking; 6 = back licking; 7 = hindlimb licking. * $p \leq 0.01$ vs. normal mice.

Concluding remarks

The overall results, though preliminary, implicate a cerebello-neocortical pathway in the completion of grooming chains, and a striato-pallido-neocortical pathway in the serial ordering of grooming chains. However, this conclusion is likely to be dependent on the method of analysis. In particular, there is evidence that the basal ganglia contribute to chain completion in paradigms with a heavy emphasis on face washing.

There is a need to complete such research attempts by lesioning specific neocortical and subcortical regions. In particular, the serial ordering of grooming chains is a prepotent rodent response. Therefore, the neural pathway underlying this response is likely to yield important clues concerning serial ordering in other types of animal behaviors, such as nest building and burrowing, as well as provide insight into serial ordering in humans, an important facet of human behavior.

Acknowledgment

This research was funded by a grant from the Natural Sciences and Engineering Research Council of Canada to RL.

References

Berntson GG and Paulicci TS (1979): Fastigial modulation of brainstem behavioral mechanisms. *Brain Res Bull* **4**:549–52.

Berntson GG and Schumacher KM (1980): Effects of cerebellar lesions on activity, social interactions, and other motivated behaviors in the rat. *J Comp Physiol Psychol* **94**:706–17.

Berntson GG and Torello MW (1982): The paleocerebellum and the integration of behavioral function. *Physiol Psychol* **10**:2–12.

Berntson GG, Potolicchio SJ Jr and Miller NE (1973): Evidence for higher functions of the cerebellum: eating and grooming elicited by cerebellar stimulation in cats (fastigial nucleus/superior cerebellar peduncle). *Proc Natl Acad Sci USA* **70**:2497–9.

Berntson GG, Jang JF and Ronca AE (1988): Brainstem systems and grooming behaviors. *Ann N Y Acad Sci* **525**:350–62.

Berridge KC (1989): Substantia nigra 6-OHDA lesions mimic striatopallidal disruption of syntactic grooming chains: a neural systems analysis of sequence control. *Psychobiology* **17**:377–85.

Berridge KC and Fentress JC (1987): Disruption of natural grooming chains after striatopallidal lesions. *Psychobiology* **15**:336–42.

Berridge KC and Whishaw IQ (1992): Cortex, striatum and cerebellum: control of serial order in a grooming sequence. *Exp Brain Res* **90**:275–90.

Blatt GJ and Eisenman LM (1985): A qualitative and quantitative light microscopic study of the inferior olivary complex of normal, *reeler*, and *weaver* mutant mice. *J Comp Neurol* **232**:117–28.

Bolivar VJ, Danilchuk W and Fentress JC (1996): Separation of activation and pattern in grooming development of *weaver* mutant mice. *Behav Brain Res* **75**:49–58.

Cannon RL, Paul DJ, Baisden RH and Woodruff ML (1992): Alterations in self-grooming sequences in the rat as a consequence of hippocampal damage. *Psychobiology* **20**:205–18.

Coscia EM and Fentress JC (1993): Neurological dysfunction expressed in the grooming behavior of developing *weaver* mutant mice. *Behav Genet* **23**:533–41.

Fornaguera J and Schwarting RKW (1999): Early behavioral changes after nigro-striatal system damage can serve as predictors of striatal dopamine depletion. *Prog Neuropsychopharmacol Biol Psychiatry* **23**:1353–68.

Hartgraves SL and Randall PK (1986): Dopamine agonist-induced stereotypic grooming and self-mutilation following striatal dopamine depletion. *Psychopharmacology* **90**:358–63.

Heckroth JA (1994): Quantitative morphological analysis of the cerebellar nuclei in normal and *Lurcher* mutant mice. I. Morphology and cell number. *J Comp Neurol* **343**:173–82.

Heckroth JA and Eisenman LM (1991): Olivary morphology and olivocerebellar atrophy in adult *Lurcher* mutant mice. *J Comp Neurol* **312**:641–51.

Herrup K and Trenkner E (1987): Regional differences in cytoarchitecture of the *weaver* cerebellum suggest a new model for *weaver* gene action. *Neuroscience* **23**:871–85.

Hirano A and Dembitzer HM (1973): Cerebellar alterations in the *weaver* mouse. *J Cell Biol* **56**:478–86.

Kalueff AV and Tuohimaa P (2004): Grooming analysis algorithm for neurobehavioral research. *Brain Res Prot* **13**:151–8.

Kolb B and Whishaw IQ (1981): Decortication of rats in infancy or adulthood produces comparable functional losses on learned and species-typical behaviors. *J Comp Physiol Psychol* **95**:468–83.

Kropf W and Kuschinsky K (1993): Effects of stimulation of dopamine D1 receptors on the cortical EEG in rats: different influences by a blockade of D2 receptors and by an activation of putative dopamine autoreceptors. *Neuropharmacology* **32**:493–500.

Lalonde R and Strazielle C (2007a): Brain regions and genes affecting postural control. *Prog Neurobiol* **81**:45–60.

Lalonde R and Strazielle C (2007b): Spontaneous and induced mouse mutations with cerebellar dysfunctions: behavior and neurochemistry. *Brain Res* **1140**:51–74.

Landsend AS, Amiry-Moghaddam M, Matsubara A *et al.* (1997): Differential localization of delta glutamate receptors in the rat cerebellum: coexpression with AMPA receptors in parallel fiber-spine synapses and absence from climbing fiber-spine synapses. *J Neurosci* **17**:834–42.

Lisander B and Martner J (1975): Integrated somatomotor and gastrointestinal adjustments induced from the cerebellar fastigial nucleus. *Acta Physiol Scand* **94**:358–67.

Maricich SM, Soha J, Trenkner E and Herrup K (1997): Failed cell migration and death of Purkinje cells and deep nuclear neurons in the *weaver* cerebellum. *J Neurosci* **17**:3675–83.

Micco DJ Jr (1974): Complex behaviors elicited by stimulation of the dorsal pontine tegmentum in rats. *Brain Res* **75**:172–6.

Patil N, Cox DR, Bhat D *et al.* (1995): A potassium channel mutation in *weaver* mice implicates membrane excitability in granule cell differentiation. *Nat Genet* **11**:126–9.

Richmond G and Sachs BD (1980): Grooming in Norway rats: the development and adult expression of a complex motor pattern. *Behaviour* **75**:82–96.

Roffler-Tarlov S and Graybiel AM (1986): Expression of the *weaver* gene in dopamine-containing neural systems is dose-dependent and affects both striatal and nonstriatal regions. *J Neurosci* **6**:3319–30.

Sachs BD (1988): The development of grooming and its expression in adult animals. *Ann N Y Acad Sci* **525**:1–17.

Schein JC, Hunter DD and Roffler-Tarlov S (1998): Girk2 expression in the ventral midbrain, cerebellum, and olfactory bulb and its relationship to the murine mutation *weaver*. *Dev Biol* **204**:432–50.

Shishliannikova LV (1980): Behavioral activation developing after cessation of electrical stimulation of the caudate nucleus. *Fiziol ZH SSR Im Sechenova* **66**:1165–70.

Snider RS (1975): A cerebellar-ceruleus pathway. *Brain Res* **88**:59–63.

Strazielle C and Lalonde R (1998): Grooming in *Lurcher* mutant mice. *Physiol Behav* **64**:57–61.

Strazielle C, Lalonde R, Amdiss F *et al.* (1998): Distribution of dopamine transporters in basal ganglia of cerebellar ataxic mice by [^{125}I]RTI-121 quantitative autoradiography. *Neurochem Int* **32**:61–8.

Takayama C, Nakagawa S, Watanabe M, Mishina M and Inoue Y (1996): Developmental changes in expression and distribution of the glutamate receptor channel delta 2 subunit according to the Purkinje cell maturation. *Dev Brain Res* **92**:147–55.

Triarhou LC, Norton J and Ghetti B (1988): Mesencephalic dopamine cell deficit involves areas A8, A9 and A10 in *weaver* mutant mice. *Exp Brain Res* **70**:256–65.

Vanderwolf CH, Kolb B and Cooley RK (1978): Behavior of the rat after removal of the neocortex and hippocampal formation. *J Comp Physiol Psychol* **92**:156–75.

Vogel MW, McInnes M, Zanjani HS and Herrup K (1991): Cerebellar Purkinje cells provide target support over a limited spatial range: evidence from *Lurcher* chimeric mice. *Dev Brain Res* **64**:87–94.

Whishaw IQ, Nonneman AJ and Kolb B (1981): Environmental constraints on motor abilities used in grooming, swimming, and eating by decorticate rats. *J Comp Physiol Psychol* **95**:792–804.

Winn P (1998): Frontal syndrome as a consequence of lesions in the pedunculopontine tegmental nucleus: a short theoretical review. *Brain Res Bull* **47**:551–63.

Zuo J, De Jager PI, Takahashi KA *et al.* (1997): Neurodegeneration in *Lurcher* mutant mice caused by mutation in the delta 2 glutamate receptor gene. *Nature* **388**:769–73.

9

Striatal implementation of action sequences and more: grooming chains, inhibitory gating, and the relative reward effect

HOWARD CASEY CROMWELL

Summary

The striatum is a subcortical structure thought to be important for higher motor functions and reward processing. It is part of a larger system called the basal ganglia (BG) and composed of multiple subregions thought to be functionally heterogeneous. This review provides information and evidence for the role of the striatum in implementing the fixed action pattern of the grooming chain in the rat. The support for the involvement of the dorsolateral striatal subregion involved in the production of this movement sequence is described, and the general functional significance of implementation by striatal circuitry is discussed. Implementation is meant to refer to the ability of local processing within striatal circuits to enable motor action plans to be completed without distraction from competing sensory or motor demands. The idea that the striatum is involved in more than motor functions is developed and evidence for detailed processing of reward outcomes is presented. We introduce the possibility that the general nature of striatal function of "implementing" chains of information crosses different functional boundaries between movement and reward information. For movement plans, the implementation includes enabling motor sequences for appropriate output and for reward plans, the implementation includes enabling reward incentive hierarchies for appropriate outcome choices. These types of functions could rely upon cross-talk among striatal subregions and reveal a possible shared integrative

Neurobiology of Grooming Behavior, eds. Allan V. Kalueff, Justin L. LaPorte, and Carisa L. Bergner. Published by Cambridge University Press.

function for the different "loops" of the BG circuitry. These ideas have clinical implications for understanding complex symptoms in traditional BG disorders such as Huntington's disease and Parkinson's disease as well as other complex impairments observed in obsessive–compulsive disorder (OCD) and other mental illnesses.

Some reasons for studying action sequencing

The fields of behavioral science and neurobiology interested in motor control have well established interests in examining the parameters of movement and the physiology of action at several different levels from reflexes to complex voluntary behavior (Kugler and Turvey 1987; Lashley 1951; Lemon 2008; Marsden 1982). In the majority of cases, these diverse interests remain largely dissociated, and the distance dividing distinct research areas makes development of a complete analysis of motor control arduous to say the least.

Action sequencing is one example of higher motor control that can involve complex processing and multiple levels of brain control. Interestingly, action sequencing is one area of motor control that has been examined in diverse fields of research including neurology, psychology, ethology, psychiatry, and endocrinology. This book highlights the fact that these different fields of interest bring something important and useful to the study of action sequencing, and each area has found the examination of serial order of movement very profitable. This current work calls attention to the need to bring together the different areas of interest to formulate a comprehensive view for the neurobiology and functional significance of action sequencing.

Previously, Fentress (1990) emphasized the need for a "synthetic neuroscience," an integrated approach that incorporated dynamic processes and multidisciplinary perspectives. There remains a need for this synthesis of information and experimentation across diverse disciplines today and the outlook is promising with examples of excellent cross-disciplinary work (Cromwell *et al.* 2008; Ijspeert 2008; Van Craenenbroeck *et al.* 2005). Several fields of research have made exciting progress recently by combining information and working together including areas such as neuropsychopharmacology and psychoneuroendocrinology. These areas are growing in popularity because they "synthesize" information and take into account the dynamic and holistic processes better than viewpoints in isolation.

Examining action sequencing holds great promise for several reasons and has been the target of interest because of the wealth of information obtained from its study. It is clear that higher order interdependencies for movement abound in the natural world (Fentress 1983; Hinde 1970) yet they can be difficult to identify because of the blending of chains of action into the background noise of other activities. It can be a tedious and difficult task to uncover serial action relationships but once they are revealed they can offer a variety of diverse insights related to the natural ordering of elements over time. An example includes the analysis of gait and locomotor movements, which has provided information about the comparative ethology of movement control (Grillner and Zangger 1975), the neural basis of walking (Ryou and Wilson 2004) and the neuropathology of motor disease (Iansek *et al.* 2006). It is clear that the study of action sequencing has offered valuable information to fields of medicine such as neurology and psychiatry. Disorders such as Parkinson's disease and Huntington's disease have been shown to be characterized by impairments in ordering actions and these alterations are nested within a more general deficit of movement initiation and termination (Folstein *et al.* 1986; Georgiou *et al.* 1995; Marsden 1982). Typically, the more general motor problems of a disorder are more obvious while deficits in higher motor processing might take extra effort in terms of the detailed examination to reveal, but they can surely be just as debilitating (Aizenstein *et al.* 2005; Cattaneo *et al.* 2007).

Finally the study of action sequencing as a topic of interest has potential for substantial impact outside of the traditional motor sciences. Fentress emphasized the ability to understand the processing of nonmotor sequencing based upon the rules and interdependencies found for action sequencing (Fentress 1988). This idea is a very important one because it fosters a broad application for the study of action sequencing that can influence the study of language, mental illness, and motivational systems. For example, Fentress and Stilwell (1973) commented that the "grammar of movement" could provide a window into the underlying mechanisms involved in language syntax. We can postulate that the neural mechanisms related to action sequencing are similar for the process of ordering thoughts and potentially for the ordering of incentive information over time. This review provides some ideas about this possibility and focuses on a particular action sequence, the grooming chain of the rat, and its implementation by a subcortical region known as the striatum.

The grooming chain and action sequencing

Rats spend a large proportion of their time grooming (Bolles 1960) and a great amount of grooming activity is composed of "flexible" grooming behavior. These "flexible" sequences of grooming action consist of ordered combinations of forelimb strokes, licking, and scratching (Berridge *et al.* 1987) and the term "flexible" denotes a less rigid character for these movements relative to the "grooming chain" observed in rodents. This grooming chain is a natural, reliable sequence,

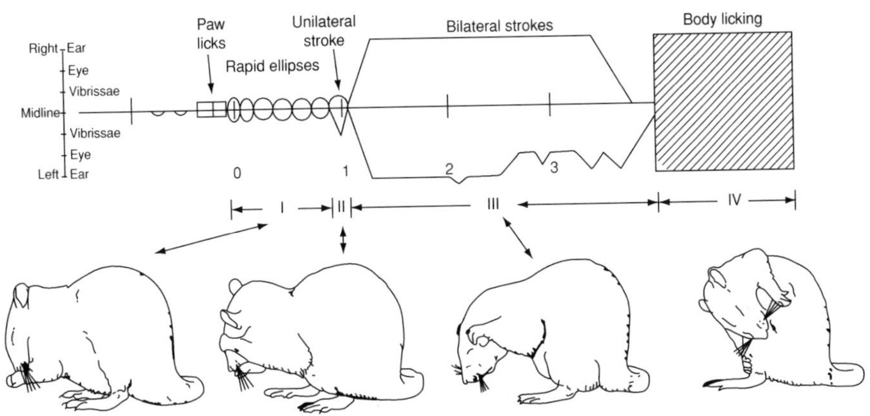

Figure 9.1 The grooming chain of the rat depicted using a drawing of a rat performing the actions of each of the four phases and by notation form. The notated version shows the actions relative to the midline with symbols representing the movements. Squares represent licking, with small = paw licking and larger = body licking. Small ovals represent the rapid ellipses and lines drawn away from the midline represent forelimb strokes from the midline of the face. Originally published in Cromwell and Berridge (1996).

which is less often observed and has been found to be sensitive to manipulations of the BG system (Aldridge and Berridge 1998; Berridge and Fentress 1987; Cromwell and Berridge 1996). The sequence does not require training and occurs naturally in a developmental trajectory (Colonese *et al.* 1996). The natural, reliable sequence is an ordered grooming chain of up to 25 actions, which occurs 13 000 times greater than chance based upon component action probabilities (see Figure 9.1; Berridge *et al.* 1987). The grooming pattern is composed of four consecutive phases:

Phase I. A set of five to nine rapid small bilateral forepaw strokes over the snout and the mystacial vibrissae at 6 to 7 Hz

Phase II. A short bout of one to four slower paw strokes, which are usually asymmetric and at differing amplitudes

Phase III. A set of larger symmetrical strokes (three to ten), which go back behind the ears and most of the head

Phase IV. A bout of body licking directed towards the lateral and ventral torso.

Once the initial components appear, the entire sequence follows with a completion rate of 80% to 95% in intact rats (Berridge *et al.* 1987).

The properties of the grooming chain have been examined in great detail using several manipulations to understand the role of contextual influences on the different types of grooming actions embedded in the different forms of the behavior (Fentress 1988). Woolridge and Fentress (Woolridge 1975) used elastic threads to pull on the forearms of the animals during grooming and found that the grooming acts outside of the chain were dramatically disrupted while those within the chain were impervious to the changes in proprioceptive load. Berridge and Fentress (1987) used deafferentiation (transection of mandibular and maxillary branches of the trigeminal nerve) of the tactile information from the rostral face and mouth to examine this contextual dependency. They found that the sequential integrity of the grooming actions remained despite the lack of sensory feedback. It was an elegant demonstration of action control and independence of tactile cues. Overall the grooming chain has been found to be an extremely useful tool to study motor sequencing due to its nonreliance on learning and having a set of similar actions that are committed outside of the chain for comparison. The grooming chain has been studied more recently to examine the details of brain mechanisms of action sequencing, and it is clear that the striatum plays a key role in the production of normal proficiency of grooming chains.

The striatum and basal ganglia system

The BG system includes the subcortical structures of the striatum, the pallidal regions, and the substantia nigra (Figure 9.2). Other structures intimately connected with the BG system include the subthalamic nucleus, amygdala, and the pedunculopontine nucleus (Kelley *et al.* 1982; Lee *et al.* 2000). Each of these brain regions has well demarcated subregions and each subregion has a set of distinct connections that make it unique from adjacent subregions. The striatum comprises three well-recognized major subregions: the caudate nucleus, putamen, and ventral striatum or nucleus accumbens. The globus pallidus has medial (i.e., internal segment), lateral (i.e., external segment), and ventral subregions; the substantia nigra contains two main subregions: the pars compacta and the pars reticulata. This set of BG structures comprises the largest subcortical system in the brain, and its complex connectivity and internal structure have made it extremely difficult to address questions about its functional nature. A relatively recent, but well-documented observation is that the system participates in numerous nonmotor functions that include perceptual, cognitive, and emotional aspects (Berridge and Cromwell 1990; Corbit and Janak 2007; Dalley *et al.* 2008; Setlow *et al.* 2003).

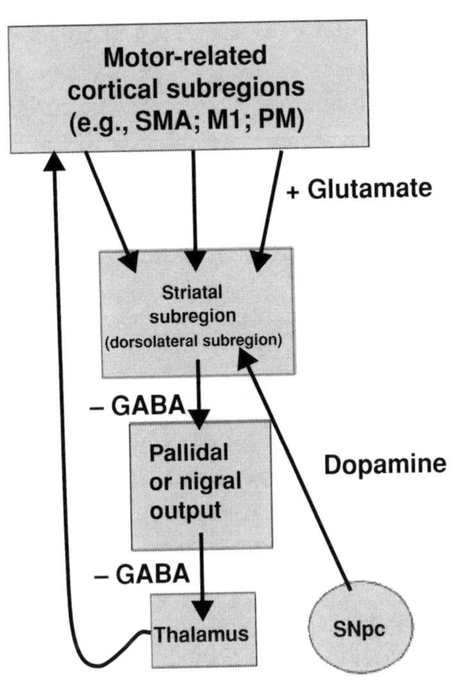

Figure 9.2 Generalized cortico–striatal loop. The BG system is characterized by connections arriving from the cortex to the BG input nuclei of the caudate nucleus and putamen (striatum). This structure then sends outputs to the BG output centers of the globus pallidus and substantia nigra. These output structures then send projections to regions in the thalamus. The thalamus has a major output to the cortex. This cortico–BG–thalamo–cortical loop has been the standard circuit in which to characterize information flow through the BG system. Most accounts of this circuitry have proposed that functionally related sets of striatal cells receive information from a related set of cortical areas. In this way, the striatum would consist of microzones of interrelated functional clusters that received inputs and funneled outputs to the next circuit location. Dopamine input from midbrain regions such as the substantia nigra pars compacta modulate striatal output and potentially amplify certain inputs relative to others. (Abbreviations: GABA, γ-aminobutyric acid; M1, primary motor cortex; PM, premotor cortex; SMA, supplementary motor area; SNpc, substantia nigra pars compacta.)

In the classic viewpoint, it was thought that the BG had a uniform motor function, mainly involved in movement generation, motor initiation, or postural control (Carpenter 1976; Martin 1977; Wilson 1914). Wilson (1914, p. 478) stressed this point in writing, "When we remember the histological simplicity and comparative structural homogeneity of the corpus striatum, in contrast with the greater dimensions, much more intricate cytoarchitectural complexity, and far wider connexions of the rolandic motor cortex, the idea of attributing all of the disturbances to striatal disease and of crowding corresponding centers into that ganglion becomes nothing short of ludicrous." This statement emphasizes the notion that a brain structure or system without obvious structural subdivisions could be thought of as having a single function, a neural solidarity that can be simply described and recognized.

In the past 30 years or so, researchers have been increasingly crowding diverse functions into the BG system and describing in detail heterogeneous loops or pathways involved in movement, sensation, emotion, motivation, and cognition (Cromwell and Berridge 1996; Haber 2003; Heimer and Van Hoesen 2006). The central location of the BG and the diverse inputs that locally converge make this system one of the most potentially integrative within the central nervous system (Kincaid *et al.* 1998; Levy *et al.* 1997; Heimer and Van Hoesen 2006). This high level of diversity of connections most likely contributes to the extraordinary level of complexity in the symptoms observed following BG dysfunction (Cromwell and King 2004; Marsden 1984). Basal ganglia disorders such as Parkinson's disease and Huntington's disease have been thought to arise from a breakdown in motor gating, sensorimotor integration, and movement sequencing (Agostino *et al.* 1992; Georgiou *et al.* 1995; Helmuth *et al.* 2000).

The striatum and the grooming chain in the rat

Lesions of the striatum using neurotoxins (kainic acid, quisqualic acid, or quinolinic acid) can produce severe impairments in motor sequencing in the rat (Pisa 1988; Sabol *et al.* 1985), cat (Cools 1980; Van Den Brecken and Cools 1982) and monkey (Delong *et al.* 1984). Most of these studies rely upon learned sequences dependent on memory processes for completion. For example, the initiation and/or completion of the pellet retrieving sequence (a series of movements the animal uses to obtain a small food pellet from a small diameter cylinder) has been impaired following lateral striatal lesions (Pisa 1988) or after 6-hydroxydopamine (6-OHDA) lesions of the nigrostriatal tract (Whishaw *et al.* 1986). However, it is difficult to distinguish between deficits in sequential learning and memory versus sequential programming per se in these types of experiments.

The striato–pallidal system seems to play a unique role compared to other forebrain structures in the sequencing of the grooming chain because lesions to other structures purported to be involved in motor sequencing do not produce a deficit in grooming-chain completion. In a previous study lesions of the primary motor cortex (Fr1, Fr3, and medial FL: agranular frontal cortex of rat where stimulation induces movement), or lesions to the secondary motor cortex (Fr2:

agranular frontal cortex where stimulation does not directly induce movement), or complete aspiration of the cerebellum did not produce a disruption of the grooming pattern (Berridge and Whishaw 1992). Only when striatum damage was added to the cortical damage (motor cortex + striatal ablation) was a sequencing deficit observed (Berridge and Whishaw 1992). A sequencing deficit is determined by calculating the proportion of chains of grooming actions performed to completion (Phase IV) between different experimental groups. More detailed analyses have included examination of the duration of the chain components and analysis of less than perfect grooming chains (e.g., those chains not ending in body licking but instead terminating in licking of other body areas – paw licking for instance) following neural manipulations. In general, this result suggested that striatopallidal damage is necessary in order to get a disruption of the grooming chain.

Another series of studies examined whether striatopallidal damage is sufficient to produce the disruption of the serial order of the grooming chain. Kainic acid lesions of the striatum (caudate nucleus and putamen) reduced the completion rate for grooming chains by at least 60% (Berridge and Fentress 1987). The 6-hydroxydopamine lesions of the substantia nigra, which depleted dopamine from the striatum, produced a similar significant reduction in chain completion compared to vehicle-injected controls (Berridge 1989). The disruption of the grooming chain seen after striatopallidal lesions is statistically equal to the disruption seen after pontine decerebration, a transection of the brain at the level of the pons, in rats (Berridge 1989). These results indicate that damage to only the striatopallidal system is needed in order to produce the sequential deficit in grooming behavior. These changes in the grooming chain following striatopallidal lesions appear sequential in nature and do not appear to be caused by an impairment in the ability to produce the individual grooming actions. Two pieces of evidence support this conclusion: (1) the individual actions are intact during variable grooming in rats following striatal lesions and (2) rats with striatal lesions can complete a certain number of chains in the appropriate order with well-formed phases.

Finally, a study was completed that examined the role of different striatal subregions in the production of the grooming chain. We targeted lesions to specific striatal and pallidal regions including: the dorsolateral region (DL-STR), ventrolateral region (VL-STR), dorsomedial region (DM-STR), ventromedial region (VM-STR), the globus pallidus (i.e., lateral GP), or the ventral pallidum (VP/SI) (Cromwell and Berridge 1996). The results of this experiment demonstrated that the DL-STR has a unique and special role in the sequencing of grooming behavior. Bilateral lesions restricted to a small region within the dorsolateral neostriatum immediately subjacent to the corpus callosum (stereotaxic coordinates: anterior/posterior, medial/lateral, and dorsal/ventral, respectively: 1.3 mm × 1.0 mm × 1.0 mm) produced deficits in grooming syntax that were comparable in magnitude to the deficits produced by large lesions of the neostriatum or even decerebration (Berridge 1989; Berridge and Fentress 1987; Berridge and Whishaw 1992). It has been shown that unilateral lesions that encapsulate this site do not produce an impairment in grooming (Cromwell and Berridge 1996). These same lesions of the DL-STR spared individual grooming actions and did not change the amount of total grooming behavior emitted.

Similar size lesions of the VL-STR, VM-STR or DM-STR, or GP or VP/SI did not produce a similar impairment in grooming-chain completion. Lesions to some of these other striatopallidal regions did produce a simple motor deficit in terms of grooming. Lesions of the VP/SI produced a severe decrease in the number of grooming actions and in the amount of grooming behavior. In some instances, rats with VP/SI lesions would appear to go through a period of "grooming arrest." In this condition described by Levitt and Teitelbaum (1975) a rat with a lateral hypothalamic lesion would slump over and appear to fall asleep. In many cases, the halting of movement would come in the middle of a grooming act hence the label "grooming arrest" (i.e., a rat would slump over in midstroke). This action was seen in rats with the VP/SI lesions but not in rats with any of the striatal lesions or GP lesions. These rats with VP/SI lesions showed a decrease in overall grooming actions and in some cases a complete "arrest" in grooming yet they completed the grooming chains at normal levels. The grooming arrest phenomenon was not seen during a chain sequence, instead these rats would complete chains to Phase IV body licking at a normal rate. In terms of the grooming behavior, the rats with VP/SI lesions are impaired in motor initiation and not impaired in motor action sequencing while the rats with DL-STR lesions are impaired in motor sequencing without an impairment in general grooming movements.

A series of electrophysiological studies have supported the idea that the DL-STR may function in a unique way to implement the grooming chain (Aldridge and Berridge 1998; Aldridge *et al.* 2004; Meyer-Luehmann *et al.* 2002). This group has performed elegant frame-by-frame analysis of the grooming chain and coupled the action pattern to the firing patterns of single neurons in the striatum and related BG structures. Firing rate increased substantially in the DL-STR during syntactic chains of action (116% above baseline). Single neurons in this subregion were more likely to be selective in their firing pattern for the grooming movements made inside of the chain compared to outside. From single neurons in the substantia nigra pars reticulata, this group has recorded from an output structure of the BG that receives input from the striatum and sends output to the thalamus and brainstem regions (Parent 1986). Activity of this structure differed from the striatum in that a majority of the firing was related to the initial phases of the grooming chain only and not the complete chain (Meyer-Luehmann *et al.* 2002). The activity changes were more intensive related to movements within the grooming chain than outside of the chain and could be more important for the initiation of the sequence and not the maintenance of the actions. The striatal neural activity, especially within the dorsolateral sector, seemed to be crucial for the entire chain of action and, hence, could be key for the enabling of the entire series of movements.

Finally, the same group has examined whether or not striatal neural activity would be related to other unlearned, sequential actions (Aldridge *et al.* 2004). They examined the neural activity during a "warm-up" sequence of actions that precede locomotion and compared activity between this "warm-up" sequence (Golani *et al.* 1981) and the grooming chain. They found a subset of striatal neurons were active during both the warm-up and grooming sequence but the strength of the activity change was weaker for the warm-up actions compared to the grooming-chain actions. The authors noted this could be due to the weaker sequential integrity of the warm-up series of movements compared to the more rigid structure of the grooming chain.

Why might the dorsolateral striatum subregion be critical for action sequencing?

Researchers have proposed that the striatopallidal system is composed of several anatomically separate circuits, and that this anatomical division could lead to functional division (see Figure 9.2; Alexander *et al.* 1986; Alheid and Heimer 1988; Dalley *et al.* 2008; Divac 1972; Levy and Dubois 2006; Nauta and Domesick 1984). In the rat and in the monkey, the DL-STR, VL-STR, DM-STR, and VM-STR each receive projections from different areas of the cortex (McGeorge and Faull 1989; Selemon and Goldman-Rakic 1985). Cortical input to the dorsolateral neostriatum comes from the primary motor cortex (M1), primary sensory cortex (S1), secondary sensory cortex (SII), and medial agranular cortex (AGm) (see Figure 9.2; McGeorge and Faull 1989). The dorsolateral sector of the rostral neostriatum receives the densest input from the motor cortex (1.2 mm to 0.7 mm anterior to bregma; McGeorge and Faull 1989). Since it is most likely motor cortex input that drives this ability to compute the combinations of sensory and motor input for appropriate motor output, it would seem likely that lesions to the motor and sensory cortical regions would produce a similar sequential impairment of grooming. Lesions to the globus pallidus (ventromedial portion), which is a member of the cortico–striato–pallido–cortical "motor" loop, do not impair the grooming chain.

These differences in lesion effects within a set of structures proposed to be a functionally continuous circuit indicate that the circuit may not be functioning in a similar manner at all processing levels. The results also indicate that there may be some functions of the striatum that are not essentially dependent upon the proposed cortico–striato–pallido–cortical loops but instead rely on crucial processing within intrinsic networks or through some other less well known pathway.

If intrinsic processing is important in the role of the dorsolateral neostriatum in sequencing the grooming chain, then it may be useful to examine its intrinsic composition more closely. Brown (1992) used deoxy-d-[^{14}C]-glucose (DG) autoradiography to show that certain regions of the dorsolateral neostriatum "light up" with maximal activity correlated with somatosensory input to either the hindlimb, trunk, or forelimb. Stroking stimuli were applied for a 45-minute period during injection of DG, and the rats were sacrificed immediately after testing. The group of rats that received that stimulation had nonoverlapping sites of activation in the primary somatosensory cortex at predictable sites according to electrophysiological studies. In the dorsolateral neostriatum, the activated portions for the different body parts "moved" as one traveled from the anterior to posterior dorsolateral neostriatum. The hindlimb, trunk, and forelimb regions were maximally nonoverlapping in the anterior region (+0.2 mm from bregma) with >1.0 mm distances between the different regions. The hindlimb was found to be represented in the most dorsomedial sector, the trunk region was ventrolateral to this region, and the forelimb was ventrolateral to the trunk region at 0.2 mm anterior to bregma. In the caudal dorsolateral neostriatum, the picture changed. The activated regions were much closer together, and the body-part representative regions switched places. The region activated by hindlimb sensory input was ventral to its anterior placement and forelimb feature was dorsal. At 0.6 mm anterior to bregma (very close to the center of the crucial site mapped in this study), the placements had moved slightly again with trunk activation sectors dorsal, followed by hindlimb and forelimb in consecutive ventral positions.

Two studies completed by West and colleagues (Carelli and West 1991; West *et al.* 1990) examined in detail a region of the dorsolateral neostriatum, which contains the crucial site identified in this study, using electrophysiological recording. In one study (West *et al.* 1990), the positions of regions within the dorsolateral neostriatum in the rat that corresponded with specific locomotor movements were determined. From +1.6 mm anterior to bregma to −1.0 mm from bregma in a ~1.0 mm wide strip just below the corpus callosum, units were found that responded specifically to hindlimb or forelimb movements. At some anterior–posterior (AP) levels relative to bregma, neurons that were related to hindlimb movements were located dorsal to neurons related to forelimb movements and at other AP levels, the relationship was converse. This region appeared to overlap with both

Brown's region of DG activation (Brown 1992), and the crucial site mapped in this study.

In the second study, West and colleagues (1990) found that specific regions within the dorsolateral neostriatum held neurons that responded to light stroking or passive manipulation of specific body parts. At 0.7 mm anterior to bregma (the AP site of the center of the crucial site found in this study) neurons were found correlated to somatosensory input of the snout, trunk, head, neck, and chin arranged from dorsal to ventral, respectively (head and neck interchanged spots at multiple recording sites). Neural units were also found that responded to forelimb and hindlimb sensation, or passive manipulation. These were located in most cases dorsal to the snout region. Responsive units to head and limb sensation were observed from 1.6 mm anterior to bregma to −0.4 mm posterior from bregma. Neurons were found that were tightly coupled to only one particular body part, but the receptive field of the individual neurons varied widely with some neurons responding to input from just a single forepaw digit and other neurons responding to the entire forelimb. A considerable percentage of units (28%) did not respond to any type of stimulation utilized (Carelli and West 1991). These results as well as results in both the cat (Malach and Graybiel 1986) and monkey (Flaherty and Graybiel 1991) show that the body map of the striatum is vastly different from the body map on the cortex. The position of regions for any particular body part shift and the different spots related to a particular body part vary in terms of the receptive field for that body part. Also, in the monkey, different regions of primary somatosensory cortex have been shown to project to the same neostriatal locus within the putamen (Flaherty and Graybiel 1991).

How do the results of these studies help us to understand the reason a sequencing impairment results from lesions to the present study's crucial site? First of all, motor and sensory processing regions overlap with intermingled forelimb, hindlimb, trunk, and face input within the crucial site. The crucial site elucidated in the present study has an AP diameter of 1.0 mm with 0.7 mm as the center. Secondly, the region where representations of body parts actually switch positions encapsulates the crucial site mapped in this study. The representation of motor processing or somatosensory input changes from rostral to caudal in this ~1.0 mm expanse. This global body representation in terms of sensation within the crucial site, and the interesting feature of shifting sites, may make the crucial region a viable location for integration of multiple sensory and motor inputs. The characteristic of the switching of processing regions in different AP levels of the dorsolateral neostriatum led Brown (1992, p. 7407) to hypothesis that this region, "provides the substrate to compute all the *permutations and combinations* of spatiotemporal input necessary for carrying out motor plans." Since this region of the dorsolateral neostriatum has intermingled sensory and motor processing subregions, it may be more accurate to state that the region contains the necessary substrate to compute *sensory and motor combinations*, which then lead to the carrying out of motor plans. This reliance upon the characteristic of shifting loci responsive to an individual body part is weak at this point because it is not known what functional attribute a shifting representation of the body contributes in sensory and motor processing.

What is the implementation function?

How might we think about implementation as a general process involved in multiple functions? Let's examine the role of the striatum in implementing an action sequence first and then we can attempt to apply similar general rules to other functions. The contribution for motor sequencing could be one of several types, which are not mutually exclusive, including: (1) the generation of the individual component grooming actions; (2) specifying the serial order of the syntactic chain; or (3) implementing the grooming chain by suppressing sensory distractions and other central action patterns. Several pieces of evidence support the idea that the DL-STR is probably **not** actually generating the component actions (e.g., face washing, forelimb stroking, and body licking). One important reason is that the total number of grooming actions actually do not change following dorsolateral neostriatal lesions. Rats with lesions and sham-injected controls emit equal numbers of total forelimb strokes during grooming and total licks to either the body or paws (even though the distribution is not the same). A second reason is that even after a decerebration that removes the whole forebrain and midbrain in a rat, grooming actions are still observed (Berridge 1989). This result indicates that the grooming component actions may actually be generated in the hindbrain.

The DL-STR could hypothetically order the component actions into a series by generating the sequential pattern. The dorsolateral neostriatum could be a central pattern generator that actually initiates the pattern from a generation site at some other location. This hypothesis seems unlikely because rats with dorsolateral neostriatal lesions still complete several grooming chains and actually initiate the same number of grooming chains as controls. Their deficit is not one of initiation or of ordering but one of a normal completion rate. A final reason that the DL-STR is probably not controlling the serial order is that complete ordered chains are still observed after a brain transection at the level of the pons and midbrain (Berridge 1989). This result indicates that even the ordering of the actions may primarily take place at some point in the hindbrain of the rat.

The dorsolateral neostriatum could implement the grooming chain of the rat by suppressing external sensory distractions and switching motor control primarily to the grooming-chain actions and away from other action sequences (Chapin and Woodward 1981; Cools 1980). This function is akin to "sensory or inhibitory gating" ideas that propose a necessary filter for information before it accesses either other sensory information or motor plans. This hypothesis postulates that the grooming actions during the grooming chain should be less susceptible to sensory distraction compared to the same actions seen outside of the chain. Experiments already discussed in this review show this to be true (see the work by Fentress and Berridge). Other work not discussed includes research using other sensory distractors including taste infusions or direct stimulation of the facial region. When a rat is given either of these stimuli as distractors, the grooming actions within the chain are significantly less likely to be disrupted (KC Berridge, personal observations) compared to the same actions outside the grooming chain. This supports the general idea that the striatal circuitry is performing essential evaluations of sensory input and enabling the motor patterns depending upon the motivational context. This idea of gating as an implementation function has been used as a descriptor for many diverse functions and many brain regions (Eccles *et al.* 1962; Freedman *et al.* 1987; Swerdlow *et al.* 2000; Woodward *et al.* 1991).

The striatum and sensorimotor gating

The BG system has been thought of as a brain system involved in inhibitory gating for many decades (Denny-Brown and Yanagisawa 1976; Cools 1980; Schneider and Lidsky 1987; Wilson 1914). Denny-Brown and Yanagisawa (1976, p. 145) noted that "the basal ganglia have all the aspects of a 'clearing house' that accumulates samples of ongoing cortical projected activity and, on a competitive basis, can facilitate any one and suppress all others." Inhibitory gating is most likely mediated by the different subpopulations of neurons in the striatum and is possibly expressed for different purposes dependent upon the striatal subregion involved. Buchwald and colleagues (1979) termed a form of this intermixing "the development of the behavioral sets" that enabled plasticity of action and ability to incorporate flexibility in stimulus–response associations. A piece of supporting evidence includes the response perseveration that ensues after lesioning regions of the striatum (Dunnett *et al.* 1999; Villablanca *et al.* 1976). Normally, motor pattern implementation would rely on rapid inhibition of stimuli and depend upon the appropriate activation of local interneuron populations housed within specific striatal compartments.

We have recently recorded neural activity from the anterior rostral striatum in the freely moving rat during an auditory gating paradigm (Cromwell *et al.* 2007). In this paradigm, two identical tones are presented with a 0.5 s interval between the tones (10 ms tone duration and a 10 s interpair interval between tone pairs). This stimulus presentation paradigm is common in neurophysiology in the clinical setting (Adler *et al.* 1982; Freedman *et al.* 1996) to test the sensory gating abilities of patients with mental illness (Adler *et al.* 1982). Typically, individuals will show a significant reduction in the neural response to the identical second tone compared to the initial tone (Freedman *et al.* 1996). This gating differs from standard habituation because the response to the initial tone remains over time while the response to the second tone is reduced. The reduction in the response to the second tone has been found to be diminished in a number of diseases including post-traumatic stress disorder (PTSD; Neylan *et al.* 1999), OCD (Olincy *et al.* 2000), gambling and addiction (Adler *et al.* 2001), and schizophrenia (Boutros *et al.* 2004). We found that single units in the striatum showed significant gating, but unlike other brain regions we have tested the proportion of units that showed gating was not as extensive (see Figure 9.3; Anstrom *et al.* 2007; Cromwell *et al.* 2005, 2007; Mears *et al.* 2006). This shows that the sensory gating in the striatum may be more selective compared to other brain areas and it demonstrates the possibility that the anterior and rostral striatum is functionally heterogeneous.

In contrast, we found that almost all the single units showed inhibitory gating within the amygdala regardless of location in the more medial or lateral regions (Cromwell *et al.* 2005). To examine the functional implications for the gating in the striatum we examined the influence of short-term or longer-term motivational shifts on the process as well as how gating in the striatum changes during locomotion. During movement compared to nonmovement, we found that the response to the first tone increased during movement and gating became stronger (closer to 0). Acute stressors such as an injection of saline led to an enhancement of gating reflected by a sharper decrease in the sensory response to the second stimulus compared to the initial stimulus. We then examined the impact of 24-hour food deprivation on gating and found the opposite effect, the prolonged, intrinsic stressor led to a reduction of gating. This latter effect appeared more similar to the types of changes observed in emotional and attentional disorders (Freedman *et al.* 1996). The difference may reflect the different contexts of extrinsic versus intrinsic information processing or the types of behaviors related to the types of perturbations. On the one hand, the acute saline led to short-term increase in locomotion during gating while the food deprivation at a 24-hour timepoint led to a decrease in exploration within an environment not associated with any past feeding experiences.

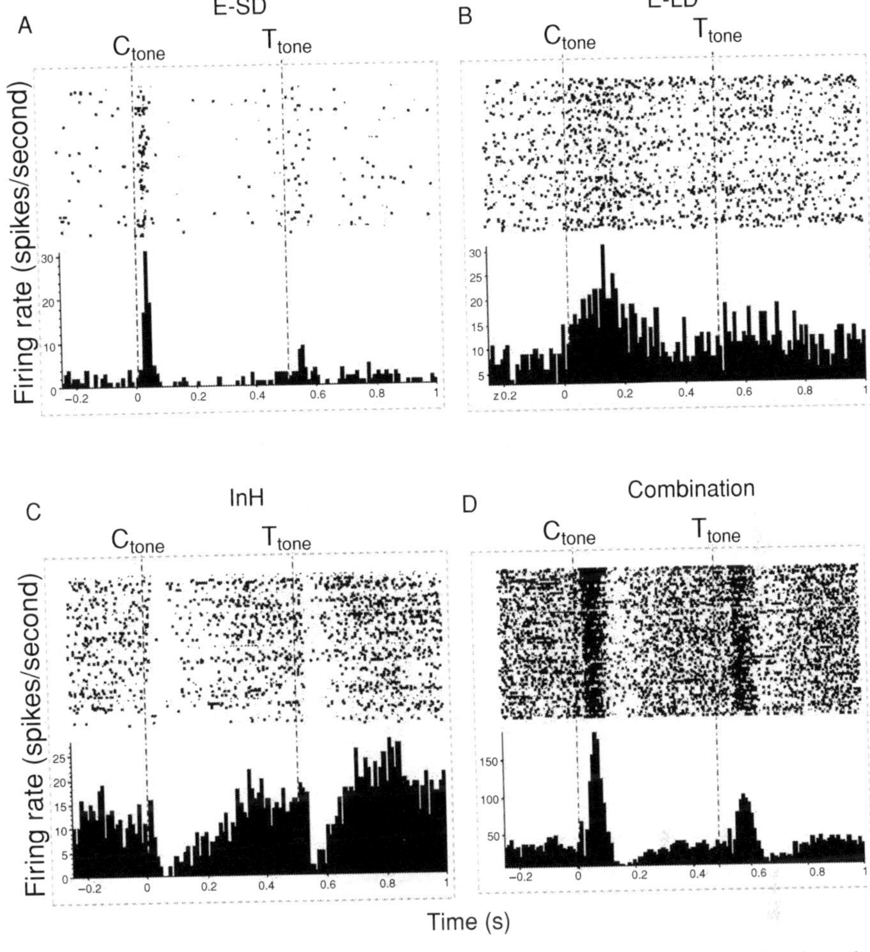

Figure 9.3 Examples of single unit responses that display inhibitory gating from the striatum: (A) excitatory short-duration response (E-SD), and (B) excitatory long-duration response (E-LD) from the anterior and central striatum. Each response displays inhibitory gating as defined by a T/C ratio of at least 0.75. Examples of responses from the striatum in which inhibitory gating is lacking (C and D). This lack of gating has not been found in other brain regions and may reflect the functional heterogeneity of intrinsic processing within a striatal subregion. C_{tone} = conditioning or initial tone and T_{tone} = test tone or second tone.

The striatum, reward, and the grooming chain

Another major function for striatal circuits involves evaluating reward outcomes and processing of reward information. The role of reward processing in the striatum has been examined thoroughly by several groups using lesion, electrophysiology, and pharmacology approaches (Cromwell *et al.* 2005; Kelley 2004;

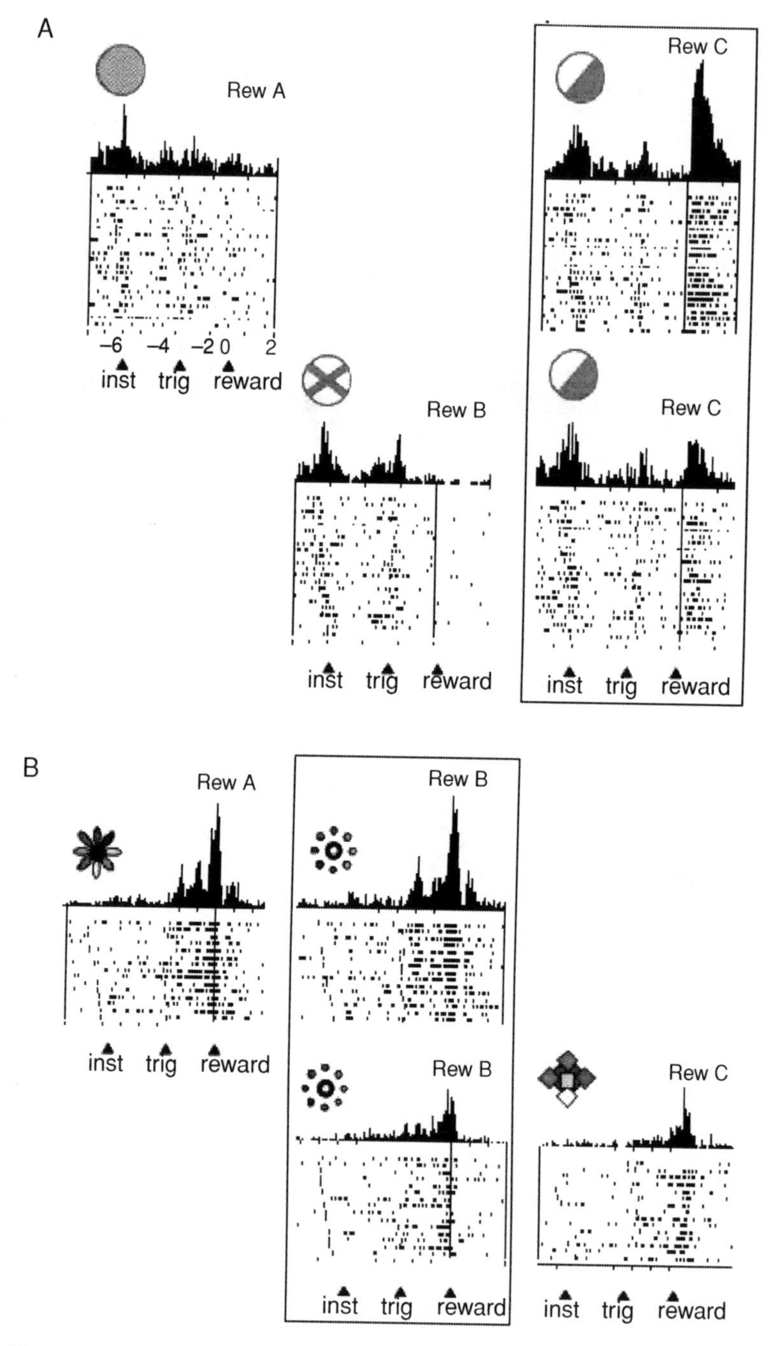

Figure 9.4 (A) The four panels show a single striatal neuron as it is being tested with three different reward outcomes. Each individual perievent and raster represents a set of trials in which the monkey is working for a reward outcome. The trial begins with the presentation of an instruction cue (inst; patterned circles or other geometric

Schultz 2006). It is clear that the striatal networks obtain information concerning the parameters of the reward outcome. Evidence has been obtained using single unit recording in the awake animal while the organism is working to obtain different outcomes. Previous work with single unit recording in the striatum in the awake primate has uncovered activations that are linked to different aspects of a delayed-response task for a reward (Apicella *et al.* 1992; Hollerman *et al.* 1998). These different elements of the task include: (1) the predictive sensory cue, (2) the delay period between the cue and movement, (3) the motor response, (4) the delay between the response and the reward, and (5) the reward itself. Hollerman's work (1998) showed that each of these elements was sensitive to the outcome of the trial and could show significant differences between trials when there was a reward and trials when there was no reward. We expanded on that work by examining how these different activations to these events would code slight differences in reward type (quality of reward) or reward amount (quantity of reward). We found that striatal single units were very sensitive to these properties of the trial (Cromwell and Schultz 2003; Hassani *et al.* 2001). For each of these events, we found single units within the striatum that showed preferential activity to one of the trial types. For the reward-quality experiment, we used different types of juice (lemon, orange, or blackcurrant juice) and found that the striatal units would code the information regarding reward type early in the sequence of the trial, seconds before the actual acquisition of the juice. The reward magnitude results were even more impressive because we found a greater proportion of neurons that were "selective" (displaying greater or lesser activity for one outcome versus another) and the differences in firing could be dramatic (see Cromwell and Schultz 2003 and Figure 9.4 for examples).

Next, we decided to examine the impact of the serial order of reward events to see whether or not one reward earlier in time could influence the neural activity of a reward trial later in time. This serial process has been labeled in different ways in different fields and for incentive-relativity work it has been called "the

designs). These different cues have been paired with specific reward outcomes. The cues are followed by the presentation of a trigger (trig) that signals the time to respond and then the delivery of the juice reward directly into the animal's mouth. The neural responses are time-locked to several events during this delayed response task but the reward contrast effect is seen to the reward response itself. The response during presentation of outcomes A and C is almost exclusively to Reward C. When Reward C is combined with the less preferred Reward B, the activity dramatically decreases. This type of response remains linked to the reward (C) but is altered due to the comparison with the other outcome. (B) The same type of influence on reward expectation activity. The activity is linked to Reward B but decreases when B is temporally linked with Reward C instead of Reward A.

relative reward effect" or contrast effect (Flaherty 1996). We found a clear influence from one reward outcome trial type to another (see Figure 9.4; Cromwell *et al.* 2005). The activation level for each of the five elements (cue, delay, movement, delay, and reward) could depend on the reward type or amount received in the previous trial. We did not examine a longer series of events, but we suspect that under certain conditions (deprivation or other motivational state shifts) a longer or more complex relationship may exist. This serial order dependency of outcome influence is similar to what has been observed in the prefrontal cortical regions (Tremblay and Schultz 1999). In this region of the brain the relative reward effect is based on preference in that the activity of the neuron is influenced by the relative preference of one reward type versus another. In the striatum, we have argued that the relative effect can be dependent upon preference so that an activation for a mildly preferred outcome can increase if this outcome becomes paired with a less preferred outcome; however, we also found activity dependencies that were independent of preference in the striatum and could reflect the sensory qualities or basic parameters (amount) of the outcomes (Cromwell *et al.* 2005).

Comparing these data to the data on action sequencing and grooming is not a direct or simple comparison to make. The relative reward influence is a basic phenomenon related to the linkage of memory and motivational processing. Yet, the striatum seems to be critical in incorporating information concerning the ordering of reward and this basic function could be enabling the appropriate response in the same way that the grooming chain can be enabled. As motivational state varies in the animal, we can expect that sensory distractors become less potent (Toates 1986) and organisms can become more efficient at procuring the desired outcome, at least up to a certain point depending upon motivational state and arousal (Yerkes and Dodson 1908).

Another related note that links the grooming chain and striatal work with our work on reward processing is the idea that grooming in general can be reinforcing part of conspecific communication (Bursten *et al.* 2000; Taira and Rolls 1996) and that this reinforcing attribute is an important part of the organismal state that maintains action sequencing amid choices of actions (Toates 1986). Grooming is expressed after feeding or drinking (Antin *et al.* 1975; Bolles 1960; Phillips *et al.* 1995) or reproductive behavior (Sachs *et al.* 1988). Certainly, allogrooming in a social context could activate reward circuitry and allow for reinforcing attributes of the behavior to be a part of future action choices (Beck 1974; Panksepp 1998). These types of processes that include "motivational gating" most likely depend upon striatal circuits to incorporate the appropriate information and enable the motor strategies to be produced (Berridge and Cromwell 1990; Mogenson *et al.*

1980). Sequencing of action becomes a vital force that the brain must invest a large computational load to in order for the production of complex behavior over time. If we include social behaviors and sequential ordering over longer time spans, then it is a remarkable ability to achieve in ordering the acts, altering the emphasis over time on particular sequences, and choosing the correct channels to express desires and needs for survival and success.

Clinical implications

Striatal control of behavioral sequencing has been an increasingly more important area of study as we begin to understand the debilitating complex motor deficits that occur following neostriatal damage. Parkinsonian patients have great difficulty in sequencing action into ordered units (Harrington and Haaland 1991; Marsden 1984; Montgomery and Buchholz 1991). In Huntington's disease, voluntary movement sequences like alternating hand movements, finger tapping, or visual saccades become impaired as the disease progresses (Folstein *et al.* 1986; Lasker *et al.* 1987; Oepen *et al.* 1985). These functions of implementation of action series dependent upon context, motivational state, and past experience could depend upon striatal networks that become disabled in these disorders. The symptoms have been shown to be conditional on these states and have a complex relationship with several variables (Marsden 1982).

To end, I would like to mention a few recent papers documenting a relationship between sequential deficits and alteration in striatal functioning. One work by Aizenstein and colleagues (2005) found altered recognition of sequences in elderly individuals with geriatric depression. The brain was imaged while these individuals were performing the task or identifying the sequences and it was found that the striatum (right caudate nucleus) had increased activity during the trials in which the sequence was violated or when the expected sequential rules were not met. Typically it might be very important to "gate" information during the sequential learning of this task and chain together the events of the task over time. The increased activation in the depressed individuals could be related to the problems these individuals are having in terms of enabling these functions. They basically lose track of the sequential process leading to decrease in accuracy.

Another interesting recent study found correlates of obsessions during implantation of deep brain stimulator electrodes into patients with OCD (Guehl *et al.* 2008). The activity of caudate nucleus cells in three patients was recorded. Firing rate frequency and variability was found to be significantly different in two patients who were highly anxious, having self-evaluated obsessions during the surgery. The one patient who seemed relaxed with a lower score on the obsessive scale did not have this abnormal striatal activity. It will be very interesting to track the effects of the deep brain stimulation on these patients and their obsessive and compulsive symptoms. Other groups have found these treatments to be effective (Abelson *et al.* 2005; Greenberg *et al.* 2008).

Finally, disruption of the BG may be involved in complex developmental disorders such as autism (Amaral *et al.* 2008; Rinehart *et al.* 2006). Recent work by Cattaneo and colleagues (2007) found an impairment in understanding and producing sequences of action in children with autism. This work was related to recent neurophysiological studies on "mirror neurons." It could be that these neurons encode the sequential nature of behavior using sensory, motor, and outcome information. Children with autism displayed significantly different electromyographic output that essentially revealed a lack of action predictability and an impairment in the ability to switch the focus of motor output over time. In conclusion, evidence is growing for a role of the striatum in action sequencing. To understand this function completely, it may be necessary to examine the broader function of the striatum involved in motor control, sensory filtering, and reward processing. These different "hats" that the striatum wears may depend upon specific subregions but they may also reflect a general mechanism that depends upon a more integrative and global process to be completed.

Acknowledgments

The author thanks the support of National Institutes of Health (Grant R03MH067136 to HCC) and the Hope for Depression Research Foundation (HCC). The author also thanks support from the Sponsored Programs and Research Office at Bowling Green State University (Research Incentive Grants in 2003 and 2005). The author is grateful for all the help and support from colleagues including Drs. Kent C. Berridge, Wolfram Schultz, and Donald J. Woodward. These mentors helped in developing these ideas and completing the experiments outlined in this review.

References

Abelson JL, Curtis GC, Sagher O *et al.* (2005): Deep brain stimulation for refractory obsessive–compulsive disorder. *Biol Psychiatry* **57**:510–16.

Adler LE, Pachtman E, Franks RD *et al.* (1982): Neurophysiological evidence for a defect in neuronal mechanisms involved in sensory gating in schizophrenia. *Biol Psychiatry* **17**:639–54.

Adler LE, Olincey A, Cawthra E *et al.* (2001): Reversal of diminished inhibitory sensory gating in cocaine addicts by a nicotinic cholinergic mechanism. *Neuropsychopharmacology* **24**:671–9.

Agostino R, Berardelli A, Formica A, Accornero N and Manfredi M (1992): Sequential arm movements in patients with Parkinson's disease, Huntington's disease and dystonia. *Brain* **115**(Pt 5):1481–95.

Aizenstein HJ, Butters MA, Figurski JL *et al.* (2005): Prefrontal and striatal activation during sequence learning in geriatric depression. *Biol Psychiatry* **58**:290–6.

Aldridge JW and Berridge KC (1998): Coding of serial order by neostriatal neurons: a "natural action" approach to movement sequence. *J Neurosci* **18**:2777–87.

Aldridge JW, Berridge KC and Rosen AR (2004): Basal ganglia neural mechanisms of natural movement sequences. *Can J Physiol Pharmacol* **82**:732–9.

Alexander GE, DeLong MR and Strick PL (1986): Parallel organization of functionally segregated circuits linking basal ganglia and cortex. *Annu Rev Neurosci* **9**: 357–81.

Alheid GF and Heimer L (1988): New perspectives in basal forebrain organization of special relevance for neuropsychiatric disorders: the striatopallidal, amygdaloid, and corticopetal components of substantia innominata. *Neuroscience* **27**:1–39.

Amaral DG, Schumann CM and Nordahl CW (2008): Neuroanatomy of autism. *Trends Neurosci* **31**:137–45.

Anstrom KK, Cromwell HC and Woodward DJ (2007): Effects of restraint and haloperidol on sensory gating in the midbrain of awake rats. *Neuroscience* **146**:515–24.

Antin J, Gibbs J, Holt J, Young RC and Smith GP (1975): Cholecystokinin elicits the complete behavioral sequence of satiety in rats. *J Comp Physiol Psychol* **89**:784–90.

Apicella P, Scarnati E, Ljungberg T and Schultz W (1992): Neuronal activity in monkey striatum related to the expectation of predictable environmental events. *J Neurophysiol* **68**:945–60.

Beck J (1974): Contact with male or female conspecifics as a reward for instrumental responses in estrus and anestrus female rats. *Acta Neurobiol Exp (Wars)* **34**:615–20.

Berridge KC (1989): Progressive degradation of serial grooming chains by descending decerebration. *Behav Brain Res* **33**:241–53.

Berridge KC and Cromwell HC (1990): Motivational-sensorimotor interaction controls aphagia and exaggerated treading after striatopallidal lesions. *Behav Neurosci* **104**:778–95.

Berridge KC and Fentress JC (1987): Deafferentation does not disrupt natural rules of action syntax. *Behav Brain Res* **23**:69–76.

Berridge KC and Whishaw IQ (1992): Cortex, striatum and cerebellum: control of serial order in a grooming sequence. *Exp Brain Res* **90**:275–90.

Berridge KC, Fentress JC and Parr H (1987): Natural syntax rules control action sequence of rats. *Behav Brain Res* **23**:59–68.

Bolles RC (1960): Grooming behavior in the rat. *J Comp Physiol Psychol* **53**:306–10.

Boutros NN, Korzyukov O, Jansen B, Feingold A and Bell M (2004): Sensory gating deficits during the mid-latency phase of information processing in medicated schizophrenia patients. *Psychiatry Res* **126**:203–15.

Brown LL (1992): Somatotopic organization in rat striatum: evidence for a combinational map. *Proc Natl Acad Sci USA* **89**:7403–7.

Buchwald NA, Hull CD and Levine MS (1979): Neuronal activity of the basal ganglia related to the development of "behavioral sets." In: Brazier MAB, ed., *Brain Mechanisms in Memory and Learning: From the Single Neuron to Man*. New York: Raven Press, pp. 93–103.

Bursten SN, Berridge KC and Owings DH (2000): Do California ground squirrels (*Spermophilus beecheyi*) use ritualized syntactic cephalocaudal grooming as an agonistic signal? *J Comp Psychol* **114**:281–90.

Carelli RM and West MO (1991): Representation of the body by single neurons in the dorsolateral striatum of the awake, unrestrained rat. *J Comp Neurol* **309**: 231–49.

Carpenter MB (1976): Anatomy of the basal ganglia and related nuclei: a review. *Adv Neurol* **14**:7–48.

Cattaneo L, Fabbri-Destro M, Boria S *et al.* (2007): Impairment of actions chains in autism and its possible role in intention understanding. *Proc Natl Acad Sci USA* **104**:17825–30.

Chapin JK and Woodward DJ (1981): Modulation of sensory responsiveness of single somatosensory cortical cells during movement and arousal behaviors. *Exp Neurol* **72**:164–78.

Colonnese MT, Stallman EL and Berridge KC (1996): Ontogeny of action syntax in altricial and precocial rodents: grooming sequences of rat and guinea pig pups. *Behaviour* **113**:1165–95.

Cools AR (1980): Role of the neostriatal dopaminergic activity in sequencing and selecting behavioural strategies: facilitation of processes involved in selecting the best strategy in a stressful situation. *Behav Brain Res* **1**:361–78.

Corbit LH and Janak PH (2007): Inactivation of the lateral but not medial dorsal striatum eliminates the excitatory impact of Pavlovian stimuli on instrumental responding. *J Neurosci* **27**:13977–81.

Cromwell HC and Berridge KC (1996): Implementation of action sequences by a neostriatal site: a lesion mapping study of grooming syntax. *J Neurosci* **16**: 3444–58.

Cromwell HC and King BH (2004): Involvement of basal ganglia in the production of self-injurious behavior in developmental disorders. *Int Rev Ment Retard Res* **29**:119–58.

Cromwell HC and Schultz W (2003): Effects of expectations for different reward magnitudes on neuronal activity in primate striatum. *J Neurophysiol* **89**: 2823–38.

Cromwell HC, Anstrom K, Azarov A and Woodward DJ (2005): Auditory inhibitory gating in the amygdala: single-unit analysis in the behaving rat. *Brain Res* **1043**:12–23.

Cromwell HC, Klein A and Mears RP (2007): Single unit and population responses during inhibitory gating of striatal activity in freely moving rats. *Neuroscience* **146**:69–85.

Cromwell HC, Mears RP, Wan L and Boutros NN (2008): Sensory gating: a translational effort from basic to clinical science. *Clin EEG Neurosci* **39**:69–72.

Dalley JW, Mar AC, Economidou D and Robbins TW (2008): Neurobehavioral mechanisms of impulsivity: fronto-striatal systems and functional neurochemistry. *Pharmacol Biochem Behav* **90**:250–60.

Delong MR, Georgopoulos AP, Crutcher MD *et al.* (1984): Functional organization of the basal ganglia: contributions of single-cell recording studies. *Ciba Found Symp* **107**:64–82.

Denny-Brown D and Yanagisawa N (1976): The role of the basal ganglia in the initiation of movement. *Res Publ Assoc Res Nerv Ment Dis* **55**:115–49.

Divac I (1972): Delayed alternation in cats with lesions of the prefrontal cortex and the caudate nucleus. *Physiol Behav* **8**:519–22.

Dunnet SB, Nathwani F and Brasted PJ (1999): Medial prefrontal and neostriatal lesions disrupt performance in an operant delayed alternation task in rats. *Behav Brain Res* **106**:13–28.

Eccles JC, Schmidt RF and Willis WD (1962): Presynaptic inhibition of the spinal monosynaptic reflex pathway. *J Physiol* **161**:282–97.

Fentress JC (1983): The analysis of behavioral networks. In: Ewert J, ed., *Advances in Vertebrate Neuroethology*. New York: Plenum Press.

Fentress JC (1988): Expressive contexts, fine structure, and central mediation of rodent grooming. *Ann N Y Acad Sci* **525**:18–26.

Fentress JC (1990): Analytical ethology and synthetic neuroscience. In: Bateson P and Marler P, eds., *The Development and Integration of Behaviour*. Cambridge: Cambridge University Press.

Fentress JC and Stilwell FP (1973): Letter: Grammar of a movement sequence in inbred mice. *Nature* **244**:52–3.

Flaherty AW and Graybiel AM (1991): Corticostriatal transformations in the primate somatosensory system. Projections from physiologically mapped body-part representations. *J Neurophysiol* **66**:1249–63.

Flaherty CF (1996): *Incentive Relativity*. Cambridge: Cambridge University Press.

Folstein SE, Leigh RJ, Parhad IM and Folstein MF (1986): The diagnosis of Huntington's disease. *Neurology* **36**:1279–83.

Freedman R, Adler LE, Gerhardt GA *et al.* (1987): Neurobiological studies of sensory gating in schizophrenia. *Schizophr Bull* **13**:669–78.

Freedman R, Adler LE, Myles-Worsley M *et al.* (1996): Inhibitory gating of an evoked response to repeated auditory stimuli in schizophrenic and normal subjects. Human recordings, computer simulation, and an animal model. *Arch Gen Psychiatry* **53**:1114–21.

Georgiou N, Bradshaw JL, Phillips JG, Chiu E and Bradshaw JA (1995): Reliance on advance information and movement sequencing in Huntington's disease. *Mov Disord* **10**:472–81.

Golani I, Bronchti G, Moualem D and Teitelbaum P (1981): "Warm-up" along dimensions of movement in the ontogeny of exploration in rats and other infant mammals. *Proc Natl Acad Sci USA* **78**:7226–9.

Greenberg BD, Askland KD and Carpenter LL (2008): The evolution of deep brain stimulation for neuropsychiatric disorders. *Front Biosci* **13**:4638–48.

Grillner S and Zangger P (1975): How detailed is the central pattern generation for locomotion? *Brain Res* **88**:367–71.

Guehl D, Benazzouz A, Aouizerate B *et al.* (2008): Neuronal correlates of obsessions in the caudate nucleus. *Biol Psychiatry* **63**:557–62.

Haber SN (2003): The primate basal ganglia: parallel and integrative networks. *J Chem Neuroanat* **26**:317–30.

Harrington DL and Haaland KY (1991): Sequencing in Parkinson's disease. Abnormalities in programming and controlling movement. *Brain* **114**(Pt 1A):99–115.

Hassani OK, Cromwell HC and Schultz W (2001): Influence of expectation of different rewards on behavior-related neuronal activity in the striatum. *J Neurophysiol* **85**:2477–89.

Heimer L and Van Hoesen GW (2006): The limbic lobe and its output channels: implications for emotional functions and adaptive behavior. *Neurosci Biobehav Rev* **30**:126–47.

Helmuth LL, Mayr U and Daum I (2000): Sequence learning in Parkinson's disease: a comparison of spatial-attention and number-response sequences. *Neuropsychologia* **38**:1443–51.

Hinde R (1970): *Animal Behavior: A Synthesis of Ethology and Comparative Psychology*. New York: McGraw-Hill.

Hollerman JR, Tremblay L and Schultz W (1998): Influence of reward expectation on behavior-related neuronal activity in primate striatum. *J Neurophysiol* **80**: 947–63.

Iansek R, Huxham F and McGinley J (2006): The sequence effect and gait festination in Parkinson disease: contributors to freezing of gait? *Mov Disord* **21**:1419–24.

Ijspeert AJ (2008): Central pattern generators for locomotion control in animals and robots: a review. *Neural Netw* **21**:642–53.

Kelley AE (2004): Ventral striatal control of appetitive motivation: role in ingestive behavior and reward-related learning. *Neurosci Biobehav Rev* **27**:765–76.

Kelley AE, Domesick VB and Nauta WJ (1982): The amygdalostriatal projection in the rat – an anatomical study by anterograde and retrograde tracing methods. *Neuroscience* **7**:615–30.

Kincaid AE, Zheng T and Wilson CJ (1998): Connectivity and convergence of single corticostriatal axons. *J Neurosci* **18**:4722–31.

Kugler PN and Turvey MT (1987): *Information, Natural Law, and the Self-Assembly of Rhythmic Movement*. Hillsdale, NJ: Lawrence Erlbaum Associates Inc.

Lashley K (1951): The problem of serial order in behavior. In: Jeffress L, ed., *Cerebral Mechanisms of Behavior*. New York: John Wiley.

Lasker AG, Zee DS, Hain TC, Folstein SE and Singer HS (1987): Saccades in Huntington's disease: initiation defects and distractibility. *Neurology* **37**:364–70.

Lee MS, Rinne JO and Marsden CD (2000): The pedunculopontine nucleus: its role in the genesis of movement disorders. *Yonsei Med J* **41**:167–84.

Lemon RN (2008): Descending pathways in motor control. *Annu Rev Neurosci* **31**: 195–218.

Levitt DR and Teitelbaum P (1975): Somnolence, akinesia, and sensory activation of motivated behavior in the lateral hypothalamic syndrome. *Proc Natl Acad Sci USA* **72**:2819–23.

Levy R and Dubois B (2006): Apathy and the functional anatomy of the prefrontal cortex-basal ganglia circuits. *Cereb Cortex* **16**:916–28.

Levy R, Hazrati LN, Herrero MT *et al.* (1997): Re-evaluation of the functional anatomy of the basal ganglia in normal and Parkinsonian states. *Neuroscience* **76**:335–43.

Malach R and Graybiel AM (1986): Mosaic architecture of the somatic sensory-recipient sector of the cat's striatum. *J Neurosci* **6**:3436–58.

Marsden CD (1982): The mysterious motor function of the basal ganglia. *Neurology* **32**:524–32.

Marsden CD (1984): The pathophysiology of movement disorders. *Neurol Clin* **2**:435–59.

Martin JP (1977): The basal ganglia and postural mechanisms. *Agressologie* **18** Spec No:75–81.

McGeorge AJ and Faull RL (1989): The organization of the projection from the cerebral cortex to the striatum in the rat. *Neuroscience* **29**:503–37.

Mears RP, Klein AC and Cromwell HC (2006): Auditory inhibitory gating in medial prefrontal cortex: single unit and local field potential analysis. *Neuroscience* **141**:47–65.

Meyer-Luehmann M, Thompson JF, Berridge KC and Aldridge JW (2002): Substantia nigra pars reticulata neurons code initiation of a serial pattern: implications for natural action sequences and sequential disorders. *Eur J Neurosci* **16**:1599–608.

Mogenson GJ, Jones DL and Yim CY (1980): From motivation to action: functional interface between the limbic system and the motor system. *Prog Neurobiol* **14**:69–97.

Montgomery EB Jr and Buchholz SR (1991): The striatum and motor cortex in motor initiation and execution. *Brain Res* **549**:222–9.

Nauta WJ and Domesick VB (1984): Afferent and efferent relationships of the basal ganglia. *Ciba Found Symp* **107**:3–29.

Neylan TC, Fletcher DJ, Lenoci M *et al.* (1999): Sensory gating in chronic posttraumatic stress disorder: reduced auditory P50 suppression in combat veterans. *Biol Psychiatry* **46**:1656–64.

Oepen G, Mohr U, Willmes K and Thoden U (1985): Huntington's disease: visuomotor disturbance in patients and offspring. *J Neurol Neurosurg Psychiatry* **48**:426–33.

Olincey A, Ross RG, Harris JG *et al.* (2000): The P50 auditory event-evoked potential in adult attention-deficit disorder: comparison with schizophrenia. *Biol Psychiatry* **47**:969–77.

Panksepp J (1998): *Affective Neuroscience*. Cambridge: Cambridge University Press.

Parent A (1986): *Comparative Neurobiology of the Basal Ganglia*. New York: John Wiley and Sons.

Phillips GD, Howes SR, Whitelaw RB, Robbins TW and Everitt BJ (1995): Analysis of the effects of intra-accumbens SKF-38393 and LY-171555 upon the behavioural satiety sequence. *Psychopharmacology (Berl)* **117**:82–90.

Pisa M (1988): Motor functions of the striatum in the rat: critical role of the lateral region in tongue and forelimb reaching. *Neuroscience* **24**:453–63.

Rinehart NJ, Tonge BJ, Bradshaw JL *et al.* (2006): Gait function in high-functioning autism and Asperger's disorder: evidence for basal-ganglia and cerebellar involvement? *Eur Child Adolesc Psychiatry* **15**:256–64.

Ryou JW and Wilson FA (2004): Making your next move: dorsolateral prefrontal cortex and planning a sequence of actions in freely moving monkeys. *Cogn Affect Behav Neurosci* **4**:430–43.

Sabol KE, Neill DB, Wages SA, Church WH and Justice JB (1985): Dopamine depletion in a striatal subregion disrupts performance of a skilled motor task in the rat. *Brain Res* **335**:33–43.

Sachs BD, Clark JT, Molloy AG, Bitran D and Holmes GM (1988): Relation of autogrooming to sexual behavior in male rats. *Physiol Behav* **43**:637–43.

Schneider JS and Lidsky TI (1987): *Basal Ganglia and Behavior: Sensory Aspects of Motor Functioning*. Amsterdam, Holland: Hogrefe Publishing.

Schultz W (2006): Behavioral theories and the neurophysiology of reward. *Annu Rev Psychol* **57**:87–115.

Selemon LD and Goldman-Rakic PS (1985): Longitudinal topography and interdigitation of corticostriatal projections in the rhesus monkey. *J Neurosci* **5**:776–94.

Setlow B, Schoenbaum G and Gallagher M (2003): Neural encoding in ventral striatum during olfactory discrimination learning. *Neuron* **38**:625–36.

Swerdlow NR, Braff DL and Geyer MA (2000): Animal models of deficient sensorimotor gating: what we know, what we think we know, and what we hope to know soon. *Behav Pharmacol* **11**:185–204.

Taira K and Rolls ET (1996): Receiving grooming as a reinforcer for the monkey. *Physiol Behav* **59**:1189–92.

Toates F (1986): *Motivational Systems*. Cambridge: Cambridge University Press.

Tremblay L and Schultz W (1999): Relative reward preference in primate orbitofrontal cortex. *Nature* **398**:704–8.

Van Craenenbroeck K, De Bosscher K, Vanden Berghe W, Vanhoenacker P and Haegeman G (2005): Role of glucocorticoids in dopamine-related neuropsychiatric disorders. *Mol Cell Endocrinol* **245**:10–22.

Van Den Bercken JH and Cools AR (1982): Evidence for a role of the caudate nucleus in the sequential organization of behavior. *Behav Brain Res* **4**:319–27.

Villablanca JR, Marcus RJ and Olmstead CE (1976): Effects of caudate nuclei or frontal cortical ablation in cats. 1. Neurology and gross behavior. *Exp Neurol* **52**: 389–420.

West MO, Carelli RM, Pomerantz M *et al.* (1990): A region in the dorsolateral striatum of the rat exhibiting single-unit correlations with specific locomotor limb movements. *J Neurophysiol* **64**:1233–46.

Whishaw IQ, O'Connor WT and Dunnett SB (1986): The contributions of motor cortex, nigrostriatal dopamine and caudate-putamen to skilled forelimb use in the rat. *Brain* **109**(Pt 5):805–43.

Wilson SAK (1914): An experimental research into the anatomy and physiology of the corpus striatum. *Brain* **36**:427–92.

Woodward DJ, Moises HC, Waterhouse BD, Yeh HH and Cheun JE (1991): The cerebellar norepinephrine system: inhibition, modulation, and gating. *Prog Brain Res* **88**:331–41.

Woolridge MW (1975): A quantitative analysis of short-term rhythmical behavior in rodents. *Animal Behavior*. Oxford: Wolfson College.

Yerkes RM and Dodson JD (1908): The relation of strength of stimulus to rapidity of habit-formation. *J Comp Neurol Psychol* **18**:459–82.

10

An ethological analysis of barbering behavior

BRETT D. DUFOUR AND JOSEPH P. GARNER

Summary

"Barbering" is an abnormal behavior in mice. Barbering mice pluck fur and/or whiskers from cage-mates and/or themselves, leaving idiosyncratic patches of hair loss. The behavior is a paradox: barbering is common in laboratory mice, but it is not seen in wild mice, it does not benefit the plucker, and it is costly to the recipient. This chapter will attempt to resolve the barbering paradox by asking how and why barbering behavior occurs. Using Tinbergen's (1963) framework for an ethological analysis, we assess barbering in terms of adaptive function, phylogeny, development, and mechanism.

The first section discusses hypotheses of adaptive function. The dominance hypothesis is refuted by several studies; the coping hypothesis remains untested; and the pathology hypothesis is supported by multiple lines of evidence. The pathology hypothesis therefore provides the best resolution to the barbering paradox. Accordingly, throughout, we compare and contrast barbering to trichotillomania (TTM) and other human disorders characterized by repetitive behavior. The second section assesses the phylogenetic underpinnings of barbering by comparing and contrasting hair-plucking behavior across species and between mouse strains. The third section reviews the developmental processes that underlie barbering behavior, particularly developmental risk factors, learning, the laboratory environment, and transgenic effects. The final section reviews the behavioral mechanisms, eliciting stimuli, and physiological mechanisms that might mediate barbering. Here, we outline the role of cortico–striatal circuitry in abnormal repetitive

Neurobiology of Grooming Behavior, eds. Allan V. Kalueff, Justin L. LaPorte, and Carisa L. Bergner. Published by Cambridge University Press.

behavior in general, how it can be used to delineate disorders, and insights it provides into barbering.

In conclusion, barbering appears to be homologous to human TTM, and distinct from other repetitive behavior disorders, particularly obsessive–compulsive disorder (OCD).

Introduction

Barbering behavior consists of an individual mouse plucking whiskers and/or fur from its cage-mates (Hauschka 1952; Sarna *et al.* 2000) and/or itself (Garner *et al.* 2004b; Long 1972). The behavior is common in laboratory colonies and has been referred to also as "whisker-eating" (Hauschka 1952), "whisker trimming" and "hair nibbling" (Long 1972), "behavior associated alopecia areata" (Militzer and Wecker 1986), and "the Dalila effect" (Sarna *et al.* 2000). Despite interest in barbering, both as a husbandry problem and as a behavioral assay in biomedical research, very little is known about how or why it occurs.

Plucking whiskers can be broken into four distinct stages (Sarna *et al.* 2000):

(1) Hold: the "barber" presses down on the back and neck of its cage-mate
(2) Grasp: the "barber" grasps a single hair from the victim with its incisors
(3) Pluck: the "barber" pulls its head away from the victim, removing the hair from the root
(4) Manipulation: the "barber" often manipulates the removed hair with its paws, sometimes ingesting the hair.

The process for plucking fur has not been described in such detail, but is presumably very similar. Over time, barbers pluck hair from focused areas on the body of the recipients (Sarna *et al.* 2000), leaving idiosyncratic patterns of alopecia (Garner *et al.* 2004b; Long 1972; Thornburg *et al.*1973) (see Figure 10.1). Each barber typically plucks a similar, matching pattern from all accessible cage-mates, referred to as the barber's "cutting style" (Sarna *et al.* 2000). For example, one barber may pluck the whiskers, while another barber may pluck only a spot on the left flank of its cage-mates. Cutting styles also differ between strains, some only plucking whiskers and from the face (e.g., C3H, A2G, and 129/S6SvEvTac strains; Hauschka 1952; Strozik and Festing 1981) while others pluck their idiosyncratic pattern from any area that is accessible (e.g., C57BL strains/hybrids; Kalueff *et al.* 2006b; Long 1972; Militzer and Wecker 1986). Alopecia caused by barbering is distinct from hair loss caused by other factors: barbering lesions are nonpruritic and leave a signature histopathology of the skin, which includes epidermal

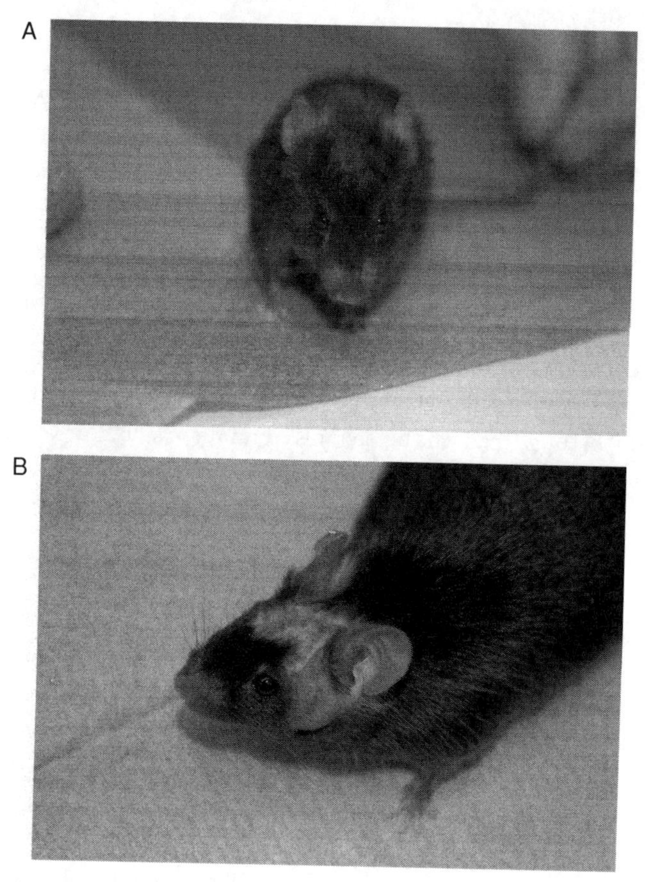

Figure 10.1 Barbering-induced alopecia. Typical patterns of barbering-induced alopecia in C57BL/6J mice, showing plucked (A) whiskers and (B) body/facial fur.

thickening, clogged/clumped follicles, and a disturbance of the normal wave pattern of mouse hair growth (Thornburg *et al.* 1973).

There is a relatively small body of literature that directly assesses barbering behavior. Various features of the behavior have been investigated, such as: the role of dominance (Garner *et al.* 2004a; Strozik and Festing 1981); associated changes in healthy grooming behavior (Militzer and Wecker 1986); social transmission (Strozik and Festing 1981); and a detailed description of the behavior itself (Sarna *et al.* 2000). Many of these studies also included epidemiological data sets detailing the incidence of barbering within the researchers' population of experimental mice (Garner *et al.* 2004b; Hauschka 1952; Kalueff *et al.* 2006b; Long 1972; Militzer and Wecker 1986; Strozik and Festing 1981). However, it has been difficult to synthesize the results of these experiments into general theories about how and why barbering behavior occurs. Firstly, results have sometimes been contradictory;

for example, in C57BL/6J mice, a male sex bias has been reported (Strozik and Festing 1981), as well as a female bias (Garner *et al.* 2004a, b; Militzer and Wecker 1986), and no bias at all (Long 1972). Secondly, little is known about the development and underlying neurophysiology of barbering behavior; thus it is perhaps premature to assume homology between strains, and developmental factors are probably not controlled for between labs. These complications may account for the discrepancies between studies, and also impede general understanding.

Recently there has been a renewed interest in barbering behavior, consisting primarily of research assessing barbering as a model for human obsessive–compulsive spectrum disorders (OCSD), including TTM and OCD; and transgenic studies that assess changes in many behavioral phenotypes, including barbering. Developing a better understanding of how and why barbering behavior occurs is integral to accurately validating barbering as a model for human psychiatric disorder, and will also increase the usefulness of barbering in transgenic studies.

Ethology is more than a simple description of an animal's behavior in its natural environment – it employs a comprehensive framework for understanding how and why animals behave in the ways that they do. An ethological analysis asks four questions (Tinbergen 1963):

(1) Adaptive value – How does the behavior increase the fitness of an individual, population, or species?
(2) Phylogeny – What is the evolutionary history of the behavior?
(3) Development – How does the behavior change over an individual's lifetime?
(4) Mechanism – What are the internal and external factors that underlie the performance of the behavior?

The goals of this review are to analyze barbering behavior in mice using this ethological framework, and provide a comprehensive review of the literature to date.

Adaptive value

The behavior of a species, and the physiology that underlies it, has been shaped by evolution to maximize the fitness of individuals (Tinbergen 1963). Thus behaviors are adaptive, allowing animals to succeed in the environment in which they have evolved. Interestingly, barbering has never been documented in wild mice (Reinhardt 2005) and therefore is presumed not to be a part of the natural behavioral repertoire of the free-living house mouse (*Mus musculus*). However, it appears in many strains of laboratory mice (Garner *et al.* 2004b; Long 1972; Militzer and Wecker 1986).

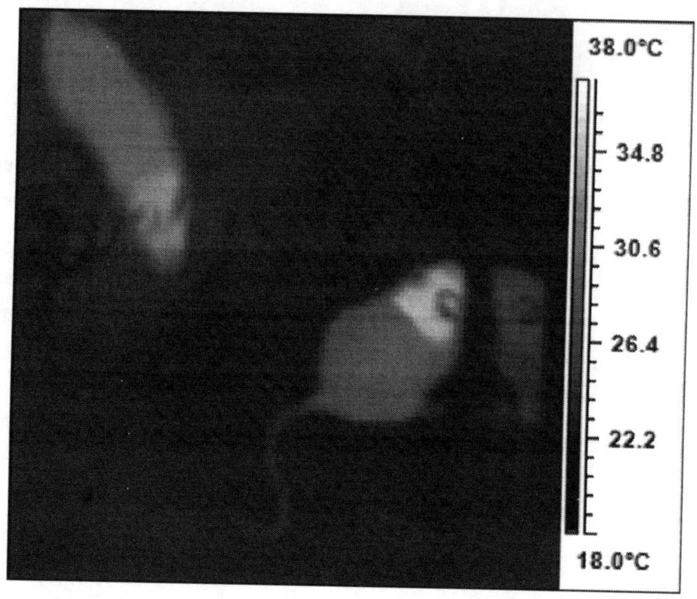

Figure 10.2 Heat loss from barbered mice. This thermal image illustrates the heat loss from barbered areas of the skin. Both mice have been barbered on the head. For the mouse in the right of the picture, the mean temperature of this area is 35.4°C, and the mean temperature of the body, where fur is present, is 29.5°C. By comparison, typical mouse core body temperatures are 36–37.5°C. Mice control heat loss primarily through vasodilation of the tail. This mouse's tail is an average of 22.6°C, indicating that it is cold stressed.

There is no evidence that hair-plucking behavior increases the fitness of barbering mice. However, it does reduce the fitness of the individuals that have been plucked. Recipients show reduced dendritic density in the barrel cortex (Sarna *et al.* 2000); lose essential sensory input from the whiskers; show disturbed hair growth patterns (Thornburg *et al.* 1973); and possibly experience reduced efficiency in thermoregulation as is seen in feather-plucked hens (Glatz 2001) and fur-chewed chinchillas (Vanjonack and Johnson 1973). Figure 10.2 illustrates similar heat loss in barbered mice. This highlights the barbering paradox: why does barbering behavior, which is never seen in the wild, frequently emerge in the laboratory when it has no obvious adaptive value for the barber and reduces the fitness of those that are plucked? Answers to this paradox are the most contentious issue within the barbering literature, to which three distinct hypotheses have been developed: the dominance hypothesis; the coping hypothesis; and the pathology hypothesis.

Regardless of adaptive function, barbering meets the criteria for abnormal behavior in animals adapted (Garner 2005) from the criteria for abnormal behavior in human psychiatry (Davison and Neale 1998). Accordingly, barbering is abnormal in that it is only seen in captivity; it is performed excessively; it affects normal social interaction; and it is peculiar to a subset of individuals. Abnormal behaviors can be further subdivided into maladaptive or malfunctional behaviors (Mills 2003). Maladaptive behaviors reflect a healthy animal behaving adaptively in an abnormal environment (such as excessive infanticide and aggression in mice in the laboratory). On the other hand, malfunctional behaviors reflect the functioning of an animal that develops abnormally (i.e., disturbed neurophysiology) in response to an abnormal environment (such as, perhaps, stereotypies). The three theories that address the barbering paradox, and thus the adaptive function of barbering, clearly align with either one of these two clusters. The dominance hypothesis and the coping hypothesis categorize barbering as a maladaptive behavior, resulting from a healthy and adaptive behavioral response to an abnormal environment. The pathology hypothesis categorizes barbering as a malfunctional behavior, resulting not as a functional adaptation to the laboratory environment but as a nonadaptive or pathological consequence to developing in it.

The dominance hypothesis

Clear dominance hierarchies develop in mice in the wild (Crowcroft 1966; Latham and Mason 2004) and the laboratory (Benton and Brain 1979; Messeri *et al.* 1975). Dominance hierarchies are essentially agreements between animals about who gains priority access to resources. Because hierarchies remove the need to fight over resources, they benefit both dominant and subordinate individuals (Tinbergen 1968). The dominance hypothesis attempts to explain the barbering paradox by claiming that mice pluck hair in order to establish their dominance over their cage-mates.

The dominance hypothesis is the most widely held view by veterinarians and mouse biologists. However, it is actually refuted by the available evidence. Thus for the dominance hypothesis to be correct, it must be supported not only by evidence of a correlational relationship between barbering and social hierarchy, but also by evidence of causality – that the performance of barbering behavior shapes the dominance hierarchy of a cage. As barbering behavior is typically expressed socially, the converse is equally possible – that the dominance hierarchy of a cage determines which individuals will become the recipients of barbering behavior. The crux of the issue depends on which one develops first (barbering or dominance) and how they influence each other from that point on.

Garner *et al.* (2004b), Long (1972) and Van de Weerd *et al.* (1992) each argue that barbering does not cause dominance, but that dominant individuals tolerate barbering least. Other researchers (Hauschka 1952; Kalueff *et al.* 2006b; Long

1972; Strozik and Festing 1981) interpreted *correlations* between the barbering and social status as barbering being causal to social dominance. However, any causal information from these studies actually supports the opposite view – that social hierarchy has a causal effect on barbering behavior, with dominant mice allowing themselves to be plucked less than subordinates.

Long (1972) is often cited as support for the dominance hypothesis (Kalueff *et al.* 2006b; Sarna *et al.* 2000; Strozik and Festing 1981). This observational paper relates the "cage histories" of barbers and their cage-mates. When caging a group of mice, Long observed that barbering behavior occurs after the development of a social hierarchy, but not before (i.e., barbering is not causal to dominance). Within a cage, the barber plucked some cage-mates more than others. While Long argues a link between social dominance and barbering, she doesn't directly claim causality. Instead, she argues that one can infer social status of victims *only*, by the relative amount of hair loss seen in cage-mates.

Hauschka (1952) identified six barbering C3H mice from his colony (four males and two females), each with fur and whiskers intact, and recaged them into three same-sex pairs. After nine days, all animals had "a deteriorated moustache," but only one individual from each cage also plucked hair from the face and orbits, which Hauschka deemed the winner of the pair. This result does not show that barbering behavior shapes the social hierarchy (in which one mouse would have ceased barbering). Rather, it is consistent with the theory that social hierarchy determines how much mice will allow themselves to be plucked.

Strozik and Festing (1981) assessed 12 cages, each consisting of 4 male mice of different strains (used to differentiate between cage-mates). However, using strains in this way is problematic: barbering is related to strain (see Table 10.1), as are other traits such as aggression and weight (The Mouse Phenome Project/The Jackson Laboratory 2008), which likely influence dominance relationships. The social hierarchy of each cage was assessed by a tube-dominance test, and the dominant mouse of each cage was identified. After 45 days, there was alopecia consistent with barbering in 4 cages out of 12, and in none of the cases had the dominant mouse been barbered. If we can overlook the fact that only four cages showed barbering, and this is insufficient for a statistically significant result, this result might seem to support the dominance hypothesis (as Strozik and Festing concluded). However, in three of the cages, at least one of the three subordinate mice had their fur and whiskers fully intact. As only alopecia scores were taken and barbering behavior was not directly assessed, it is possible that in those three cases the dominant mouse was not the barber and lacked hair loss because it resisted barbering by a subordinate.

Table 10.1 *Hair-plucking behavior in mammals. A comparison of abnormal plucking behaviors across mammalian species. Variables that are unknown for a particular species are indicated with a cross (X)*

Species	Order	Behavior name	Sex bias	Plucking target (self or social)	Mode of plucking (oral or manual)	Post-plucking manipulation	Post-plucking ingestion
Humans	Primates	Trichotillomania	Female[4,6,23]	Self[5]	Manual[3,23]	Yes[5]	Yes[5]
Nonhuman primates	Primates	Hair pulling	Female[20]	Both[21]	Both[21]	Yes[21]	Yes[21]
Mice	Rodentia	Barbering	Female[9,12]	Both[19]	Oral[22]	Yes[22]	Yes[22]
Chinchillas	Rodentia	Fur chewing	Female[16]	Both[16]	Oral[17]	×	No[24]
Guinea pigs	Rodentia	Hair pulling	Female[10]	Social[10]	Oral[21]	×	Yes[21]
Rabbits	Lagomorpha	Hair pulling	×	Both[1]	Oral[11]	×	Yes[11,2]
Sheep	Artiodactyla	Wool pulling	×	Social[8,13]	Oral[21]	×	×
Muskox	Artiodactyla	Wool pulling	Female[20]	Social[21]	Oral[22]	×	Yes[18]
Cats	Carnivora	Psychogenic alopecia	Female[15]	Self[21]	Oral/licking[15]	×	×
Dogs	Carnivora	Acral lick dermatitis	×	Self[21]	Oral/licking[15]	×	×
Mink	Carnivora	Fur chewing	Female[14]	Self[14]	Oral/licking[7]	×	×

1. Blackmore *et al.* 1986; **2.** Boers *et al.* 2002; **3.** Christenson *et al.* 1991a; **4.** Christenson *et al.* 1991c; **5.** Christenson and Mansueto 1999; **6.** Cohen *et al.* 1995; **7.** de Jonge 1988; **8.** Done-Currie *et al.* 1984; **9.** Garner *et al.* 2004b; **10.** Gerold *et al.* 1997; **11.** Gunn and Morton 1995; **12.** Long 1972; **13.** Marsden and Woodgush 1986; **14.** Mason 1994; **15.** Moon-Fanelli *et al.* 1999; **16.** Ponzio *et al.* 2007; **17.** Rees 1962; **18.** Reinhardt and Flood 1983; **19.** Reinhardt 1984; **20.** Reinhardt *et al.* 1986; **21.** Reinhardt 2005; **22.** Sarna *et al.* 2000; **23.** Swedo and Leonard 1992; **24.** Tisljar *et al.* 2002.

Using aggression marks as indicators of dominance, Kalueff *et al.* (2006b) concluded that "dominant" NMRI strain mice were never barbered, while their cage-mates often were. Kalueff *et al.* suggested that the dominant mouse should have few aggression marks, the subordinate mouse or mice have more. However, aggression marks have not been validated as a measure of dominance (by for instance, correlating aggression marks with formal observations of submissive behavior by subordinates). Furthermore, wounding aggression typically indicates that animals are evenly matched, and uncertain of their dominance relationships, and are thus forced to fight (Howerton *et al.* 2008). Thus aggression marks are typically indicative of cages where no clear dominance hierarchy exists (Howerton *et al.* 2008). Nevertheless, taken at face value these results support the theory that dominant mice often do not allow themselves to be barbered, not that mice express dominance through barbering behavior.

Experiments designed to test causality (Garner *et al.* 2004a; Van de Weerd *et al.* 1992) directly contradict the dominance hypothesis, showing that barbering behavior does not shape social hierarchy. Van de Weerd *et al.* removed the vibrissae of mice with scissors, caged them in pairs with an intact mouse, and then assessed the social hierarchy between them using a tube-dominance test (such manipulations are a classic ethological technique for confirming the role of a behavior, signal, or ornamentation in social interactions). Removal of vibrissae was found not to have any effect on the social status of the mice. Using a tube-dominance test, we found that the dominant mice in the cages were no more likely to be barbers than the subordinates (Garner *et al.* 2004a). However, there was a clear relationship between dominance and barbering: the amount of hair plucked from individuals was directly correlated to their social status within the cage, with subordinate mice having more hair removed and dominant mice having less removed (Garner *et al.* 2004a).

There are also specific features of barbering behavior that are inconsistent with the dominance hypothesis. Barbering mice don't only pluck from their cage-mates, but also from themselves (Garner *et al.* 2001; Long 1972). In breeding pairs, it is more common for the female to pluck from the male (Garner *et al.* 2001; Long 1972; Militzer and Wecker 1986) although females are never dominant to male mice (Crowcroft 1966). Sometimes mice also pluck hair from their young, before they are weaned (Pinkus 1964; Thornburg *et al.* 1973) and barbering mice have been shown to pluck from rats twice their size, when caged together (Hauschka 1952). Also, barbering is also performed after the development of social hierarchy, but not before (Long 1972). In all of these cases, it is doubtful or impossible for the mouse to be exerting dominance through hair-plucking behavior.

Thus, there is no evidence that barbering behavior plays a role in the initiation or maintenance of social hierarchy. Instead, available evidence indicates that social hierarchy influences the severity of hair loss in the victims of barbering behavior; dominant mice will submit themselves to the plucking of a barber less than subordinate mice do.

The coping hypothesis

Given the problems with the dominance hypothesis as an explanation for the function of barbering behavior, alternatives have been proposed. Van den Broek (1993) concluded that "social ranking alone does not determine whether a mouse actively trims or is being trimmed," and McMahon and Sundberg (1994) argued that "barbering is probably far more than just an expression of social dominance." Van den Broek instead suggested that barbering "is a form of aberrant behavior developed to cope with inadequate housing conditions" (Van den Broek *et al.* 1993), without elaborating which features were inadequate, and what coping entails. Although this view is largely unsubstantiated, Van den Broek raised an important issue that had largely been ignored in the barbering literature – that barbering behavior may develop as a consequence of the characteristic barren and stressful elements of the mouse housing, as discussed below.

Coping is often invoked as a functional explanation for other abnormal behaviors seen in captivity, such as stereotypy (Mason 1991). The coping hypothesis attempts to explain the barbering paradox by claiming that mice pluck hair in order to reduce stress that is induced by captivity. It implies that the adaptive value of plucking hair is to reduce stress in individuals that pluck, which should improve the health and survival of laboratory mice. However, there is no evidence that the performance of barbering behavior provides any anxiolytic or stress-reducing effects for barbers.

The pathology hypothesis

The pathology hypothesis attempts to explain the barbering paradox by arguing that laboratory mice pluck hair as a result of abnormal brain function, which is induced by the unnatural environment in which they develop. In contrast to the other hypotheses, it implies that barbering behavior has no adaptive value, but instead occurs as a symptom of disturbed neurophysiology. There is circumstantial evidence for the pathology hypothesis: barbering behavior has never been documented in wild mice, but is seen often in captive mice that live in unnatural conditions (which are known to disturb neurobiological development) (e.g., Würbel 2001).

A growing body of evidence suggests that abnormal repetitive behavior in both humans and animals involves dysfunction in the mechanisms that regulate the selection and sequencing of behavior, which are found in a series of circuits running between the frontal cortex and the basal ganglia, called the cortico–striatal loops (Albin and Mink 2006; Albin *et al.* 1995; Aouizerate *et al.* 2004; Chamberlain *et al.* 2005; Crider 1997; Garner 2006a; Greer and Capecchi 2002; Hill *et al.* 2007; Lewis and Bodfish 1998; Mink 2003; Rosenberg and Keshavan 1998; Stein *et al.* 1999; Welch *et al.* 2007). Several lines of evidence implicate these mechanisms in barbering. For instance, barbering mice show the same pattern of neuropsychological deficits as trichotillomania patients, which uniquely implicate deficits in this circuitry (Garner 2006a). Similarly, transgenic mice with alterations in this circuitry show increased levels of barbering behavior (e.g., Greer and Capecchi 2002). These results are discussed in detail in the section titled "Mechanism."

Thus, in summary, there is strong preliminary evidence that supports the pathology hypothesis, while other hypotheses have either not been substantiated (the coping hypothesis) or clearly refuted (the dominance hypothesis).

The barbering paradox essentially asks "Why should barbering be so common in the laboratory, when it has no adaptive value?" and in doing so it assumes that the behavior must have adaptive value. The pathology hypothesis resolves the barbering paradox by questioning this assumption – if barbering is pathological, then there is no paradox to explain.

Phylogeny

The behavior patterns of a species, and the physiological capacities to perform them, have developed and been shaped through evolutionary processes (Tinbergen 1963). Although there is no fossil record for behavior, the comparative method is often employed to assess the homology of normal and abnormal behavioral processes and their physiological underpinnings across living species (e.g., Clubb and Mason 2004). This section will assess the homology of hair-plucking behavior across various species, with particular attention to the hair-plucking behaviors of mice and humans, as well as homology across different genetic lines within *Mus musculus*.

Species occurrence

In captivity, plucking behaviors are a common occurrence, particularly among birds (e.g., Garner *et al.* 2006a; Savory 1995) and mammals (Reinhardt 2005), and are often considered to be abnormal (see Table 10.1 for a species comparison). In mammals, hair-plucking behavior has been documented in humans (e.g., Christenson *et al.* 1991c), six nonhuman primate species (Reinhardt 2005), mice, guinea pigs (Gerold *et al.* 1997), chinchillas (Tisljar *et al.* 2002), rabbits (Gunn and Morton 1995), sheep (Done-Currie *et al.* 1984), muskoxen (Reinhardt and Flood 1983), dogs and cats (Moon-Fanelli *et al.* 1999), mink (de Jonge 1988), and beech martins (Hansen 1992; as cited by Mason 2006). There are many similarities across species: hair-plucking behaviors occur more frequently in females than males (Christenson and Mansueto 1999; Garner *et al.* 2004b; Long 1972; Reinhardt 2005); onset at/following reproductive maturity (Christenson 1995; Garner *et al.* 2004b); immediately after plucking, hairs are often manipulated manually/orally and sometimes ingested (Christenson and Mansueto 1999; Reinhardt 2005; Reinhardt and Flood 1983; Sarna *et al.* 2000); and while the behaviors frequently appear in various nonhuman species developing in artificial (i.e., captive) environments, they have never been documented for any of the same species that have developed in the wild (Reinhardt 2005). The differences in hair-plucking behaviors between these species largely reflect general differences in healthy grooming patterns characteristic of each. Plucking usually is carried out with the same anatomical part that is used for grooming by each species – for example, mice pluck orally (Sarna *et al.* 2000), nonhuman primates pluck both orally and manually (Reinhardt 2005), and humans pluck with their hands. Species that tend to groom primarily themselves, such as humans, dogs, and cats, also tend to pluck only from their own bodies (Reinhardt 2005). Species that tend to groom both themselves and conspecifics, such as mice, rabbits, and nonhuman primates, also tend to pluck from both themselves and cage-mates (Reinhardt 2005).

Plucking behaviors in other species are widely regarded as abnormal (Latham and Mason 2004; Reinhardt 2005; Walsh 1982): they are repetitive, goal-directed behaviors that are performed without any known adaptive benefit to the plucker, and occur at a cost to the recipient. The negative impacts of plucking have been best shown in chinchillas: self-plucking "fur-chewers" show elevated corticosterone and thyroid hormones, reduced core body temperature, and adrenocortical hypertrophy (Vanjonack and Johnson 1973). Very little is known about the neurophysiological mechanisms that are different for individuals that pluck versus those who do not.

As very little is known about the mechanisms that underlie the performance of hair-plucking behavior, it is impossible directly to assess the homology of mechanism across species at this time. However, the overwhelming similarities across species indicate that homology is a likely possibility, particularly when considering that most differences in performing the behavior simply reflect species-specific adaptations that dictate how an individual grooms.

Human insights

While the subjective psychological states of animals are difficult to assess (Dawkins 1993), the subjective psychological states of human hair pluckers can be assessed through self-reporting and structured interviews. In this way, human research may provide important insight into the hair-plucking behavior in nonhuman animals, particularly the subjective experience of performing the behavior, as well as highlight issues relevant to the development and performance of the behavior across species.

Human hair-plucking behavior, called trichotillomania (TTM) (American Psychiatric Association 1994), is one of the oldest documented psychiatric disorders (Christenson and Mansueto 1999). Although the prevalence of TTM varies between studies (Christenson and Mansueto 1999), Christenson *et al.*'s (1991c) estimate of 3.4% of women and 1.5% of men is probably the best, and broadly agrees with other studies. Several studies show that TTM has a female bias (e.g., Cohen *et al.* 1995; Swedo and Leonard 1992). Thus, like barbering behavior, TTM occurs in a small but significant subpopulation and shows a female bias. Also similar to barbering, TTM develops primarily during sexual maturity – one sample estimated the average age of onset at 13.1 years (Christenson 1995). Trichotillomania is largely believed to be a chronic condition (Christenson 1995; Christenson and Mansueto 1999) – in one sample, 59% of TTM patients had never stopped plucking hair for a month or longer. Although very little is known about the etiology of TTM, a childhood history of psychosocial loss or stress has been associated with TTM onset, such as death/illness of a parent, parental divorce, alienation from friends, childhood illness/injury, or violence/abuse (Boughn and Holdom 2003; Christenson and Mansueto 1999).

Trichotillomania patients habitually pluck hairs, typically one by one, from preferential regions of their body (often from the scalp, eyelashes, eyebrows, and pubic regions) (Christenson *et al.* 1991a; Swedo and Leonard 1992; Woods *et al.* 2006). While they are aware of the alopecia that results, they are unable to stop performing the behavior (Christenson and Mansueto 1999). The subjective experience of hair plucking may differ between individuals, and also differ between bouts within the same individual (Christenson and Crow 1996; Christenson and Mansueto 1999). The process is sometimes carried out without a conscious motivation (called automatic plucking), and other times occurs under the conscious control of the plucker (called focused plucking) in response to the specific urge to pluck (Azrin 1977; Christenson *et al.* 1994; Christenson and Mansueto 1999). Cues differentially affect the two types of plucking: negative affect (such as embarrassment, anxiety, depression, etc.) precipitates bouts of focused plucking, and sedentary/contemplative circumstances increase the incidence of automatic plucking in patients. Circumstances that are perceived to be either stressful or "boring" are known to trigger bouts of hair-plucking behavior (Christenson and Mansueto 1999). Plucking behaviors are also sensitive to menstrual cycles in some female patients, in which both the urge to pluck and hair-plucking frequency increases during the week prior to menstruation (Keuthen *et al.* 1997).

Trichotillomania is often compared with OCD (see Stanley and Cohen 1999 for a review). However, the two disorders are clearly distinct. Trichotillomania and

OCD differ in terms of behavioral mechanism, etiological profile, pharmacological treatment response (Stanley and Cohen 1999), and thus most likely differ in terms of underlying pathophysiology. The compulsive behaviors seen in OCD largely develop as a healthy response to a pathological anxiety that manifests as an obsession (Graybiel and Rauch 2000). For example, an individual who has a pathological fear of (i.e., an obsession with) contamination may compulsively wash his/her hands in an attempt to remove disease-causing bacteria or viruses. Hair plucking in TTM is not performed as a result of an anxiety-driven obsession: patients often pluck hair without realizing that they are doing it, some even pluck while asleep (Altman *et al.* 1982; Deluca and Holborn 1984). Neuropsychological and imaging studies indicate different deficits in OCD and TTM patients. While both groups show deficits in set-shifting, only OCD patients show deficits in cognitive inhibition (Bohne *et al.* 2005c) and reversal learning (Remijnse *et al.* 2006); also, orbitofrontal dysfunction is associated with OCD (Insel 1992), but not with TTM (Stanley and Cohen 1999). Currently there is no known efficacious pharmacotherapy for TTM; unlike OCD, serotonergic agents have been shown to be largely ineffective in treating TTM symptoms in controlled trials (Christenson *et al.* 1991b; Streichenwein and Thornby 1995; van Minnen *et al.* 2003). Assessing the efficacy of serotonergic therapies in treating barbering behavior in mice will be an important step in assessing predictive validity with the human disorder; accordingly, negative results would provide one piece of evidence for pathophysiological homology.

Trichotillomania has been classified with clusters of other human psychiatric disorders, according to behavioral similarities. These groupings include: OCSD, which involve the performance of repetitive and/or impulsive behaviors such as OCD, eating disorders, and Tourette's syndrome; impulse control disorders (ICD), which involve deficits in impulse control, such as kleptomania, pyromania, and impulsive aggression (APA 1994); and body-focused repetitive behaviors (BFRB), which involve the expression of repetitive and compulsive grooming behaviors, such as skin picking and nail biting (Bohne *et al.* 2005b). While there are similarities with TTM and a variety of other disorders, these groupings that are often used are not exclusive, are sometimes used inconsistently, and are not necessarily based on either homologous pathophysiological mechanisms or pharmacological treatment profiles. Accordingly, using hair-plucking behaviors to model any particular one of these groups of disorders may be misleading. Inference from barbering studies should be considered tentative until a better understanding of their relationships is elucidated (Garner 2006a; Stein *et al.* 2007).

Considering the many similarities, barbering may very well serve as an important model for human TTM – providing the many advantages that mice offer as a model species, particularly in terms of understanding pathophysiological mechanisms and establishing successful pharmacological treatment strategies. The converse is also true: TTM may also provide insights into why mice pluck hair, implying perhaps a role for stress in the disorders etiology, as well as showing that the behavior may take place without any conscious motivation or obsession.

A comparison of mouse strains

Barbering behavior has been documented in a variety of mouse strains, and there is considerable variation reported in the occurrence of barbering behavior among them. Most of these differences have been demonstrated through epidemiological studies, in which the data were independently collected and reported by various labs. It is difficult to compare the data from these studies, given the variation in husbandry specific to each lab. This problem has been discussed in the literature, particularly inconsistencies regarding the sex bias in the occurrence of barbering (Militzer and Wecker 1986). Taking into consideration the inherent variability of labs and their husbandry practices, Table 10.2 summarizes the findings from the epidemiological studies, highlighting occurrence in different strains and sexes, while accounting for age. In the rest of this section, we will highlight the variation in strains, with particular emphasis on the C57BL/6J and 129 strains. We focus on these strains because most of the known features of barbering have been elucidated through work with the C57BL/6J strain; both strains show a high prevalence of barbering behavior; and these are the two most commonly used mouse strains, particularly in transgenic studies.

The C57BL/6J is the most commonly used inbred strain of mouse (The Jackson Laboratory 2008). They tend to perform barbering behavior in a relatively high frequency; about 20% of these mice are estimated to barber in laboratory conditions (Garner *et al.* 2004b; Long 1972; Sarna *et al.* 2000). Barbering develops more frequently in females than males (Long 1972; Militzer and Wecker 1986) in both stock cages and breeding cages. While mice of other strains have only been documented to pluck from each other, C57BL/6J barbers have also been documented to pluck hair from their pups (Pinkus 1964; Thornburg *et al.* 1973) and themselves (Garner *et al.* 2004b; Long 1972). The cutting styles characteristic of C57BL/6J barbers are also quite unique: they will pluck idiosyncratic patterns of hair, from cage-mates and/or themselves, in any area that is accessible. From cage-mates, whiskers are commonly plucked, as well as the face, between the ears, and often patches appear in various regions of the back and flanks (Garner *et al.* 2004b; Long 1972). In self-barbering (Figure 10.3A), the mouse will pluck from its ventral surface, forelimbs, and the anogenital and tail areas (Garner *et al.* 2004b). The C57BL/6J

Table 10.2 *Strain differences in barbering behavior. This table is a summary of the existing epidemiological data showing the prevalence of barbering behavior in different strains of mice. Here, "n" refers to the overall sample described (either individuals or cages). Prevalence was calculated as the number of individual mice or cages (depending on the study, as indicated in the "Sample" column) for which barbering was detected out of the total sample*

Population characteristics				Overall		Males		Females			Citation	
Strain	Age	Housing system	Sample	n	Prevalence	n	Prevalence	n	Prevalence	Sex bias	Author	Year
129S6/SvEvTac	2–3 mo	Stock (pairs)	# mice	100	45.00%						Lijam *et al.*	1997
129S6/SvEvTac	6–7 mo	Stock (pairs)	# mice	100	75.00%						Lijam *et al.*	1997
129S6/SvEvTac	15 wks	Stock (pairs)	# mice	92	76.09%	44	77.27%	48	75.00%	No (NS)	Dufour & Garner	unpub
A/J		Breeding + Stock	# cages	62	0.00%						Long	1972
BALB/c		Breeder pairs	# mice	100	0.00%	50	0.00%	50	0.00%	No (NS)	Militzer & Wecker	1986
C3H	Adults	Stock	# mice	869	2.88%						Hauschka	1952
C3H		Breeding + Stock	# cages	16	6.25%						Long	1972
C3H		Breeder pairs	# mice	78	0.00%	39	0.00%	39	0.00%	No (NS)	Militzer & Wecker	1986
C57/10		Breeding + Stock	# cages	110	4.55%						Long	1972
C57BL/6J	Adults	Stock	# mice	461	9.98%						Hauschka	1952
C57BL/6J		Breeding + Stock	# cages	83	16.87%						Long	1972
C57BL/6J	15 wks	Stock	# mice	152	34.21%	74	21.62%	76	47.37%	Female	Dufour & Garner	unpub
C57BL/6J		Breeder pairs	# mice	252	24.21%	126	10.32%	1	38.10%	Female	Militzer & Wecker	1986
CBA		Breeder pairs	# mice	14	0.00%	7	0.00%	7	0.00%	No (NS)	Militzer & Wecker	1986
Webster Swiss	Adults	Stock	# mice	632	1.90%						Hauschka	1952
Across Strains	Adults	Stock (pairs)	# mice	1962	4.23%	1001	4.00%	961	4.47%	No (NS)	Hauschka	1952

Abbreviations: Stock = mice housed in same-sex cages; Breeder pairs = cages of pairs of mice, one male and one female; NS = nonsignificant difference; Unpub = unpublished.

A

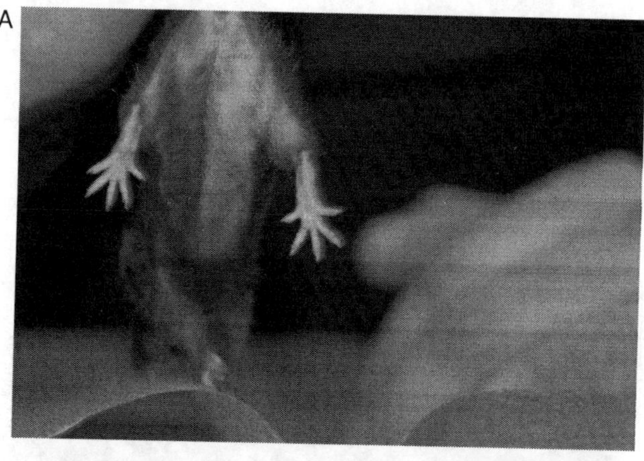

B

Figure 10.3 Barbering-induced alopecia. Less common patterns of barbering-induced hair loss in C57BL/6J mice are shown here, illustrating (A) ventral alopecia induced by self-plucking, and (B) posterior alopecia in pups plucked by their parent.

mice also barber their pups (Figure 10.3B), typically plucking the posterior dorsal surface clean (Pinkus 1964), as well as whiskers (personal observation).

Like the C57BL/6J strain, 129S6/SvEvTac mice are also a workhorse of transgenic research. There is a high incidence of barbering in the wild type, with 50% estimated to be pluckers (Greer and Capecchi 2002); this high prevalence makes 129S6/SvEvTac mice a potentially useful strain for further study of barbering behavior. The cutting style of 129S6/SvEvTac barbers is quite homogenous: frequently whiskers and facial fur are plucked, and in more extreme cases fur from the neck and back is also plucked (Greer and Capecchi 2002, and personal observations). While only the barbering of cage-mates has been documented in the literature, 129S6/SvEvTac mice will often pluck the whiskers from their pups and sometimes (but rarely) self-pluck from the dorsal surface of their forelimbs (unpubl. data). No sex bias has been documented in 129S6/SvEvTac, but a female bias was found in a related strain, 129S1 (Kalueff *et al.* 2006b).

There are some other notable strains, both pertinent to the general phenomenology of barbering and the history of barbering research. The NMRI strain has been found to have a very high prevalence of barbering behavior: in only one investigation barbering was found in 100% of male cages and 87% of female cages (Kalueff *et al.* 2006b). While a male bias was observed, the sample was quite small (total of 13 cages) and the difference is not statistically significant (Kalueff *et al.* 2006b). However, due to the high prevalence of the behavior, NMRIs may make a useful subject for barbering studies. In contrast to NMRI mice, certain strains (BALB/c and A/J) have been consistently shown to never perform barbering behavior (Garner *et al.* 2004b; Kalueff *et al.* 2006b; Long 1972; Militzer and Wecker 1986). The A2G and C3H strains were used in early barbering studies, but haven't been used since Strozik and Festing (1981).

Development

Questions of behavioral development address how an animal acquires the internal "behavioral machinery" (i.e., mechanism) to perform novel behaviors, even when the environment is held constant (Tinbergen 1963). One aspect of barbering essential to understanding its development is that it likely does not develop in the wild, but only appears in mice developing under laboratory conditions. However, even within the laboratory, barbering behavior develops differentially in individual mice: it normally appears only within a subset of cages in a colony (e.g., Long 1972; Militzer and Wecker 1986), and typically in a subset of individuals within those cages (e.g., Long 1972; Strozik and Festing 1981). Thus, while some mice never express hair-plucking behaviors, in others it develops and is performed habitually. This raises an interesting question: why do mice of similar biological backgrounds (especially inbred strains that are, in theory, genetically identical; Silver 1995), which develop in supposedly identical environments, differentially develop the behavior?

This section will address this question regarding differential development in the laboratory, by assessing known risk factors for barbering onset; the role of learning; the role of captive environments; and knockout studies. While genes, learning, and the environment all may play a role in barbering development, it is important to recognize that barbering is an abnormal behavior, likely malfunctional, and accordingly may develop very differently from ways in which normal behaviors develop. While it will be important to assess how biological and environmental factors may lead to its development, it is possible that there are also factors that may prevent it. Considering that little is known about barbering development from direct study, we will also integrate into our discussion comparative evidence that illustrates how hair-plucking behaviors develop in other species, as well as theoretical models that assess ways in which the laboratory environment is believed to induce abnormal behaviors.

Known risk factors

Particular "risk factors" for barbering development have been identified through assessing the prevalence of the behavior across a large sample of mice (Garner *et al.* 2004a, b). Mice differed in their expression (and by inference, development) of the behavior according to age, sex, husbandry, reproductive events, and genetic background (i.e., mouse strain).

As mice age the incidence of barbering behavior has been reported to increase in a number of studies (Garner *et al.* 2004b; Hauschka 1952; Strozik and Festing 1981). While there is no clear critical window to the development of the behavior – it seems to continue to develop as mice age – it does only develop after mice reach reproductive maturity (Garner *et al.* 2004b). Kalueff *et al.* (2006b) argued that barbering is performed by young weanlings plucking from their mothers; however, the alopecia patterns shown in this study were consistent with locations typically plucked by self-barbers, as seen in mice housed alone (Garner *et al.* 2004b). As the performance of barbering was not directly observed in this study, it is more likely that the mothers were plucking themselves.

Barbering behavior develops more frequently in females than in males. While this may not hold true for all strains, our epidemiological analysis across many strains clearly showed that there is a significant female bias to the behavior (Garner *et al.* 2004b). Although the mechanisms responsible for this sex bias are unknown, it is a consistent risk factor across species (Reinhardt 2005).

Reproductive events confer risk in barbering behavior onset in two ways: barbering behavior is first seen in mice when they reach reproductive maturity (but never before); and the incidence of barbering behavior tends to be higher in breeding cages than in stock (same-sex) cages (Garner *et al.* 2004b). Although unknown for mice, the symptoms are exacerbated in premenstrual female TTM patients (Keuthen *et al.* 1997). Also, symptoms sometimes change in TTM patients during pregnancy; however, this finding was inconsistent among patients: some increased, some decreased, some didn't change at all (Keuthen *et al.* 1997). The mechanisms that link reproductive events with hair plucking are not known; however, they strongly implicate reproductive physiology in the onset and activation of the behavior.

The fourth factor, the one that confers the highest risk, is genetic background – mice of different strains vary significantly in the expression of hair-plucking behaviors (Garner *et al.* 2004b), as discussed above. While genetics confers significant risk for, and perhaps protection against, barbering development, the role of specific genes in its "spontaneous" development in the laboratory is unknown. Through a backcrossing experiment, Hauschka (1952) demonstrated that barbering does not represent the phenotypic expression of a single gene/allele. Even in most high-incidence inbred strains it only appears in a small subset of cages (Garner *et al.* 2004b), although individuals are genetically identical. Accordingly, while an individual's genetics may increase risk for spontaneous barbering development, it is not enough; there must be some type of gene by environment interaction in the behavior's etiology.

Role of learning

Learning processes are integral to the development of many behaviors, and could provide a necessary complement to the other known risk factors. However, the only direct assessment of learning's role in barbering development comes from a cross-fostering experiment (Carruthers *et al.* 1998), presented only in a conference abstract, in which pups of a high-incidence strain were cross-fostered with a nonbarbering strain parents, and vice versa. The pups of the barbering strain, raised by nonbarbering parents, still performed barbering behavior. Thus, exposure to barbering (and by inference learning) is not necessary for its development. Interestingly, the pups of the nonbarbering strain, raised by the barbering parents, did barber. While it is tempting to conclude from this case that barbering behavior can be learned, the effects of learning are clearly confounded with any potential stresses resulting from the cross as well as the likely stress of being barbered.

While there is little evidence that learning facilitates the development (i.e., onset) of barbering behavior itself, it does shape the cutting style of individuals that have already started to pluck. Cutting styles are more highly correlated between barbers within the same cage than between barbers in different cages (Garner *et al.* 2004a), likely implicating social transmission of cutting styles.

Role of the laboratory environment

An empirical understanding of environmental mechanisms that lead to barbering behavior comes from only two studies, which show that enrichment may reduce the severity of hair-plucking behavior in established barbers (DeLuca 1997), and specific social and husbandry factors, such as being caged with relatives (i.e., siblings) or in metal cages, may increase the incidence of the behavior (Garner *et al.* 2004a). Considering the lack of understanding from direct study, in this section we will address ways in which the laboratory environment is known to induce abnormal behavior in captive animals, generally, and specifically assess these potential mechanisms for barbering behavior onset. We will also integrate known mechanisms essential to the onset of hair-plucking behaviors in other species.

The physiology and behavior of animals is altered in animals developing in barren laboratory conditions (Presti and Lewis 2005; Würbel and Stauffacher 1997). We hypothesize that barbering is a malfunctional behavior, resulting from these developmental disturbances in neurophysiology induced by the laboratory environment. Standard rodent housing interferes with brain and behavioral development through a variety of mechanisms, including: early environmental deprivation, thwarting of behavioral response rules, and disruption of habitat-dependent adaptation processes (Würbel 2001). Würbel argues that these mechanisms are responsible for the emergence of abnormal behaviors in the laboratory, and below we set out how they may help explain the onset of barbering behavior.

Early environmental deprivation, particularly early social deprivation, is associated with altered neurobiology and behavior. Early weaning increases tail biting (fur chewing/sucking) in mink (Mason 1994) and stereotypic behaviors in mice (Würbel and Stauffacher 1997). In rats, isolation rearing disturbs dopamine systems in the prefrontal cortex and the striatum (Hall *et al.* 1998; Robbins *et al.* 1996) and disrupts cognitive processes, particularly inhibitory control in attentional selection (Schrijver and Würbel 2001). While isolation rearing has not been identified as a risk factor for barbering mice, they do show similar deficits of inhibitory control in attentional selection (i.e., disturbed set-shifting, indicative of dorsolateral prefrontal cortex [DLPFC] loop dysfunction). In other species, hair-plucking behaviors are associated with social stress, due to isolation housing in nonhuman primates (Tully *et al.* 2002) and rabbits (Kraus *et al.* 1994), and crowding in nonhuman primates (Elton 1979; Reinhardt *et al.* 1986) and sheep (Fraser 1995). Also, in humans, the onset of hair pulling is often associated with early psychosocial loss or trauma (Christenson and Mansueto 1999). Although the role of social stress in the hair-plucking behaviors of mice is largely unknown from direct study, converging evidence from other species implicates this type of early environmental deprivation as a mechanism through which the behavior may emerge. In particular, the set-shifting deficits shown in barbering mice (Garner 2006a), shown elsewhere to reflect deficits in cortico–striatal function (Dias *et al.* 1996), parallel the findings in isolation-reared rats and accordingly illustrate that isolation rearing, or some other type of social stress, may be sufficient to induce barbering behavior. It will be important to assess which aspects of current husbandry practices and the laboratory environment are socially stressful for mice, such as unnaturally early weaning ages, crowding, and isolation housing. However, as barbering is typically expressed socially, isolation housing may not be as important as other types of social stress in the onset of barbering behavior.

Behavioral response rules (i.e., innate responses to the laboratory environment, such as escape attempts) may be prevented by confinement. Captive environments also disrupt habitat-dependent adaptation processes (i.e., innately motivated behaviors that are adaptive in the wild but unnecessary in captivity, such as foraging). Both of these mechanisms by which highly motivated behaviors are prevented are associated with the performance of maladaptive behaviors (i.e., infanticide and excessive fighting in mice) and stereotypies, another type of abnormal repetitive behavior (Lewis *et al.* 2007; Odberg 1986; Würbel and Stauffacher 1998; Würbel *et al.* 1996). Thwarting of highly motivated behaviors may induce deficits in brain function that underlie the development of abnormal behaviors (Garner 2005). While the role of thwarted motivational systems is unknown in barbering onset (it has never been investigated), they are a likely candidate for how the environment may bring about the development of barbering behavior.

While barbering could emerge specifically from either the thwarting of motivated behaviors or disturbances in social structure, it is also possible that barbering emerges simply as a result of the chronic stress associated with them. The Moberg (2000) model of animal stress defines stress as "the biological response elicited when an individual perceives a threat to its homeostasis" and emphasizes that small innocuous stressors can eventually deplete an animal's ability to respond to the point of distress (defined as stress that compromises function in a second biological system). In this way, distress may disturb developmental processes and/or directly affect neurophysiological function.

Thus, multiple risk factors may bring about the development of barbering behavior by increasing the stress load and/or by increasing the stress reactivity of each individual. In this way, the increased onset of barbering as mice age may reflect how an accumulation of stress may ultimately result in barbering onset. Biological resources are much more quickly exhausted during the adolescent period of high growth and physiological reorganization, marking a period in which individuals are more susceptible to distress (Moberg 2000), and thus may explain the pubertal onset of barbering. Specific strain backgrounds are also known to differ in stress reactivity – for example, the high-barbering NMRI strain is also listed as one of the most stress-reactive strains (The Mouse Phenome Project/The Jackson Laboratory 2008), according to open-field and light-dark box transition measurements. Strains also vary across a multitude of idiosyncratic biological characteristics, such as having a reduced or absent corpus callosum in 129 strains (Wahlsten *et al.* 2001) or an increased tendency to obesity and hyperglycemia in C57s. These idiosyncrasies may directly relate to their increased or decreased risk of barbering development among subgroups: as biological currency may be expended due to these conditions, mice of certain strains may have a higher tendency towards distress.

Insights from transgenic studies

Alterations in barbering behavior have been induced in a variety of transgenic studies. However, these studies are difficult to interpret. While changes in barbering are clearly correlated with the particular "knockout" at hand in these studies, the mechanism by which the modified genotypes influence the barbering phenotype is rarely clear. Given that barbering development is multifactorial, genetic manipulations may alter barbering prevalence by influencing any one of the many processes that lead to barbering behavior. This section will review these experiments, describing the changes in barbering behavior observed and discussing possible mechanisms (developmental, behavioral, and physiological) in which the genetic manipulation may have altered the prevalence of the behavior. Wild type function of knocked out alleles, and their influence on barbering behavior are summarized in Tables 10.3 and 10.4, respectively.

However, it is important to keep in mind that knockout mice often suffer from limits on their interpretation that are rarely discussed in the literature, particularly: the flanking gene problem (Wolfer *et al.* 2002; Wolfer and Lipp 2000), the spider's knees fallacy, and an overreliance on convergent over discriminant validity. Because transgenic mice are typically congenic between 129 and C57BL/6J strains, phenotypes in a transgenic can (and often are) due to the insertion of 129 genes onto the C57BL/6J background flanking the manipulated gene. This problem is easily resolved with a variety of controls (Wolfer *et al.* 2002; Wolfer and Lipp 2000). In the case of barbering this is particularly problematic as both these strains are very prone to the behavior. Greer and Capecchi (2002) include such controls, but the other mouse models described below do not. The spider's knees fallacy is committed when a treatment that disrupts a secondary function necessary for the expression of the phenotype is misinterpreted as evidence that the treatment is the mechanism of the phenotype. For instance, strain differences in the Morris water maze are traditionally interpreted as reflecting differences in learning. However, strain differences in blindness are actually a major predictor of performance in the maze (Brown and Wong 2007). Thus if a gene essential for retinal function was knocked out, and the mouse performs poorly on the water maze, we would commit the spider's knees fallacy if we concluded that the retinal gene was involved in memory. Hill *et al.* (2007) address this issue experimentally, Greer and Capecchi (2002) acknowledge it, but the other mouse models do not. Finally, knockout phenotypes are typically interpreted on the basis of convergent validity (that two things look the same), and this interpretation is rarely tested

Table 10.3 *Transgenic studies: wild type allele function*

Author	Year	Gene	Product	Function	Effect	Tissues expressed (WT)
Lijam *et al.*	1997	DVL1	WNT signal (disheveled 1)	Organizational effects on brain development	Cell fate determination	Cerebellum, hippocampus, olfactory bulb
Greer & Capecchi	2002	HoxB8	Homeobox gene B8	Organizational effects on brain development	Unknown (in adults)	Olfactory bulb, basal ganglia, hippocampus, cortex
Kassed & Herkenham	2004	NF-kB p50	Transcription factor	Immune system activation expressed in response to stress/pathogens	Neuro-protective	Hippocampus, cortex, amygdala, hypothalamus, brainstem
Kalueff *et al.*	2006a	VDR	Vitamin D receptor	Neurosteroid hormone system: cell differentiation, homeostasis	Varied	Widespread in brain and spinal cord
Hill *et al.*	2007	Ar	Aromatase enzyme	Converts androgen to estrogen	Varied	Brain, bone, adipose tissue, and gonads
Welch *et al.*	2007	Sapap-3	Post-synaptic scaffolding protein	Regulates trafficking of post-synaptic receptors (AMPA/NMDA) and signaling	Signaling integrity	Exitatory synapses, striatum

Table 10.4 *Transgenic studies: knockout effects on behavior*

Author	Year	Gene	Barbering effects	WT	KO	Grooming effects	Social interaction	Anxiety
Lijam *et al.*	1997	DVL1	Eliminated barbering	50–75%	0%	*	Large reduction (elimination)	*
Greer & Capecchi	2002	HoxB8	Increased hair loss and skin lesioning	50%	100%	Increased (initiation, duration and elicitation)	*	*
Kassed & Herkenham	2004	NF-kB p50	Decreased	69%	10%	Slight reduction (93% to 72%)	*	Reduced
Kalueff *et al.*	2006a	VDR	Eliminated barbering	53%	0%	Increased self-grooming, eliminated allogrooming	Reduced (not eliminated)	Increased
Hill *et al.*	2007	Ar	Males – increased	–	–	Males – increased	*	*
			Females – no effect	–	–	Females – no effect		
Welch *et al.*	2007	Sapap-3	Induced hairless lesions	0%	100%	Large increase	*	Increased

by excluding alternative explanations (discriminant validity). For instance, a wide range of repetitive behaviors in mice have been interpreted as "OCD" or "OCD-like," when in fact their equivalent in humans would preclude a diagnosis of OCD (Garner 2006a). This fallacy is sometimes referred to as "all dogs bark, but not all that barks is a dog," and is committed by all the mouse models listed below. Barbering is an excellent example – as we will explain later.

The *Hoxb8* mutant mice generated by Greer and Capecchi (2002) showed a 100% penetrant barbering phenotype. Greer and Capecchi speculate that this disturbance is mediated by pathology of the "OCD-circuit," which is the limbic cortico–striatal loop of the basal ganglia. However, other areas of the brain were also affected by this mutation, including other areas of the basal ganglia, the hippocampus, and cortical regions. Alternatively, it is possible that deficits associated with any of these other regions mediate the barbering phenotype.

Aromatase is an enzyme expressed in neural, gonadal, and other tissues, which converts androgen to estrogen, and is integral for brain masculinization (Lephart 1996). Male aromatase knockout (ArKO) mice show increased barbering, although the behavior is unaffected in females (Hill *et al.* 2007). As the knockout increases all grooming behaviors in males, it is possible that there is a general increase in both healthy and abnormal (i.e., barbering) grooming, without necessarily affecting the idiosyncratic neural circuitry specifically associated with barbering. The prevalence of barbering in male ArKO mice is similar to both female ArKO and wild type (WT) mice in this study. As the process of brain masculinization is dependent on aromatase, the male knockout individuals have feminized brains. Thus, an alternative explanation to this study is that the knockout renders males similar to females in their risk for barbering onset.

Three other knockout studies reduced the prevalence of barbering behavior. In all of these cases the mutant mice also have behavioral abnormalities in elements likely critical to the development and/or performance of barbering behavior, such as anxiety/stress and social interaction. Knockout of the gene for NF-kB p50, eliminated the incidence of barbering behavior (Kassed and Herkenham 2004). The NF-kB p50 is a transcription factor that is activated by physical stress and pathogens and has neuroprotective properties (Kassed and Herkenham 2004). Behaviorally, the knockout not only reduced barbering behavior, but also reduced measures of anxiety in the mutants relative to WT. It is probable that the mechanism drastically reduces stress reactivity in individuals. If this is the case, it may be that NF-kB p50 knockout reduces barbering by alleviating stress in individuals, reducing the likelihood of experiencing distress, and thus reducing the likelihood of distress-induced basal ganglia dysfunction. The incidence of barbering behavior was also reduced to zero in mutant mice without a functional copy of the Dvl1 gene (Lijam *et al.* 1997). While the direct effects of this knockout on neurophysiology are largely unknown at this time, the gene product encoded plays a role in the WNT signaling pathway, which is integral to cell fate determination (Lijam *et al.* 1997). These mutant mice demonstrated severe social deficits that reduced incidence of huddling and allogrooming, and thus reduce the opportunity for barbering. Similarly, knockout of the vitamin D receptor (VDR) also eliminated social grooming behavior, both in terms of healthy allogrooming, as well as barbering (Kalueff *et al.* 2006a). As barbering is typically expressed socially (particularly in the 129S6/SvEvTac background strains used in both experiments), it is likely that the social deficits prevented barbering behavior from developing at all.

Mechanism

In the last section we discussed ways in which mice differentially develop the tendency to pluck hair; in this section, we will ask the question: of those mice that do become barbers, how is the behavior expressed and what are the processes that mediate its performance?

Behavioral mechanism

In most studies, the patterns of alopecia resulting from barbering have been used as a direct measure of the behavior (e.g., Garner *et al.* 2004a, b; Kalueff *et al.* 2006b; Long 1972), with little regard to the actual behavioral processes that cause the hair loss. The process by which mice pluck has been systematically described only once (i.e., Sarna *et al.* 2000) and is limited to the behavioral processes of whisker plucking. Others have also discussed the plucking process (i.e., Hauschka 1952), but have only referred to anecdotal observations without any type of systematic study or description of postures, movements, or behavioral sequences.

During a single bout of whisker plucking, barbers do not seem to specifically approach cage-mates to pluck out hair; in fact barbers are just as likely to approach their “victim” (recipient), as to be approached by them (Sarna *et al.* 2000). When together, the barber will mount or climb the recipient and press down on the body with his/her front paws, grasp a single shaft of hair with their teeth, then proceed to pluck it out from the root by pulling back (Hauschka 1952; Sarna *et al.* 2000). After plucking, the barber often rears on its haunches, transfers the removed hair to its forepaws, and then begins to orally manipulate it (Sarna *et al.* 2000); thus completing the behavioral sequence. Presumably, fur plucking also consists of the same process; however, it is not clear if hairs are plucked one by one, or in clumps. Over time bouts of hair plucking are repeated, with individual barbers preferentially plucking from specific regions of the victim’s body, resulting in the patterns of alopecia that reflect each barber's cutting style (Sarna *et al.* 2000). Typically these patterns have smooth and well-defined borders (see Figure 10.1), which is more consistent with plucking a few or single hairs at each location, than with the plucking of large clumps.

Allogrooming, or social grooming, occurs in mice and many other species and helps to maintain social relationships and provides care of body surfaces (Spruijt *et al.* 1992). As barbering is often expressed socially, it could be considered a modified or disturbed form of allogrooming. However, barbering is actually counterproductive to the adaptive benefit of allogrooming: removed hairs expose body surfaces and thus make skin more susceptible to damage, and the process of plucking may disturb social relationships. The exact relationship of barbering and allogrooming is not clear. Bouts of barbering and allogrooming do begin similarly, with the groomer pressing down on the cage-mate's body (Grant and Mackintosh 1963; Sarna *et al.* 2000), but are presumably divergent past this point. In allogrooming, the mouse proceeds to mouth and lick the body surface, often in specific regions, and sometimes even gently biting the fur without pulling any hair out (Grant and Mackintosh 1963). In whisker plucking, some mouthing and licking may take place, but often hairs are plucked out in the absence of these normal grooming behaviors after the cage-mate has been pressed down (Sarna *et al.* 2000).

Overall, barbering mice do not spend more time grooming than do nonbarbering mice; however, compared to nonbarbers, cage-mate barbers have significantly higher levels of allogrooming and significantly lower levels of self-grooming (Militzer and Wecker 1986). In Militzer and Wecker's study, plucking was not directly assessed and accordingly the increase in allogrooming may reflect time spent plucking hair. Their findings are interesting in two ways: hair plucking may appear similar enough to allogrooming that they can be mistaken for each other; and even though barbers increase time spent allogrooming, they do not increase the overall amount of grooming that they perform (this increase is at the expense of self-grooming). Thus, there is a qualitative difference in the way that barbers groom, but not necessarily a quantitative difference in grooming behavior, overall.

Drawing from the limited data available directly assessing barbering behavior, we can conclude that:

(1) barbering is a goal-directed behavior – mice grasp hairs with their teeth in order to remove, manipulate, and ingest them
(2) this goal is not functionally adaptive and accordingly is not a part of the natural behavioral repertoire of mice
(3) barbering is repetitive in two ways – bouts of plucking are repeated frequently and specific regions of the victims are repeatedly visited.

Considering that hairs are typically plucked one at a time, alopecia patterns (particularly the amount of hair plucked and the severity of the lesions) provide an objective measure of the frequency of the behavior. However, inference from hair-loss measures should be limited to the frequency of the behavior, and not necessarily used as an indicator of an individual's "motivation" or "inherent tendency" to pluck; social and environment factors also influence the performance of plucking behaviors, and accordingly affect the severity of these lesions without necessarily affecting an individual's "tendency" to pluck (e.g., Garner *et al.* 2004a).

Cues and constraints

Taking into account the behavioral pattern that constitutes barbering, in this section we will assess factors that may precipitate or constrain its performance.

In humans, both stress and boredom are known to precipitate bouts of hair-plucking behavior in affected individuals (Christenson and Mansueto 1999). While laboratory mice are widely believed to experience both stress and boredom as a consequence of the barren artificial environments that they typically inhabit, to date the effect of the environment on precipitating the behavior has not been investigated. The ability of environmental enrichment to reduce the severity of hair-plucking behavior (DeLuca 1997) may have been mediated by reducing stressful factors (through providing shelters and nesting material) and by reducing boredom (through adding novel objects that the mice interacted with). Stress and isolation housing have also been identified as precipitating factors in nonhuman primates, guinea pigs, sheep, cats, and dogs (Reinhardt 2005)

There are two major ways in which the performance of barbering behavior is constrained: social factors have been shown empirically to alter the performance of the behavior; and being plucked is likely painful and should alter the performance as well.

Barbering is largely performed in a social context, with barbering individuals plucking from cage-mates that may or may not be barbers themselves. Van den Broek (1993) showed that the victims of the barbering will participate in the behavior, even when given the opportunity to escape from it. This is the likely mechanism to explain the alteration in barbering phenomena resulting from dominance hierarchies. As discussed earlier, dominance hierarchies form a natural part of murine social structure, both in the wild and in the laboratory. Barbering is not a dominance behavior: it is performed by both dominant and subordinate mice (Garner *et al.* 2004a; Hauschka 1952; Long 1972). However, dominant mice have smaller barbering lesions relative to subordinate mice (Garner *et al.* 2004a;

Hauschka 1952). In the light of Van den Broek's work, it is likely that dominant mice will participate less than will subordinate mice.

Sarna *et al.* (2000) described that mice often winced when hair was plucked out; they concluded that the process is likely painful. However, victims will facilitate their cage-mate's behavior, even when given the ability to escape it (Van den Broek *et al.* 1993). The most logical rationale for this phenomenon is that mice are motivated to allow themselves to be groomed by cage-mates. The painful nature of hair plucking also likely explains why barbering mice preferentially pluck from cage-mates, rather than from themselves.

Neurophysiological mechanism

In this section we introduce a general theoretical framework for abnormal repetitive behaviors (ARBs) in humans and animals. For detailed discussion see Albin and Mink (2006), Aouizerate *et al.* (2004), and Garner (2006a). We then present evidence that barbering fits in this general framework, showing both discriminant and convergent validity as a model of TTM.

The organization of behavior has long been viewed as hierarchical – although the three major behavioral traditions (ethology, comparative psychology, and neurobiology) (Grier 1984) differ in terminology, each has identified four basic layers of control (Table 10.5). Functional behavior involves selection and inhibition – i.e., sequencing – at each level. For instance, for reproductive motivation to be selected, currently active motivations (e.g., feeding) must be inhibited. Then suitable reproductive goals must be selected in the correct order (e.g., there is no point in displaying to a mate before one has been found) before particular motor programs within each goal are enacted. Within each motor program muscle movements must be coordinated in turn. Although control further down the hierarchy is directed by upstream selections, processing continues in higher levels as lower aspects of behavior are executed. Thus, motor patterns in feeding behavior can be interrupted by a higher change in motivation (e.g., if a predator appears – stop eating and run away) (Grier 1984; Norman and Shallice 1986). Disinhibition, overactivation, or alterations in hysteresis potentially predispose the system to inappropriate repetitive or chaotic behavioral transitions (Albin *et al.* 1989; Aouizerate *et al.* 2004; Berridge *et al.* 2005; Garner 2006a; Lyon and Robbins 1975; Mink 2003; Norman and Shallice 1986).

The control of behavior is effected by cortico–striatal loops (Alexander and Crutcher 1990; Alexander *et al.* 1986, 1990; Rolls 1994) that are broadly conserved throughout the vertebrates (Reiner *et al.* 1998; Steiner and Gerfen 1998). Each loop involves the same basic circuitry, comprised of a "direct pathway" that activates behavioral elements, and an "indirect pathway" that inhibits them (McHaffie *et al.*

Table 10.5 *The hierarchical organization of behavior, and its correlates*

Level of behavioral organization	Cortico–striatal loop	Neuropsychological deficit/neurological sign	Spontaneous symptoms
Motivational state	"Limbic"	Affective shifting serial reversal learning	Obsessions? Compulsive behaviors?
Goal-directed behavior/ attentional set	"Prefrontal"	Set shifting stuck-in-set perseveration	Compulsive and impulsive behaviors
Fixed action patterns/ motor programs	"Motor" Premotor cortex	Response shifting recurrent perseveration	Stereotypies
Isolated muscle movements	"Motor"	Continuous perseveration	Single-muscle tics
References	McHaffie *et al.* 2005; Redgrave *et al.* 1999.	Dias *et al.* 1996; Garner 2006a, b; Luria 1965; Norman and Shallice 1986.	Albin and Mink 2006; Chamberlain *et al.* 2005; Garner 2006a, b.

2005; Redgrave *et al.* 1999). Separate loops subserve separate functions, such that key loops may select and sequence each hierarchy of behavioral control discussed above (McHaffie *et al.* 2005; Redgrave *et al.* 1999). These loops have been repeatedly implicated in ARBs across a diverse range of disorders (Albin and Mink 2006; Albin *et al.* 1995; Aouizerate *et al.* 2004; Chamberlain *et al.* 2005; Crider 1997; Lewis and Bodfish 1998; Mink 2003; Rosenberg and Keshavan 1998; Stein *et al.* 1999). Importantly, dysfunction in different loops may be responsible for fundamentally distinct symptoms (Albin and Mink 2006; Alexander *et al.* 1990; Aouizerate *et al.* 2004; Chamberlain *et al.* 2005; Garner 2006a; Turner 1997), although some disagreement exists on the particulars. The consistent theme in these accounts is that ARB is the result of disinhibition of the indirect pathway, and the form of repetition seen in spontaneous symptoms reflects the component of behavior each loop sequences (Albin and Mink 2006; Aouizerate *et al.* 2004; Chamberlain *et al.* 2005; Garner 2006a, b; Sheppard *et al.* 1999; Turner 1997). Each loop can be associated with a neuropsychological deficit or neurological sign, some of which have associated lesion-validated tasks (Dias *et al.* 1996). Some contradictory evidence does exist to the scheme in Table 10.5 (e.g., Peterson *et al.* 2003).

This scheme allows us to assess discriminant validity. Thus TTM is associated with neuropsychological deficits indicative of altered prefrontal cortico–striatal loop function (e.g., Stanley *et al.* 1997), but not of the limbic loop (e.g., Bohne *et al.* 2005a, c). In contrast OCD involves neuropsychological deficits indicative of altered function in both limbic and prefrontal cortico–striatal loops (e.g., Bohne *et al.* 2005a, c; Lucey *et al.* 1997; Remijnse *et al.* 2006). Conversely, stereotypies in humans and animals are consistently associated with deficits in "response shifting," which is not associated with either loop, and possibly indicates motor cortico–striatal loop dysfunction (Garner 2006a). Thus, if barbering is homologous to OCD it should be associated with deficits indicative of limbic and prefrontal loops; if homologous to TTM with deficits indicative of the prefrontal loop; and if a stereotypy, with response-shifting deficits.

Accordingly, we recently developed mouse versions of the neuropsychological tasks used to measure these functions in humans (e.g., Garner *et al.* 2006b) and administered this battery of tasks to a population of C57BL/6J mice. As predicted, barbering mice showed a selective deficit only in "set shifting," which assays prefrontal loop function (Dias *et al.* 1996; Garner 2006a; Garner *et al.* 2006b), and the severity of their deficit correlated with the severity of their barbering (Garner 2006a). They were unimpaired on "affective shifting," which assays limbic loop function (Dias *et al.* 1996; Garner 2006a; Garner *et al.* 2006b), or in "response shifting." Conversely, stereotypies in the same mice were correlated only with deficits in "response shifting" (Garner 2006a). Thus these data show, with discriminant validity, that barbering is neither homologous to OCD, nor is it a stereotypy; but that it is consistent with TTM.

Conclusion

Barbering behavior is striking: it is strange, conspicuous, occurs relatively frequently, and does so without any clear function. It is a behavior defined more by its effects than its actual expression; however, the patterns of alopecia it leaves offer a unique and objective indicator of the behavior's performance. Although barbering was first described over 50 years ago, it has been studied only rarely. Findings have historically been inconsistent due in part to the variability in husbandry and strains employed in these studies, as well as the inability to control factors key to its development that largely remain unknown. There are many misconceptions about barbering, based more on tradition than empirical evidence. Barbering is not a dominance behavior – it in no way dictates the social hierarchy within mouse cages. There is also no known functional difference between plucking from whiskers or from the body. Although barbering is expressed socially, it is not a healthy social behavior – it is an abnormal repetitive behavior, which is not selected for evolutionarily, and is very likely related to the abnormal plucking behaviors that unnatural environments induce across a multitude of species, reflecting behavioral and neurophysiological pathology.

A better understanding about how and why barbering behavior occurs is emerging through recent research. Barbering behavior has no adaptive function: it is performed repetitively and inappropriately (e.g., the behavior is outside of the behavioral repertoire of free-living *Mus musculus* and is performed in the laboratory with a considerable fitness cost). In this way, barbering behavior is best classified as a type of ARB, related to other types of ARB such as stereotypies or obsessive–compulsive behaviors, although unique and functionally distinct. Abnormal plucking behaviors are not exclusive to mice; they appear frequently in phylogenetically related mammalian and avian species. Considering that these behaviors are only found in individuals developing in unnatural environments, plucking behaviors likely represent similar developmental disturbances and malfunction of adaptive neural structures that are phylogenetically conserved. Although very little is known about the developmental mechanisms responsible for barbering onset, it is clear that this process is complex and multifactorial. Early evidence in mice coincides with the more robust comparative evidence, implicating a role for stress and behavioral frustration. Likewise, while the neural machinery that generates the behavior remain largely unknown, as physiological differences between "spontaneous" barbers and nonbarbers have never been investigated, comparative evidence points to dysfunction in the cortico–striatal loops of the basal ganglia as the likely source of ARB. Knockout studies disturbing these structures can induce hair-plucking behavior. Our finding of set-shifting deficits provides further support for this hypothesis, specifically implicating dysfunction in the dorsolateral prefrontal cortico–striatal loops, and clearly demonstrating that barbering is not a stereotypy. Considering that the cortico–striatal loops structures are highly conserved across mammalian and avian species, and that unnatural environments can clearly disturb their development and function, the cortico–striatal loops hypothesis of ARB generally, and barbering in particular, is not only plausible, but is increasingly supported by empirical evidence.

There are still many questions about barbering behavior that should be addressed. Developmental and pathophysiological mechanisms have been the least studied, and direct evidence is not only needed, but will be essential to understanding the behavior. By identifying specific environmental features key to its development, researchers may finally elucidate the developmental mechanism; additionally enabling a refinement in barbering research by controlling for these factors. Although indirect evidence implicates disturbed DLPFC cortico–striatal loops function as the pathophysiological underpinnings of barbering, it is not understood how this circuitry is altered in barbering mice, or by what physiological mechanism these alterations come about. Understanding the pathophysiological substrate for the behavior will enable more accurate comparison of the behavior with other ARBs; it may also illuminate the healthy behavioral function of the basal ganglia and its cortico–striatal loops projections.

Refinement in barbering studies is essential, taking into account the numerous inconsistencies and inaccurate folklore that has been generated. Husbandry conditions, such as stocking density (number of mice per cage) as well as gender and caging system (breeding or same-sex pairs) all may influence its development and performance. Housing mice in either same-sex or mixed-sex breeding pairs only, would enable better comparison among studies. Developmental questions are best addressed by looking at measures of barbering incidence, while mechanistic questions are better served by assessing changes (or differences) in alopecia severity. Alopecia patterns represent an unbiased indicator of the frequency of hair-plucking behavior. However, the alopecia left by barbering represents a combination of compulsive traits, the frequency of cues that elicit the behavior, and the constraints on the performance of the behavior. When assessing changes in alopecia patterns, all three of these factors must be accounted for. Cues are not well understood from direct study, but comparative evidence points to multiple factors: stress (particularly social stress), lack of complexity or boredom, reproductive hormones (particularly oscillations), and factors that induce grooming may also increase barbering behavior. Most constraints are likely social factors: social hierarchies or social isolation may alter the performance of the behavior, simply by providing constraints on its performance, without influencing the compulsive mechanism.

Barbering behavior represents a research opportunity for human modeling and transgenic studies. In particular, the behavior bears a striking resemblance to human TTM. Considering its prevalence, debilitating impact on patients, and lack of effective treatments, a better understanding of the pathophysiology and pharmacological treatment profile of TTM are urgently needed. Barbering may represent the best opportunity to make these leaps in understanding.

References

Albin RL and Mink JW (2006): Recent advances in Tourette syndrome research. *Trends Neurosci* **29**:175–82.

Albin RL, Young AB and Penney JB (1989): The functional anatomy of basal ganglia disorders. *Trends Neurosci* **12**:366–75.

Albin RL, Young AB and Penney JB (1995): The functional anatomy of disorders of the basal ganglia. *Trends Neurosci* **18**:63–4.

Alexander GE and Crutcher MD (1990): Functional architecture of basal ganglia circuits: neural substrates of parallel processing. *Trends Neurosci*, **13**:266–71.

Alexander GE, Delong MR and Strick PL (1986): Parallel organization of functionally segregated circuits linking basal ganglia and cortex. *Annu Rev Neurosci* **9**: 357–81.

Alexander GE, Crutcher MD and Delong MR (1990): Basal ganglia-thalamocortical circuits: parallel substrates for motor, oculomotor, prefrontal and limbic functions. *Prog Brain Res* **85**:119–46.

Altman K, Grahs C and Friman P (1982): Treatment of unobserved trichotillomania by attention reflection and punishment of an apparent covariant. *J Behav Ther Exp Psychiatry* **13**:337–40.

American Psychiatric Association (1994): *Diagnostic and Statistical Manual of Mental Disorders*, 4th edn. Washington, DC: APA.

Aouizerate B, Guehl D, Cuny E *et al.* (2004): Pathophysiology of obsessive–compulsive disorder: a necessary link between phenomenology, neuropsychology, imagery and physiology. *Prog Neurobiol* **72**:195–221.

Azrin NH (1977): *Habit Control in a Day*. New York: Simon and Schuster.

Benton D and Brain PF (1979): Behavioral-comparisons of isolated, dominant and subordinate mice. *Behav Processes* **4**:211–19.

Berridge KC, Aldridge JW, Houchard KR and Zhuang XX (2005): Sequential super-stereotypy of an instinctive fixed action pattern in hyper-dopaminergic mutant mice: a model of obsessive compulsive disorder and Tourette's. *BMC Biol* **3**:4.

Blackmore DK, Schultze WH and Absolon GC (1986): Light-intensity and fur-chewing in rabbits. *N Z Vet J*, **34**:158.

Boers K, Gray G and Love J (2002): Comfortable quarters for rabbits in research institutions. In: Reinhardt A and Reinhardt V, eds., *Comfortable Quarters for Laboratory Animals*. Washington, DC: Animal Welfare Institute, pp. 44–50.

Bohne A, Keuthen NJ, Tuschen-Caffier B and Wilhelm S (2005a): Cognitive inhibition in trichotillomania and obsessive-compulsive disorder. *Behav Res Ther* **43**:923–42.

Bohne A, Keuthen NJ and Wilhelm S (2005b): Pathologic hairpulling, skin picking, and nail biting. *Ann Clin Psychiatry* **17**:227–32.

Bohne A, Savage CR, Deckersbach T, *et al.* (2005c): Visuospatial abilities, memory, and executive functioning in trichotillomania and obsessive-compulsive disorder. *J Clin Exp Neuropsychol*, **27**:385–99.

Boughn S and Holdom JJ (2003): The relationship of violence and trichotillomania. *J Nurs Scholarsh* **35**:165–70.

Brown RE and Wong AA (2007): The influence of visual ability on learning and memory performance in 13 strains of mice. *Learn Mem*, **14**:134–44.

Carruthers EL, Halkin L and King TR (1998): Mouse barbering: investigations of genetic and experiential control. Poster presented by Edward Carruthers at Animal Behavior Society national meeting. Carbondale, IL.

Chamberlain SR, Blackwell AD, Fineberg NA, Robbins TW and Sahakian BJ (2005): The neuropsychology of obsessive compulsive disorder: the importance of failures in cognitive and behavioural inhibition as candidate endophenotypic markers. *Neurosci Biobehav Rev* **29**:399–419.

Christenson GA (1995): Trichotillomania: from prevalence to comorbidity. *Psychiatr Times* **12**:44–8.

Christenson GA and Crow SJ (1996): The characterization and treatment of trichotillomania. *J Clin Psychiatry* **57**:42–9.

Christenson GA and Mansueto CS (1999): Trichotillomania: descriptive characteristic and phenomenology. In: Stein DJ, Christenson GA and Hollander E, eds., *Trichotillamania*. Washington, DC: American Psychiatric Press, Inc., pp. 1–41.

Christenson GA, Mackenzie TB and Mitchell JE (1991a): Characteristics of 60 adult chronic hair pullers. *Am J Psychiatry* **148**:365–70.

Christenson GA, Mackenzie TB, Mitchell JE and Callies AL (1991b): A placebo-controlled, double-blind crossover study of fluoxetine in trichotillomania. *Am J Psychiatry* **148**:1566–71.

Christenson GA, Pyle RL and Mitchell JE (1991c): Estimated lifetime prevalence of trichotillomania in college-students. *J Clin Psychiatry* **52**:415–17.

Christenson GA, Mackenzie TB and Mitchell JE (1994): Adult men and women with trichotillomania – a comparison of male and female characteristics. *Psychosomatics*, **35**:142–9.

Clubb R and Mason G (2004): Pacing polar bears and stoical sheep: testing ecological and evolutionary hypotheses about animal welfare. *Anim Welf* **13**:S33–S40.

Cohen LJ, Stein DJ, Simeon D *et al.* (1995): Clinical profile, comorbidity, and treatment history in 123 hair pullers: a survey study. *J Clin Psychiatry* **56**:319–26.

Crider A (1997): Perseveration in schizophrenia. *Schizophr Bull* **23**:63–74.

Crowcroft P (1966): *Mice All Over*. Great Britain: GT Foulis & Co.

Davison GC and Neale JM (1998): *Abnormal Psychology*. New York: John Wiley & Sons, Inc.

Dawkins MS (1993): *Through Our Eyes Only*? Oxford: WH Freeman/Spektrum.

de Jonge G (1988): Genetics and evolution of tail-biting by farmed mink. *Behav Genet* **18**:713–14.

DeLuca AM (1997): Environmental enrichment: does it reduce barbering in mice? *AWIC Newsletter* **8**:7–8.

Deluca RV and Holborn SW (1984): A comparison of relaxation training and competing response training to eliminate hair pulling and nail biting. *J Behav Ther Exp Psychiatry* **15**:67–70.

Dias R, Robbins TW and Roberts AC (1996): Dissociation in prefrontal cortex of affective and attentional shifts. *Nature* **380**:69–72.

Done-Currie JR, Hecker JF and Wodzickatomaszewska M (1984): Behavior of sheep transferred from pasture to an animal house. *Appl Anim Behav Sci* **12**:121–30.

Elton RH (1979): Baboon behaviour under crowded conditions. In: Erwin J, Maple T and Mitchell G, eds., *Captivity and Behaviour*. New York: Van Nostrand Reinhold, pp. 125–39.

Fraser AF (1995): Sheep. In: Rollin BE and Kessel ML, eds., *The Experimental Animal in Biomedical Research*. Boca Raton, FL: CRC Press, pp. 87–118.

Garner JP (2005): Stereotypies and other abnormal repetitive behaviors: potential impact on validity, reliability, and replicability of scientific outcomes. *ILAR J* **46**:106–17.

Garner JP (2006a): Perseveration and stereotypy. In: Mason G and Rushen J, eds., *Stereotypic Animal Behaviour: Fundamentals and Applications to Welfare*. Wallingford, England, UK: CABI.

Garner JP (2006b): Box 10.2. Implications of recognizing mechanistic differences in abnormal repetitive behaviour. In: Mason G and Rushen J, eds., *Stereotypic Animal Behaviour: Fundamentals and Applications to Welfare*. Wallingford, England, UK: CABI, pp. 293–4.

Garner JP, Weisker SM, Dufour B, Gregg LE and Mench JA (2001): The epidemiology of barbering (whisker trimming) in laboratory mice. In: Garner JP, Mench JA and Heekin S, eds., *Proceedings of the 35th International Congress of the International Society for Applied Ethology*. Davis, CA, USA: Center For Animal Welfare, UC Davis, p. 129.

Garner JP, Dufour B, Gregg LE, Weisker SM and Mench JA (2004a): Social and husbandry factors affecting the prevalence and severity of barbering ('whisker trimming') by laboratory mice. *Appl Anim Behav Sci* **89**:263–82.

Garner JP, Weisker SM, Dufour B and Mench JA (2004b): Barbering (fur and whisker trimming) by laboratory mice as a model of human trichotillomania and obsessive–compulsive spectrum disorders. *Comp Med* **54**:216–24.

Garner JP, Meehan CL, Famula TR and Mench JA (2006a): Genetic, environmental, and neighbor effects on the severity of stereotypies and feather picking in Orange-winged Amazon parrots (*Amazona amazonica*): an epidemiological study. *Appl Anim Behav Sci* **96**:153–68.

Garner JP, Thogerson CM, Würbel H, Murray JD and Mench JA (2006b): Animal neuropsychology: validation of the intra-dimensional extra-dimensional set shifting task in mice. *Behav Brain Res* **173**:53–61.

Gerold S, Huisinga E, Iglauer F *et al.* (1997): Influence of feeding hay on the alopecia of breeding guinea pigs. *J Vet Med A Physiol Pathol Clin Med* **44**:341–8.

Glatz PC (2001): Effect of poor feather cover on feed intake and production of aged laying hens. *Asian-Australas J Anim Sci* **14**:553–8.

Grant EC and Mackintosh JH (1963): A comparison of the social postures of some common laboratory rodents. *Behaviour* **21**:246–59.

Graybiel AM and Rauch SL (2000): Toward a neurobiology of obsessive–compulsive disorder. *Neuron* **28**:343–7.

Greer JM and Capecchi MR (2002): Hoxb8 is required for normal grooming behavior in mice. *Neuron* **33**:23–34.

Grier JW (1984): *Biology of Animal Behavior*. St. Louis: Times Mirror/Mosby College Pub.

Gunn D and Morton DB (1995): Inventory of the behavior of New-Zealand White rabbits in laboratory cages. *Appl Anim Behav Sci* **45**:277–92.

Hall FS, Wilkinson LS, Humby T *et al.* (1998): Isolation rearing in rats: pre- and postsynaptic changes in striatal dopaminergic systems. *Pharmacol Biochem Behav* **59**:859–72.

Hansen SW (1992): Stress reactions in farm mink and beech marten in relation to housing and domestication. Unpublished PhD thesis, Cøpenhagen: Cøpenhagen University.

Hauschka TS (1952): Whisker-eating mice. *J Hered* **43**:77–80.

Hill RA, McInnes KJ, Gong ECH *et al.* (2007): Estrogen deficient male mice develop compulsive behavior. *Biol Psychiatry* **61**:359–66.

Howerton CL, Garner JP and Mench JA (2008): Effects of a running wheel-igloo enrichment on aggression, hierarchy linearity, and stereotypy in group-housed male CD-1 (ICR) mice. *Appl Anim Behav Sci* **115**:90–103.

Insel TR (1992): Toward a neuroanatomy of obsessive–compulsive disorder. *Arch Gen Psychiatry* **49**:739–44.

Kalueff AV, Keisala T, Minasyan A *et al.* (2006a): Behavioural anomalies in mice evoked by "Tokyo" disruption of the Vitamin D receptor gene. *Neurosci Res* **54**:254–60.

Kalueff AV, Minasyan A, Keisala T, Shah ZH and Tuohimaa P (2006b): Hair barbering in mice: implications for neurobehavioural research. *Behav Processes* **71**:8–15.

Kassed CA and Herkenham M (2004): NF-kappa B p50-deficient mice show reduced anxiety-like behaviors in tests of exploratory drive and anxiety. *Behav Brain Res* **154**:577–84.

Keuthen NJ, O'Sullivan RL, Hayday CF *et al.* (1997): The relationship of menstrual cycle and pregnancy to compulsive hairpulling. *Psychother Psychosom* **66**:33–7.

Kraus AL, Weisbroth SH and Flatt RE (1994): Biology and disease of rabbits. In: Fox JG, Cohen BJ and Loew FM, eds., *Laboratory Animal Medicine*. Orlando: Academic Press, pp. 207–40.

Latham N and Mason G (2004): From house mouse to mouse house: the behavioural biology of free-living *Mus musculus* and its implications in the laboratory. *Appl Anim Behav Sci* **86**:261–89.

Lephart ED (1996): A review of brain aromatase cytochrome P450. *Brain Res Rev* **22**:1–26.

Lewis MH and Bodfish JW (1998): Repetitive behavior disorders in autism. *Ment Retard Dev Disabil Res Rev* **4**:80–9.

Lewis MH, Tanimura Y, Lee LW and Bodfish JW (2007): Animal models of restricted repetitive behavior in autism. *Behav Brain Res* **176**:66–74.

Lijam N, Paylor R, McDonald MP *et al.* (1997): Social interaction and sensorimotor gating abnormalities in mice lacking Dvl1. *Cell* **90**:895–905.

Long SY (1972): Hair-nibbling and whisker-trimming as indicators of social hierarchy in mice. *Anim Behav* **20**:10–12.

Lucey JV, Burness CE, Costa DC *et al.* (1997): Wisconsin Card Sorting Test (WCST) errors and cerebral blood flow in obsessive–compulsive disorder (OCD). *Br J Med Psychol* **70**:403–11.

Luria AR (1965): Two kinds of motor perseveration in massive injury of the frontal lobes. *Brain* **88**:1–11.

Lyon M and Robbins T (1975): The action of central nervous system stimulant drugs: a general theory concerning amphetamine effects. *Curr Dev Psychopharmacol* **2**:79–163.

Marsden D and Woodgush DGM (1986): A note on the behavior of individually-penned sheep regarding their use for research purposes. *Anim Prod* **42**:157–9.

Mason GJ (1991): Stereotypies: a critical review. *Anim Behav* **41**:1015–37.

Mason GJ (1994): Tail-biting in mink (*Mustela Vison*) is influenced by age at removal from the mother. *Anim Welf* **3**:305–11.

Mason GJ (2006): Box 7.1. Are wild-born animals 'protected' from stereotypy when placed in captivity? In: Mason G and Rushen J, eds., *Stereotypic Animal Behaviour*. Wallingford, England, UK: CABI, p. 196.

McHaffie JG, Stanford TR, Stein BE, Coizet W and Redgrave P (2005): Subcortical loops through the basal ganglia. *Trends Neurosci* **28**:401–7.

McMahon WM and Sundberg JP (1994): Barbering behavioral abnormalities in inbred laboratory mice. In: Sundberg JP, ed., *Handbook of Mouse Mutations with Skin and Hair Abnormalities: Animal Models and Biomedical Tools*. Boca Raton: CRC Press, pp. 493–7.

Messeri P, Eleftheriou BE and Oliverio A (1975): Dominance behavior: phylogenetic analysis in mouse. *Physiol Behav* **14**:53–8.

Militzer K and Wecker E (1986): Behavior-associated alopecia-areata in mice. *Lab Anim* **20**:9–13.

Mills DS (2003): Medical paradigms for the study of problem behaviour: a critical review. *Appl Anim Behav Sci* **81**:265–77.

Mink JW (2003): The basal ganglia and involuntary movements: impaired inhibition of competing motor patterns. *Arch Neurol* **60**:1365–8.

Moberg GP (2000): Biological response to stress: implications for animal welfare. In: Moberg GP and Mench GP, eds., *The Biology of Animal Stress*. New York: CABI Publishing, pp. 1–21.

Moon-Fanelli AA, Dodman NH and O'Sullivan RL (1999): Veterinary models of compulsive self-grooming: parallels with trichotillomania. In: Stein DJ, Christenson GA and Hollander E, eds., *Trichotillomania*. Washington, DC: American Psychiatric Press, pp. 63–92.

Norman DA and Shallice T (1986): Attention to action: willed and automatic control of behaviour. In: Davidson RJ, Schwartz GE and Shapiro D, eds., *Consciousness and Self-Regulation: Advances in Research and Theory*. New York: Plenum Press, pp. 1–18.

Odberg FO (1986): The jumping stereotypy in the bank vole (*Clethrionomys glareolus*). *Biol Behav* **11**:130–43.

Peterson BS, Thomas P, Kane MJ *et al.* (2003): Basal ganglia volumes in patients with Gilles de la Tourette syndrome. *Arch Gen Psychiatry* **60**:415–24.

Pinkus H (1964): Transient alopecia in weanling BD mice (Trichomalacia). 743–53.

Ponzio MF, Busso JM, Ruiz RD and de Cuneo MF (2007): A survey assessment of the incidence of fur-chewing in commercial chinchilla (*Chinchilla lanigera*) farms. *Anim Welf* **16**:471–9.

Presti MF and Lewis MH (2005): Striatal opioid peptide content in an animal model of spontaneous stereotypic behavior. *Behav Brain Res* **157**:363–8.

Redgrave P, Prescott TJ and Gurney K (1999): The basal ganglia: a vertebrate solution to the selection problem? *Neuroscience* **89**:1009–23.

Rees RG (1962): Fur-chewing. *National Chinchilla Breeders* **18**:20–2.

Reiner A, Medina L and Veenman CL (1998): Structural and functional evolution of the basal ganglia in vertebrates. *Brain Res Rev* **28**:235–85.

Reinhardt V (1984): Behavioural sex differences in muskox calves. *Biological Papers of the University of Alaska* **4**:110–17.

Reinhardt V (2005): Hair pulling: a review. *Lab Anim* **39**:361–9.

Reinhardt V and Flood PF (1983): Behavioral-assessment in Muskox calves. *Behaviour* **87**:1–21.

Reinhardt V, Reinhardt A and Houser D (1986): Hair pulling and eating in captive rhesus-monkey troops. *Folia Primatol* **47**:158–64.

Remijnse PL, Nielen MMA, van Balkom AJLM *et al.* (2006): Reduced orbitofrontal-striatal activity on a reversal learning task in obsessive–compulsive disorder. *Arch Gen Psychiatry* **63**:1225–36.

Robbins TW, Jones GH and Wilkinson LS (1996): Behavioural and neurochemical effects of early social deprivation in the rat. *J Psychopharmacol* **10**:39–47.

Rolls ET (1994): Neurophysiology and cognitive functions of the striatum. *Rev Neurol (Paris)* **150**:648–60.

Rosenberg DR and Keshavan MS (1998): Toward a neurodevelopmental model of obsessive–compulsive disorder. *Biol Psychiatry* **43**:623–40.

Sarna JR, Dyck RH and Whishaw IQ (2000): The Dalila effect: C57BL6 mice barber whiskers by plucking. *Behav Brain Res* **108**:39–45.

Savory CJ (1995): Feather pecking and cannibalism. *Worlds Poult Sci J* **51**:215–19.

Schrijver NCA and Würbel H (2001): Early social deprivation disrupts attentional, but not affective, shifts in rats. *Behav Neurosci* **115**:437–42.

Sheppard DM, Bradshaw JL, Purcell R and Pantelis C (1999): Tourette's and comorbid syndromes: obsessive compulsive and attention deficit hyperactivity disorder. A common etiology? *Clin Psychol Rev* **19**:531–52.

Silver LM (1995): *Mouse Genetics: Concepts and Applications*. Oxford: Oxford University Press.

Spruijt BM, Vanhooff JARAM and Gispen WH (1992): Ethology and neurobiology of grooming behavior. *Physiol Rev* **72**:825–52.

Stanley MA and Cohen LJ (1999): Trichotillomania and obsessive–compulsive disorder. In: Stein DJ, Christenson GA and Hollander E, eds., *Trichotillomania*. Washington, DC: American Psychiatric Press, Inc., pp. 225–61.

Stanley MA, Hannay HJ and Breckenridge JK (1997): The neuropsychology of trichotillomania. *J Anxiety Disord* **11**:473–88.

Stein DJ, O'Sullivan RL and Hollander E (1999): The neurobiology of trichotillomania. In: Stein DJ, Christenson GA and Hollander E, eds., *Trichotillomania*. Washington, DC: American Psychiatric Press, Inc., pp. 43–61.

Stein DJ, Garner JP, Keuthen NJ *et al.* (2007): Trichotillomania, stereotypic movement disorder, and related disorders. *Curr Psychiatry Rep* **9**:301–2.

Steiner H and Gerfen CR (1998): Role of dynorphin and enkephalin in the regulation of striatal output pathways and behavior. *Exp Brain Res* **123**:60–76.

Streichenwein SM and Thornby JI (1995): A long-term, double-blind, placebo-controlled crossover trial of the efficacy of fluoxetine for trichotillomania. *Am J Psychiatry* **152**:1192–6.

Strozik E and Festing MFW (1981): Whisker trimming in mice. *Lab Anim* **15**:309–12.

Swedo SE and Leonard HL (1992): Trichotillomania: an obsessive–compulsive spectrum disorder. *Psychiatr Clin North Am* **15**:777–90.

The Jackson Laboratory (2008): JAX mice database (http://jaxmice.jax.org/strain/000664.html).

The Mouse Phenome Project/The Jackson Laboratory (2008): MPD: Mouse phenome database (http://phenome.jax.org/pub-cgi/phenome/mpdcgi).

Thornburg LP, Stowe HD and Pick JR (1973): Pathogenesis of alopecia due to hair chewing in mice. *Lab Anim Sci* **23**:843–50.

Tinbergen N (1963): On aims and methods of ethology. *Z Tierpsychol* **20**:410–33.

Tinbergen N (1968): On war and peace in animals and man: an ethologists approach to biology of aggression. *Science* **160**:1411–18.

Tisljar M, Janic D, Grabarevic Z *et al.* (2002): Stress-induced Cushing's syndrome in fur-chewing chinchillas. *Acta Vet Hung* **50**:133–42.

Tully LA, Jenne M and Coleman K (2002): Paint roller and grooming boards as treatment for over-grooming rhesus macaques. *Contemp Top Lab Anim Sci* **41**:75.

Turner M (1997): Towards an executive dysfunction account of repetitive behavior in autism. In: Russell J, ed., *Autism as an Executive Disorder*. New York: Oxford University Press.

Van de Weerd HA, Van den Broek FAR and Beynen AC (1992): Removal of vibrissae in male mice does not influence social dominance. *Behav Processes* **27**:205–8.

Van den Broek FAR, Omtzigt CM and Beynen AC (1993): Whisker trimming behavior in A2g mice is not prevented by offering means of withdrawal from it. *Lab Anim* **27**:270–2.

van Minnen A, Hoogduin KAL, Keijsers GPJ, Hellenbrand I and Hendriks GJ (2003): Treatment of trichotillomania with behavioral therapy or fluoxetine: a randomized, waiting-list controlled study. *Arch Gen Psychiatry* **60**:517–22.

Vanjonack WJ and Johnson HD (1973): Relationship of thyroid and adrenal function to fur-chewing in chinchilla. *Comp Biochem Physiol* **45**:115–20.

Wahlsten D, Crabbe JC and Dudek BC (2001): Behavioural testing of standard inbred and 5HT(IB) knockout mice: implications of absent corpus callosum. *Behav Brain Res* **125**:23–32.

Walsh S (1982): A vocabulary of abnormal-behavior for restrictive-reared chimpanzees (Pan troglodytes). *Am J Phys Anthropol* **57**:239.

Welch JM, Lu J, Rodriguiz RM *et al.* (2007): Cortico-striatal synaptic defects and OCD-like behaviours in Sapap3-mutant mice. *Nature* **448**:894–900.

Wolfer DP and Lipp HP (2000): Dissecting the behaviour of transgenic mice: is it the mutation, the genetic background, or the environment? *Exp Physiol* **85**:627–34.

Wolfer DP, Crusio WE and Lipp HP (2002): Knockout mice: simple solutions to the problems of genetic background and flanking genes. *Trends Neurosci* **25**:336–40.

Woods DW, Flessner CA, Franklin ME *et al.* (2006): The Trichotillomania Impact Project (TIP): exploring phenomenology, functional impairment, and treatment utilization. *J Clin Psychiatry* **67**:1877–88.

Würbel H (2001): Ideal homes? Housing effects on rodent brain and behaviour. *Trends Neurosci* **24**:207–11.

Würbel H and Stauffacher M (1997): Age and weight at weaning affect corticosterone level and development of stereotypies in ICR-mice. *Anim Behav* **53**:891–900.

Würbel H and Stauffacher M (1998): Physical condition at weaning affects exploratory behaviour and stereotypy development in laboratory mice. *Behav Processes* **43**: 61–9.

Würbel H, Stauffacher M and von Holst D (1996): Stereotypies in laboratory mice: quantitative and qualitative description of the ontogeny of 'wire-gnawing' and 'jumping' in Zur:ICR and Zur:ICR nu. *Ethology* **102**:371–85.

11

Should there be a category: "grooming disorders?"

LARA J. HOPPE, JONATHAN IPSER, CHRISTINE LOCHNER, KEVIN G. F. THOMAS, AND DAN J. STEIN

Summary

Grooming disorders, such as trichotillomania (TTM), nail biting, and skin picking, are receiving increasing attention from researchers and clinicians. This is in part due to their possible link to obsessive–compulsive disorder (OCD). It is imperative that these disorders are categorized correctly in order to facilitate research and aid diagnosis and treatment. However, within current psychiatric classification systems TTM is currently conceptualized as an impulse control disorder and nail biting, lip biting, and skin picking are not yet included in the official nomenclature. It is therefore unclear whether these grooming disorders should form a category on their own (i.e., "grooming disorders" or "pathological grooming behaviors"), or whether they should be classified as obsessive–compulsive spectrum disorders (OCSD), impulse control disorders, or as body-focused repetitive behaviors. This chapter will discuss these diagnostic and taxonomic issues particularly as they pertain to clinical practice, fostering further discussion on the psychopathology of aberrant grooming behaviors.

Introduction

The positioning of TTM and other conditions characterized by self-directed repetitive behaviors (e.g., nail biting and pathological skin picking [PSP]) within existing psychiatric classification systems has recently been debated (e.g., Stein *et al.* 2007). Trichotillomania is classified in the most recent edition of the *Diagnostic and Statistical Manual of Mental Disorders* (DSM-IV-TR; APA, 2000) as an impulse

Neurobiology of Grooming Behavior, eds. Allan V. Kalueff, Justin L. LaPorte, and Carisa L. Bergner. Published by Cambridge University Press.

control disorder. However, some researchers suggest that TTM is an OCSD (Swedo and Leonard 1992). More recently some researchers have argued that these disorders (TTM, PSP, lip biting, and nail biting) should be classified as body-focused repetitive behaviors (Stein *et al.* 2007) or that grooming disorders (Bienvenu *et al.* 2000; Fineberg *et al.* 2007) may be a more appropriate term. In this chapter, we will attempt to identify the source of the dispute between researchers proposing divergent taxonomies, and determine whether the classification of disorders characterized by self-directed repetitive behaviors as a separate psychiatric entity not only has clinical utility, but is also warranted by the existing evidence database. Although this chapter will cover the whole range of putative "grooming disorders," there will be an emphasis on TTM, as the majority of research conducted thus far has focused on this particular disorder.

There are a number of conceptual challenges involved in recategorizing disorders within the DSM. It is unclear, for example, how much data (e.g., on phenomenology, genetics, imaging, treatment response) is required before decisions to restructure are made (Phillips *et al.* 2003a, b). There are three main approaches to psychiatric categorization: the classical approach, the critical approach, and an integrated approach (Stein 2008: see Table 11.1). This chapter will use the integrated approach to explore the relationship of TTM, nail biting, lip biting, and PSP to each other and to the other psychiatric diagnoses. By combining explanations at the level of clinical practice and psychobiology, an integrated approach helps make categorization practically useful (Stein, 2008).

Phenomenology, psychobiology and treatment of "grooming disorders"

Trichotillomania

Trichotillomania (from the Greek *thrix* – hair, *tillein* – pulling out, and *mania* – hysteria or madness) is a distressing and debilitating condition characterized by significant hair loss due to repeated hair pulling from different bodily sites (Diefenbach 2005). Common sites of hair pulling include the scalp, eyebrows, and eyelashes (Woods *et al.* 2006a). Researchers have suggested that there may be two subtypes of TTM, namely focused pulling and automatic pulling subtypes (Du Toit *et al.* 2001; Flessner *et al.* 2008b). Focused pullers engage in pulling as a premeditated behaviour in order to stop unpleasant urges, bodily feelings, or thoughts, whereas automatic pullers are not always aware of their behavior and usually engage in hair pulling while in a relaxed state (Flessner *et al.* 2008b). Empirical evidence demonstrates that TTM is associated with psychological and social dysfunction, and possible serious medical sequelae (Christenson *et al.* 1991a; O'Sullivan *et al.* 1997a, 1996).

Table 11.1 *Classical, critical, and integrated approaches (from Stein 2008, p. 2)*

	Classical	Critical	Integrated
Philosophical influences	Plato, logical positivists, early Wittgenstein	Vico, Herder, much continental philosophy	Aristotle, Bhaskar, Lakoff
View of categories	Categories can be defined using necessary and sufficient criteria	Categories reflect human practices	Categories reflect both human practices and real/underlying structures/mechanisms
View of psychiatric disorders	Disorders can be defined using necessary and sufficient criteria	Psychiatric classifications reflect human practices	Nosologies reflect both human practices and real/underlying structures/mechanisms
Is disorder X in category or spectrum Y?	Research the relevant necessary and sufficient components	Answers reflect practices inside human medicine (clinical utility) and outside	Debate both the relevant underlying psychobiology and the clinical utility

Trichotillomania was included in the DSM-III-R (APA 1987) as an impulse control disorder, and by the International Statistical Classification of Diseases and Related Health Problems 10th Revision (ICD-10) (WHO 1992) as a habit and impulse control disorder. Although relatively consistent with the diagnostic criteria in the DSM, the ICD-10 does not mention the associated distress or impairment in functioning that are highlighted by DSM-IV and DSM-IV-TR (WHO 1992; APA 1994; APA 2000). (See Tables 11.2 and 11.3 for the diagnostic criteria.)

Trichotillomania may be more prevalent than originally assumed, with lifetime prevalence rates close to 0.6% in a survey of 2579 American college students who were asked questions based on the DSM-III-R criteria (Christenson *et al.* 1991c). This increased to 1.5% of men and 3.4% of women when the criteria of clinically significant hair pulling was considered on its own.[1] In a smaller study of two American college samples, 0.9% of the 711 students who participated described

[1] The large discrepancy between these figures has fueled debate surrounding whether the DSM-IV-TR diagnostic criteria are too restrictive, especially with regards to the inclusion of criteria B and C (see Table 11.2: Woods *et al.* 2006a). Indeed, these criteria might more appropriately be considered proxies for symptom severity, as there is evidence that they are more frequently met as hair pulling becomes more severe (Woods *et al.* 2006b).

Table 11.2 *DSM-IV-TR criteria (APA 2000)*[a]

	Trichotillomania
A	Ongoing pulling out of one's own hair that causes visible hair loss
B	Rising tension before pulling out one's own hair or when attempting to stop pulling out one's own hair
C	Happiness or pleasure when pulling out the hair
D	This hair pulling is not better explained by another mental disorder, nor does it result from a medical condition
E	This condition causes distress and significant functional impairment (e.g., in the social domain)

[a] The criteria in Tables 11.2 and 11.3 have been slightly modified for improved clarity.

Table 11.3 *ICD-10 diagnostic criteria (WHO 1992)*

	Trichotillomania
	The disorder causes visible hair loss due to the person being unable to restrain themselves from pulling out their own hair. Increasing anxiety is felt before engaging in hair pulling and pleasure is felt after hair pulling. This condition is not better described by a dermatological problem, nor is it occurring during a delusion or hallucination
Excludes	Stereotyped movement disorder with hair plucking

baldness and significant distress as a consequence of hair pulling (Rothbaum *et al.* 1993).

Christenson *et al.* (1991a) studied the phenomenological features of 60 hair pullers. They found that the majority of the sample was female (15:1), that hair pulling started at approximately age 13 and that hair was mainly pulled from the scalp (67% of hair pullers), but also from other areas on the face and body. Ninety-five percent of subjects said that they felt rising tension before hair pulling, while 88% said that they felt pleasure or relief after hair pulling. Most subjects (98%) tried to stop themselves from pulling out their hair, using many different methods (e.g., wearing gloves). Nail biting, knuckle cracking, nose picking, thumb sucking, tongue chewing, cheek chewing, lip biting, head banging, body rocking, scab picking, and picking at acne, were other behaviors that subjects reported engaging in at some point (85% of subjects). Moreover, people with TTM had other comorbid psychiatric disorders (e.g., 10% had OCD).

Little is known about the mechanisms implicated in the pathogenesis of TTM. Global cerebral glucose hypermetabolism and increased regional blood flow in the right parietal region and bilaterally in the cerebellum have been noted in women with TTM (Swedo *et al.* 1991). There is also evidence of bilateral reductions in the volume of the cerebellum in females with TTM (Keuthen *et al.* 2007b), part of the brain known to be involved in controlling complex, coordinated motor sequences. Dysfunctional sensorimotor pathways are also implicated in the discovery that more severe symptoms of TTM are associated with reduced volumes of the left primary sensorimotor regions (Keuthen *et al.* 2007b), as well as findings of reduced volume and activation of the putamen (Keuthen *et al.* 2007b; O'Sullivan *et al.* 1997b; Stein *et al.* 2002).

Interventions for TTM may normalize activation or volumetric differences in sensorimotor regions of the brain implicated in its mediation. An early treatment study of (n = 10) TTM patients provided evidence of a reduction in blood flow in the left putamen, as well as in frontal and temporal regions of the brain, during a symptom provocation task following 12 weeks of treatment with the selective serotonin reuptake inhibitor (SSRI) citalopram (Stein *et al.* 2002). These findings are consistent with the idea that TTM may be mediated by cortico–striatal circuits (Stein *et al.* 2002). Findings from studies of cognitive function in TTM have consistently reported across a number of cognitive domains, including executive functioning (Bohne *et al.* 2005; Chamberlain *et al.* 2006; Keuthen *et al.* 1996), nonverbal memory (Keuthen *et al.* 1996; Rettew *et al.* 1991), and divided attention (Stanley *et al.* 1997). The failure of a recent study to detect a deficit in implicit learning in a TTM sample also found no evidence of cortico–striatal dysfunction (Rauch *et al.* 2007).

A number of different interventions have been used in treating TTM (Swedo *et al.* 1989). The greatest number of controlled treatment trials for TTM to date have investigated the efficacy of cognitive behavioral therapy or SSRIs. Open-label SSRI studies have sometimes, but not always, suggested that these agents may be useful (Gadde *et al.* 2007; Iancu *et al.* 1996; Koran *et al.* 1992; Stanley *et al.* 1991, 1997; Van Minnen *et al.* 2003; Winchel *et al.* 1992). Similarly, small open-label trials of certain tricyclic antidepressants, antipsychotic, and anticonvulsant agents, as well as the serotonin–noradrenaline reuptake inhibitor, venlafaxine, have indicated some promise in treating TTM (Black and Blum 1992; Epperson *et al.* 1999; Lochner *et al.* 2006; Ninan *et al.* 1998; Pollard *et al.* 1991; Stewart and Nejtek 2003; Van Ameringen *et al.* 1999). However, few placebo-controlled randomized trials of medication for TTM have been carried out (Christenson *et al.* 1991b; Dougherty *et al.* 2006; Epperson *et al.* 1996; Ninan *et al.* 2000; Streichenwein and Thornby 1995; Swedo *et al.* 1989). Two small cross-over randomized controlled trials (RCTs) of the SSRI, fluoxetine (n = 21; n = 23), did not find this agent to be effective in the treatment of TTM (Christenson *et al.* 1991b; Streichenwein and Thornby 1995). An RCT comparing placebo, sertraline (SSRI), and habit reversal therapy (n = 42)

Table 11.4 *Diagnostic criteria for pathological skin picking proposed by Odlaug and Grant (2008a, p. 62). Reprinted with the authors' permission*

A	Recurrent skin picking resulting in noticeable skin damage
B	Preoccupation with impulses or urges to pick skin, which is experienced as intrusive
C	Feelings of tension, anxiety, or agitation immediately prior to picking
D	Feelings of pleasure, relief, or satisfaction while picking
E	The picking is not accounted for by another medical or mental disorder (e.g., cocaine or amphetamine use disorders, scabies)
F	The individual suffers significant distress, or social or occupational impairment

demonstrated that combining these therapies increased their efficacy relative to their use in isolation (Dougherty *et al.* 2006). In terms of tricyclic antidepressants, a cross-over RCT of desipramine and clomipramine (n = 14) found clomipramine to be effective in the short-term treatment of TTM (Swedo *et al.* 1989). A placebo-controlled randomized comparison of clomipramine and cognitive behavioral therapy (n = 23) found that although clomipramine was significantly more effective in the treatment of TTM than placebo, cognitive behavioral therapy (habit reversal therapy) was more effective than the antidepressant (Ninan *et al.* 2000). One cross-over RCT examined intranasal oxytocin and demonstrated that it was ineffective in treating TTM (Epperson *et al.* 1996).

Three controlled trials have assessed the effectiveness of habit reversal therapy (Ninan *et al.* 2000; Van Minnen *et al.* 2003; Woods *et al.* 2006b). A recent meta-analysis, which included these studies, compared the effectiveness of habit reversal therapy (a form of cognitive behavioral therapy), SSRIs, and clomipramine (tricyclic antidepressant: Bloch *et al.* 2007). There was no evidence from this meta-analysis that SSRIs are effective in treating TTM, although the relevant trials have limited power (Chamberlain *et al.* 2008).

Pathological skin picking

Pathological skin picking (also known as psychogenic excoriation, neurotic excoriation, compulsive skin picking, and dermatotillomania) is not classified as a psychiatric disorder in the DSM-IV (Arnold *et al.* 1998).

However, diagnostic criteria for PSP have been independently proposed by two different research teams (Arnold *et al.* 2001; Odlaug and Grant 2008a) (see Table 11.4 for one such criteria set). The authors of these criteria agree that PSP is characterized by pathological skin picking that causes distress and results in psychosocial dysfunction. Pathological skin picking has also been divided into subtypes, which are reminiscent of automatic vs. focused hair pulling (Arnold *et al.* 2001) (see Table 11.5).

Table 11.5 *Subtypes of pathological skin picking proposed by Odlaug and Grant (2008a, p. 62). Reprinted with the authors' permission*

Subtypes
Compulsive type
Skin excoriation is performed to avoid increased anxiety or to prevent a dreaded event or situation and/or is elicited by an obsession (e.g., obsession about contamination of the skin)
It is performed in full awareness
It is associated with some resistance to performing the behavior
There is some insight into its senselessness or harmfulness
Impulsive type
Skin excoriation is associated with arousal, pleasure, or reduction of tension
It is performed at times with minimal awareness (e.g., automatically)
It is associated with little resistance to performing the behavior
There is little insight into its senselessness or harmfulness
Mixed type
Skin excoriation has both compulsive and impulsive features

Using the Habit Questionnaire (Teng *et al.* 2002 based on a survey used by Woods *et al.* 1996), which assesses, among other things, how frequently a person picks their skin, the duration of skin-picking behavior, functional impairment suffered, and if the person has received medical help for this behavior, one study found that PSP had a prevalence of 2.7% in a sample of (n = 439) North American undergraduate psychology students (Teng *et al.* 2002). A slightly higher prevalence rate of 4.6% (n = 133) was observed in a study of German students, using the Skin Picking Inventory (Keuthen *et al.* 1996) to assess PSP (Bohne *et al.* 2002). Studies of people who suffer from PSP symptoms reveal that they cause much anguish and functional impairment, and that PSP is frequently comorbid with other psychiatric conditions (e.g., Arnold *et al.* 1998; Flessner and Woods 2006; Odlaug and Grant 2008a). Multiple picking sites appear to be the norm, with the face the most common site (Arnold *et al.* 1998; Bohne *et al.* 2002).

In a study of 34 adults with PSP all participants were diagnosed with a comorbid psychiatric condition (Arnold *et al.* 1998). The most common comorbid conditions in this predominantly female sample were bipolar disorder II, major depression disorder, and generalized anxiety disorder. Calikusu *et al.* (2002) found a high prevalence of OCD (45.2%) and major depressive disorder (58.1%) in their sample (n = 31). Recently, Odlaug and Grant (2008a) reported that 38.3% of their participants with PSP presented with current Axis I comorbidity.

The use of patterns of psychiatric comorbidity in informing the classification of PSP is complicated by differences in comorbid diagnoses across studies. For instance, Odlaug and Grant (2008a) reported a greater proportion of comorbid disorders lying on the putative obsessive–compulsive spectrum (e.g., TTM, nail biting, and OCD), a finding duplicated to some extent by Wilhelm *et al.* (1999) in their sample of 31 patients (e.g., OCD: 52% of the sample) and to a lesser extent by Lochner *et al.* (2002) in their sample of 21 patients (e.g., OCD: 19% of the sample). Additionally, it was found that the number of PSP symptoms was significantly correlated with the number of OCD symptoms in a German student population (Bohne *et al.* 2002). Wilhelm *et al.* (1999) and Lochner *et al.* (2002) also reported high rates of comorbid Axis II disorders, such as obsessive–compulsive and borderline personality disorders in their sample of patients with PSP (n = 31 and 21, respectively).

A comparative study between PSP and TTM patients revealed that TTM and PSP are very similar in terms of demographic characteristics, psychiatric comorbidity, and personality dimensions (Lochner *et al.* 2002). They may also have similar subtypes (focused versus automatic). However, dissociative symptoms are more frequent in TTM than in PSP. Recent research has also found the phenomenology of PSP, TTM, and comorbid TTM and PSP to be very similar (Odlaug and Grant 2008b). However, these authors found differences between the diagnostic groups in terms of the amount of time spent picking/pulling, the triggers for this behavior, the number of comorbid depressive disorders, and number of family members with lifetime diagnoses of PSP. Additionally, PSP was associated with pathological worry in another study, whereas hair pulling was linked to increased levels of anxiety, obsessive–compulsive symptoms, and stress reactivity (follow-up study of college students that frequently engaged in skin picking or hair pulling: sample: n = 72; comparison group: n = 221: Hajcak *et al.* 2006).

Grant and Christenson (2007) examined the presentation of TTM and PSP across genders. They found that there were many similarities between men and women with these grooming disorders. However, in men these disorders started later in life, were associated with more psychosocial dysfunction, and occurred alongside more comorbid anxiety disorders.

The average age of onset of PSP symptoms differs across studies, with Arnold *et al.* (1998) reporting onset age of 38 years compared to a much younger age of onset reported by other more recent studies (Odlaug and Grant 2008a: 12.3 years; Wilhelm *et al.* 1999: 15 years). The finding in a recent study that adults (n = 19) with PSP symptoms that first appeared in childhood possessed similar clinical features to the participants (n = 21) who only developed PSP symptoms in adulthood (Odlaug and Grant 2007a) suggests that age of onset of symptoms may not be that clinically relevant. However, in this study people with childhood-onset PSP symptoms were significantly less likely to have ever received medication to treat these symptoms.

Little is known about the mechanisms of the brain that are implicated in PSP. To the authors' best knowledge, no neuroimaging or neuropsychological studies have been conducted with people who have PSP. The involvement of serotonergic circuits is perhaps implicated by the partial success of selective SSRIs in treating PSP described below (e.g., Arnold *et al.* 1999; Bloch *et al.* 2001; Keuthen *et al.* 2007a; Simeon *et al.* 1997).

Only two RCTs of PSP have been conducted to date (Bloch *et al.* 2001; Simeon *et al.* 1997). Both of these assessed the efficacy of the SSRI fluoxetine, and in contrast to the findings for this agent in TTM, reported positive results. In a small relapse-prevention study (n = 15) of six weeks of open-label treatment with fluoxetine, the severity of skin-picking symptoms in four of the eight responders who were subsequently randomized to placebo treatment returned to their pretreatment levels (20 mg/d–60 mg/d: Bloch *et al.* 2001). The four treatment responders who continued with fluoxetine (six weeks) maintained their treatment gains for the six weeks that they were receiving treatment. The finding that fluoxetine is effective in treating PSP symptoms was replicated in a randomized controlled short-term trial of 20 mg/d–80 mg/d administered over ten weeks in a sample of 21 patients with PSP (Simeon *et al.* 1997). The effect of treatment was independent of the participants' other comorbid conditions.

The efficacy of other SSRIs in treating PSP has also been investigated using less rigorous designs. An open-label study (n = 14) of 12 weeks of treatment with the SSRI fluvoxamine 25 mg/d–300 mg/d observed significant reduction of skin-picking symptoms (14 participants with PSP) (Arnold *et al.* 1999). Moreover, sertraline (another SSRI) was found to be effective in improving dermatosis in 21 of 31 patients who presented to a psychodermatology clinic with acne excoriée, neurotic excoriations, and intractable pruritus ani (Kalivas *et al.* 1996). Finally the SSRI escitalopram has also been used with some success in treating PSP in an open-label study (n = 29: Keuthen *et al.* 2007a) and in a case study (10 mg/d: Pukadan *et al.* 2008). In contrast, an increase in skin-picking behavior was noted in two cases with OCD following SSRI treatment (Denys *et al.* 2003). In addition, two subjects dropped out of the study as their skin picking worsened following treatment with fluoxetine (Simeon *et al.* 1997).

A range of anecdotal reports and open-label studies suggest that other medications may also be useful for treating PSP. A case study found that riluzole (a glutamate-modulating agent) was effective in reducing skin-picking behavior in a patient with comorbid depression, OCD, and disordered eating behavior symptoms (Sasso *et al.* 2006). Moreover, two-thirds of 24 participants who were administered lamotrigine (a glutamatergic agent) in an open-label fashion responded to treatment, as assessed by a number of treatment response scales (Grant *et al.* 2007). Additionally, inositol (a simple glucose isomer) was successful in the treatment of two patients with PSP (one patient took 18 g/d in addition to citalopram 40 mg/d, the other patient took 18 g/d: Seedat *et al.* 2001). Finally, there is evidence from a case study for a possible role for aripiprazole (an antipsychotic) in the augmentation of venlaflaxine and other serotonergic medications (such as SSRIs) in the treatment of psychogenic excoriation (Carter and Shillcutt 2006). A case study also demonstrated that olanzapine (an atypical antipsychotic: 5 mg/d) augmentation of fluoxetine (40 mg/d) was effective in reducing PSP symptoms (Christenson 2004). Finally, another case study demonstrated that N-acetylcysteine (an antioxidant) may be an effective treatment for PSP (Odlaug and Grant, 2007b).

In terms of psychotherapy treatment, two uncontrolled studies of habit reversal therapy, one with four patients with PSP and the other with three, observed reductions in skin-picking behavior in these small samples (Deckersbach *et al.* 2002; Rosenbaum and Ayllon 1981). This finding was replicated by Twohig and Woods (2001) who demonstrated the effectiveness of habit reversal therapy (a three-session intervention) in two adults with a mild, chronic skin-picking problem using a nonconcurrent multiple baseline design. When compared with a wait-list control group (Teng *et al.* 2006), three sessions (one 1-hour treatment session followed by two separate 30-minute booster sessions) of habit reversal therapy appeared to improve skin-picking behavior in a sample of 19 adults (all female).

Acceptance and commitment therapy has shown some promise in the treatment of TTM (Twohig *et al.* 2006). However, treatment gains were not maintained at follow-up in a case study of five people with PSP (Twohig *et al.* 2006). The efficacy of acceptance and commitment therapy may be enhanced by combining it with habit reversal therapy in the form of acceptance-enhanced behavior therapy. Indeed, ten, 50-minute sessions (over the course of 12 weeks) of acceptance-enhanced therapy reduced the PSP symptoms of all participants ($n = 5$) (Flessner *et al.* 2008a). The same study found that the order in which acceptance and commitment therapy and habit reversal therapy are administered does not affect these treatment gains. A recent trial of an internet self-help intervention, based on cognitive behavioral therapy, reported significant reductions in symptoms among the 373 PSP patients who participated following treatment (Flessner *et al.* 2007).

Nail biting

Relatively little research has been done on onychophagia (nail biting). One small epidemiological study reported that 13.7% (60) of a sample of 439 undergraduate psychology students reported at least one body-focused repetitive behavior (i.e., nail biting, skin picking or scratching, and mouth chewing); the most frequently reported behavior was nail biting (6.4%: Teng *et al.* 2002).

Two studies have examined the treatment of nail biting. A cross-over double-blind study examined clomipramine and desipramine treatment in 25 adults (14 completers) with nail biting (Leonard *et al.* 1991). Inclusion criteria were: severe nail biting, which was defined as "fingernails bitten beyond the free edge" and "nail margin below the soft tissue border" (Malone and Massler 1952, p. 94). This study found clomipramine to be significantly more effective than desipramine in reducing nail-biting symptoms, as measured by the nail length measure and nail photographs (Leonard *et al.* 1991). Two participants stopped biting their nails and a >30% decrease in incidences and severity of nail biting was observed in five participants (on clomipramine). These findings must be viewed with caution as there was a high drop out rate in this study (11 of 25 participants dropped out).

An RCT of habit reversal therapy versus placebo in 30 adults with nail biting demonstrated that habit reversal therapy may be an effective treatment for pathological nail biting: the habit reversal therapy group reported a 22% increase in nail length after a period of three weeks, whereas those in the placebo group reported a 3% increase in nail length (Twohig *et al.* 2003). At follow-up (i.e., five months post treatment) the habit reversal therapy group still had a 19% increase in nail length in contrast to no increase in nail length in the placebo group. Inclusion criteria in this study was that participants had to report nail biting five or more times a day almost every day for a minimum of a month, and that nail biting caused them psychosocial impairment or physical injury.

Lip biting

There is a paucity of research on lip biting. A study of Croatian children (n = 1025) presenting for their annual dental check-up revealed that 33.4% (342) had oral habits, 16.96% (58) of these children with oral habits engaged in lip or cheek biting (Bosnjak *et al.* 2002). One case study described the effective treatment, using behavioral methods (self-recording, aversive response substitution, and relaxation), of a 12-year-old boy who presented with compulsive lip biting (Lyon 1983). Another case study further demonstrated the effectiveness of behavioral techniques (use of a star chart) in the treatment of lip biting in a 12-year-old boy (Ellis 1986). Moreover, dental appliances were successfully used to protect the lips and prevent the future lip biting of two patients who presented with lip biting (Turley and Henson 1983). Pharmacotherapy, specifically the use of escitalopram (SSRI: 10 mg/d), was also found to be effective in treating a 24-year-old man with chronic lip biting (Pukadan *et al.* 2008).

Similarities between TTM, PSP, lip biting and nail biting

Table 11.6 depicts the similarities between TTM, PSP, lip biting, and nail biting. There is some evidence, partially inconsistent, that habit reversal therapy may be effective in treating all four of these conditions. In addition, the phenomenology of these disorders are similar, in that many TTM and PSP patients report mounting tension prior to "grooming" (either skin picking or hair pulling) and relief or satisfaction following their "grooming" behavior. Further research will determine whether similar parts of the brain are implicated in these conditions. The evidence presented thus far would seem to provide a partial basis for the argument that these disorders should be placed in the same category. In the next section we will present alternative perspectives on the positioning of these disorders in the psychiatric taxonomy. The strengths of these different conceptualizations will be discussed in the final section of the chapter, with a view towards determining how these disorders might best be categorized.

If TTM is not a grooming disorder, what is it?

Impulse control disorder

As noted earlier in the chapter, TTM was first included in the DSM system in the third edition of the manual (DSM-III; APA 1980). The justification for its inclusion in a discrete DSM section titled "Impulse Control Disorders Not Elsewhere Classified" (along with intermittent explosive disorder, kleptomania, pyromania, and pathological gambling) appears largely to have been justified by historical descriptions of individuals who performed harmful behaviors in response to irresistible, uncontrollable, "morbid" impulses (e.g., Geller *et al.* 1986; Lewis and Yarndell 1951). What McElroy *et al.* (1992, p. 323) describe as "the similar core disturbance of impulsivity and compulsivity," then, is perhaps the reason why TTM and the other four impulse control disorders (ICDs)[2] mentioned above continue to be grouped together and housed in a separate section of the DSM-IV-TR (APA 2000). Currently, therefore, TTM and the other ICDs are described in the DSM nomenclature as the kinds of psychiatric disorders where "the essential feature... is the failure to resist an impulse, drive, or temptation to perform an act that is harmful to the person [self] or to others" (APA 2000, p. 663). A recent review found that response inhibition deficits are seen in many neuropsychiatric disorders that are linked to impulsivity; one of which is TTM (Chamberlain and Sahakian 2007). Furthermore, as noted earlier in the chapter, the manual proposes that other defining features of the ICDs are that the impulsive, harmful act is preceded by

[2] Here, the acronym refers only to the impulse control disorders as classified in the DSM-IV-TR.

Table 11.6 *Summary table of phenomenology, psychobiology, and treatment of grooming disorders*

Diagnostic category	Symptoms	Psychobiology	Treatment RCTs pharmacotherapy	Treatment RCTs psychotherapy
Trichotillomania	repeated hair pulling significant hair loss feeling of tension prior to hair pulling feeling of relief/ satisfaction after hair pulling functional impairment common sites: scalp, eyebrows, eyelashes mainly scalp comorbid: obsessive–compulsive disorder onset: 13 years old	cortico–striatal circuits right parietal region cerebellum sensorimotor regions putamen frontal and temporal regions	**Tricyclic antidepressants** *Clomipramine:* Ninan *et al.* 2000 (+) Swedo *et al.* 1989 (+) *Desipramine:* Swedo *et al.* 1989 (−) **SSRIs** *Fluoxetine:* Christenson *et al.* 1991b (−) Streichenwein and Thornby 1995 *Intranasal oxytocin:* Epperson *et al.* 1996 (−)	*Habit reversal therapy:* Ninan *et al.* 2000 (+) Van Minnen *et al.* 2003 (+) Woods *et al.* 2006b (+) *Habit reversal therapy and sertraline:* Dougherty *et al.* 2006 (+)
Pathological skin picking	feeling of tension prior to skin picking feeling of relief after skin picking distress impairment noticeable skin damage	possibly serotonergic circuits	**SSRIs** *Fluoxetine:* Bloch *et al.* 2001 (+) Simeon *et al.* 1997 (+) *Fluvoxamine:* Arnold *et al.* 1999 (+)	

Table 11.6 (*cont.*)

Diagnostic category	Symptoms	Psychobiology	Treatment RCTs pharmacotherapy	Treatment RCTs psychotherapy
	multiple sites picked common site: face comorbid: many Axis I disorders, including OCD; body dysmorphic disorder, and obsessive–compulsive personality disorder			
Nail biting	repeated nail biting	–	*Clomipramine:* Leonard *et al.* 1991 (+) *Desipramine:* Leonard *et al.* 1991 (–)	*Habit reversal therapy:* Twohig *et al.* 2003 (+)
Lip biting	acute lip biting	–	–	–

SSRIs: selective serotonin reuptake inhibitors
+/– indicates whether this drug demonstrated efficacy in the particular study

"an increasing sense of tension or arousal" (criteria B), accompanied by "pleasure, gratification, or relief at the time of committing the act" (criteria C), which is then followed by "regret, self-reproach, or guilt" (APA 2000, p. 663).

As noted above, there may be legitimate grounds for grouping TTM with the ICDs in a diagnostic taxonomy. Nevertheless, researchers have stated that "little is known about them [ICDs]... nor is it clear why these individual conditions are grouped together" (e.g., McElroy *et al.* 1992, p. 319). More than 15 years later, research is still limited with regards to the prevalence, onset, course, and comorbidity of most ICDs (although as noted in the opening sections of the chapter,

TTM has been studied in more detail than most). While TTM may have some overlap with ICDs/behavioral addictions such as pathological gambling (Lochner *et al.* 2005), there are little data to support that it belongs in this category. For instance, although deficiencies in the same dopaminergic reward systems appear in TTM and pathological gambling (Bergh *et al.* 1997; Blum *et al.* 2000; Seedat *et al.* 2000), these deficiencies appear in numerous other disorders, including hypersexual disorder and Tourette's disorder, but not in the other ICDs.

More recently, some researchers have argued that TTM may be better considered as part of an obsessive–compulsive spectrum of disorders (OCSD), featuring at its core OCD but also including a range of compulsive–impulsive (CI) disorders such as CI sexual behaviors and CI shopping (e.g., Lochner *et al.* 2005).

In summary, there is little firm empirical evidence that TTM is more similar to the ICDs with which it shares space in the DSM than it is to the OCSDs. In the next section, we review neurobiological, pharmacological, and other evidence clearly suggesting that TTM might be a part of the obsessive–compulsive spectrum.

Obsessive–compulsive spectrum disorder

It is only since the 1990s that TTM, a disorder characterized by repetitive hair pulling, has been systemically investigated. This increase in TTM research was encouraged by the evidence that patients with OCD – a condition also characterized by repetitive behaviors or rituals – responded selectively to treatment with SSRIs. The proposed construct of a spectrum of OCD-related disorders has in recent years received much support as a useful clinical and research heuristic (Stein *et al.* 1995). The next section will consider whether TTM should be considered part of the OCSD, by discussing commonalities and differences between TTM and OCD.

It has been proposed that TTM and certain other ICDs may be conceptualized as part of the spectrum of OCD-related disorders.[3] These proposals were based on their comparable clinical characteristics, familial transmission, and response to pharmacotherapy and psychotherapy (Hollander 1993; Hollander *et al.* 1992; McElroy *et al.* 1994). Similarities between OCD and TTM have also been widely recognized. For example, both TTM and OCD patients describe compulsive behaviors (Stein *et al.* 1995; Swedo 1993). The repetitive hair pulling is indeed reminiscent of the compulsions seen in OCD (Stein *et al.* 1995; Tukel *et al.* 2001). Most patients with OCD have insight into the excessiveness or irrationality of their symptoms.

[3] In the following discussion we will refer to TTM as describing a homogenous cluster of symptoms. However, it could be argued that the focused subtype is more similar to OCD than the automatic subtype (the same argument may also hold with respect to the compulsive versus impulsive subtypes of pathological skin picking, respectively).

Similarly, insight in patients with hair-pulling and related behaviors usually remains intact, and they usually recognize these symptoms as senseless and undesirable. Comorbidity patterns are also similar, with depressive and anxiety disorders being highly prevalent in both OCD and TTM (e.g., Lochner *et al.* 2005; Stein *et al.* 1995).

The exact causes of TTM and OCD are still largely unknown but there are some important overlapping features in their etiology. For example, there is evidence to support the role of a genetic and biological basis for both OCD and TTM. In one study, altering the *Hoxb8* gene of mice led to pathological grooming, resulting in hair loss and wounds (Greer and Capecchi 2002). This supports the conceptualization of TTM as an OCSD, as the *HoxB8* gene is expressed in elements of a so-called "OCD circuit" (e.g., orbitofrontal cortex: Greer and Capecchi 2002). Additionally, a recent study of mice without the *Sapap-3* gene demonstrated that this gene has an important function with regards to cortico–striatal synapses (Welch *et al.* 2007). These mice had less cortico–striatal synaptic transmission and displayed pathological grooming behaviors and anxiety that were successfully treated with fluoxetine (SSRI). Research into the stereotypies ("abnormal, repetitive, unvarying, and apparently functionless behaviours": Garner and Mason 2002) of rodents and parrots implicates basal ganglia dysfunction and associated striatal disinhibition of response selection (Garner and Mason 2002; Garner *et al.* 2003). Further investigation of the link between animal stereotypies and repetitive human behaviors may provide additional insight into the relationship between OCD and TTM.

Obsessive–compulsive symptoms or hair pulling may also present in response to significant environmental risk factors or life stressors (e.g., streptococcal infections and childhood interpersonal trauma: De Oliveira 2007; Lochner *et al.* 2001). These factors may interact with the underlying genetic susceptibility, and may subsequently lead to the development of conditions such as OCD or TTM. In terms of treatment, SRIs are effective in OCD (e.g., Jenike 1992; Leonard *et al.* 1989; Pigott *et al.* 1990; Zohar and Insel 1987). Evidence suggesting that these agents may be effective in treating TTM has already been reviewed in this chapter. In addition to relatively comparable response to SRIs, there currently also is evidence for the usefulness of behavior therapy in both OCD (Baer 1993) and TTM (Friman *et al.* 1987). In apparent contrast to these similarities, there also are important phenomenological and psychobiological differences that separate TTM, OCD, and the other OCSDs. For example, hair pulling in TTM is rarely in response to obsessions, and TTM is not typically accompanied by the characteristics of classical compulsions (i.e., changing over time in terms of focus and severity) of OCD. Trichotillomania is also much more prevalent in females; most surveys have suggested a mean female:male ratio approaching 10:1 in TTM (Christenson *et al.* 1994) and of 1.5:1.0

in OCD (Castle *et al.* 1995; Rasmussen and Tsuang 1984; Weissman *et al.* 1994). The mean age of onset also differs between the two conditions: TTM typically presents in early adolescence, with age of onset of hair pulling later in males than in females (Christenson *et al.* 1994; Du Toit *et al.* 2001). Conversely, OCD has been shown to present at any age, from childhood through to early adulthood (Rettew *et al.* 1992), but with males reporting an earlier onset compared to females (Bogetto *et al.* 1999). Although comorbid depressive and anxiety disorders are highly prevalent in both OCD and TTM, these conditions appear to be significantly more prevalent in OCD (Himle *et al.* 1995; Stanley *et al.* 1992; Tukel *et al.* 2001).

Indeed, a recent phenomenological comparison study between TTM and OCD demonstrated that comorbidity in OCD is much greater across a range of different diagnostic categories, including mood (major depressive disorder [MDD], dysthymia), anxiety (panic disorder), OCD-related (hypochondriasis), and personality disorders (obsessive–compulsive personality disorder [OCPD]) (Lochner *et al.* 2005). Although selective response to treatment with SRIs have been suggested in some studies for both TTM and OCD, there is good evidence that response to SRIs is sustained in OCD, whereas anecdotal evidence suggests that a positive response in patients with TTM is limited to fewer of the SRIs, and that the effectiveness of these agents may wane with time (Swedo 1993). In addition, whereas the usefulness of behavior therapy in both OCD (Baer 1993) and TTM (Friman *et al.* 1987) has been demonstrated, the focus and implementation of the treatment differs significantly in the two disorders (i.e., exposure and response prevention in OCD versus habit reversal in TTM). Etiological research, discussed above, is limited (compared to genetics studies depicting the close relationship between other OCSDs such as Tourette's syndrome and OCD: Lochner *et al.* 2005), and there are questions regarding whether the behaviors observed in mice correspond to grooming behaviors in humans. Yet, despite the differences described above between TTM and OCD, this research does suggest that there may still be some basis for conceptualizing TTM as an OCSD.

In conclusion, despite differences between TTM and OCD, there is some phenomenological and psychobiological overlap between these two conditions. Thus, for diagnostic clarity, conceptualization, and management purposes it may be useful to view TTM as related to OCD, and to include this condition in the category of OCSDs (Lochner and Stein 2006).

What is the right name?

Trichotillomania and OCD share some common phenomenological, genetic, and psychobiological features. However, it must be noted that research in this area is limited and inconsistent. Perhaps more importantly, TTM, PSP, lip

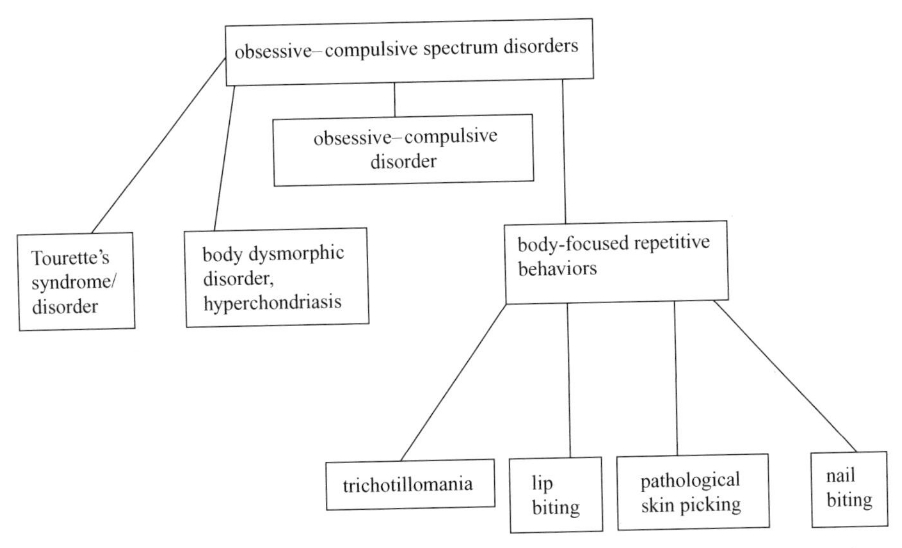

Figure 11.1 Obsessive–compulsive spectrum disorders (adapted from Lochner *et al.* 2006). A putative depiction of how TTM, PSP, lip biting, and nail biting could be classified.

biting, and nail biting have a comparable phenomenology and follow similar treatment approaches (see Table 11.6). Thus, it may be useful to classify TTM, PSP, lip biting, and nail biting together within the category of OCSDs. However, what to name these disorders within this category remains an issue.

Some researchers have proposed that it is useful to categorize these disorders as body-focused repetitive behaviors (Bohne *et al.* 2002; Teng *et al.* 2002; Stein *et al.* 2007). "The term 'body-focused, repetitive behavioural disorders' is theoretically neutral" (Stein *et al.* 2007, p. 2), which leaves scope for research to be conducted into these disorders. The label of "grooming disorders," may be restrictive in its emphasis on a neuroethological perspective of grooming (Stein *et al.* 1995). On the other hand it fits a growing emphasis on psychiatry as translational neuroscience (Cohen and Insel 2008). A diagram illustrating how the body-focused repetitive behaviors may fall within the OCSDs has been provided (Figure 11.1). Placing all these disorders within one category serves to emphasize that repetitive, nonfunctional motoric behaviors are neither limited to subjects with mental retardation nor restricted to hair pulling (Stein *et al.* 2007, p. 2).

Acknowledgments

The authors are supported by the MRC of South Africa.

References

American Psychiatric Association (1980): *Diagnostic and Statistical Manual of Mental Disorders*, 3rd edn. Washington, DC: APA.

American Psychiatric Association (1987): *Diagnostic and Statistical Manual of Mental Disorders*, Text Revised, 3rd edn. Washington, DC: APA.

American Psychiatric Association (1994): *Diagnostic and Statistical Manual of Mental Disorders*, 4th edn. Washington, DC: APA.

American Psychiatric Association (2000): *Diagnostic and Statistical Manual of Mental Disorders*, Text Revised, 4th edn. Washington, DC: APA.

Arnold LM, McElroy SL, Mutasim DF *et al.* (1998): Characteristics of 34 adults with psychogenic excoriation. *J Clin Psychiatry* **59**:509–14.

Arnold LM, Mutasim DF and Dwight MM (1999): An open clinical trial of fluvoxamine treatment of psychogenic excoriation. *J Clin Psychopharm* **19**:15–18.

Arnold LM, Auchenbach MB and McElroy SL (2001): Psychogenic excoriation: clinical features, proposed diagnostic criteria, epidemiology and approaches to treatment. *CNS Drugs* **15**:351–9.

Baer L (1993): Behavior therapy for obsessive–compulsive disorder in the office-based practice. *J Clin Psychiatry* **54**(6 Suppl-10).

Benarroche CL (1990): Trichotillomania symptoms and fluoxetine response. New Research Program and Abstracts, 143rd meeting of the APA, NY (NR327), 173.

Bergh C, Eklund T, Sodersten P and Nordin C (1997): Altered dopamine functioning in pathological gambling. *Psychol Med* **27**:473–5.

Bienvenu OJ, Samuels JF, Riddle MA *et al.* (2000): The relationship of obsessive–compulsive disorder to possible spectrum disorders: results from a family study. *Biol Psychiatry* **48**:287–93.

Black DW and Blum N (1992): Trichotillomania treated with clomipramine and a topical steroid. *Am J Psychiatry* **149**:842–3.

Bloch MR, Elliott M, Thompson H and Koran LM (2001): Fluoxetine in pathologic skin-picking: open-label and double-blind results. *Psychosomatics* **42**:314–19.

Bloch MH, Landeros-Weisenberger A, Dombrowski P *et al.* (2007): Systematic review: pharmacological and behavioural treatment for trichotillomania. *Biol Psychiatry* **62**:839–46.

Blum K, Braverman ER, Holder JM *et al.* (2000): Reward deficiency syndrome: a biogenetic model for the diagnosis and treatment of impulsive, addictive, and compulsive behaviors. *J Psychoact Drugs* **32** Suppl:i–112.

Bogetto F, Venturello S, Albert U, Maina G and Ravizza L (1999): Gender-related clinical differences in obsessive–compulsive disorder. *Eur Psychiatry* **14**:434–41.

Bohne A, Wilhelm S, Keuthen NJ, Baer L and Jenike MA (2002): Skin picking in German students: prevalence, phenomenology and associated characteristics. *Behav Modif* **26**:320–39.

Bohne A, Keuthen NJ, Tuschen-Caffier, B and Wilhelm S (2005): Cognitive inhibition in trichotillomania and obsessive–compulsive disorder. *Behav Res Ther* **43**: 923–42.

Bosnjak A, Vucicevic-Boras V and Miletic I (2002): Incidence of oral habits in children with mixed dentition. *J Oral Rehab* **29**:902–5.

Calikusu C, Yucel B, Polat A and Baykal C (2002): The relationship between psychogenic excoriation and psychiatric disorders: a comparative study. *Turk Psikiyatri Derg* **13**:282–9.

Carter WG and Shillcutt SD (2006): Aripiprazole augmentation of venlaflaxine in the treatment of psychogenic excoriation. *J Clin Psychiatry* **67**:1311.

Castle DJ, Deale A and Marks IM (1995): Gender differences in obsessive compulsive disorder. *Aust N Z J Psychiatry* **29**:114–17.

Chamberlain S and Sahakian BJ (2007): The neuropsychiatry of impulsivity. *Curr Opin Psychiatry* **20**:255–61.

Chamberlain SR, Fineberg NA, Blackwell AD, Robbins TW and Sahakian BJ (2006): Motor inhibition and cognitive flexibility in obsessive–compulsive disorder and trichotillomania. *Am J Psychiatry* **163**:1282–4.

Chamberlain S, Ipser J, Stein D and Fineberg N (2008): Regarding "Systematic review: pharmacological and behavioral treatment for trichotillomania." *Biol Psychiatry* **63**:e33.

Christenson GA, Mackenzie TB and Mitchell JE (1991a): Characteristics of 60 chronic hair pullers. *Am J Psychiatry* **148**:365–70.

Christenson GA, Mackenzie TB, Mitchell JE and Callies AL (1991b): A placebo-controlled, double-blind crossover study of fluoxetine in trichotillomania. *Am J Psychiatry* **148**:1566–71.

Christenson GA, Pyle RL and Mitchell JE (1991c): Estimated lifetime prevalence of trichotillomania in college students. *J Clin Psychiatry* **52**:415–17.

Christenson GA, Mackenzie TB and Mitchell JE (1994): Adult men and women with trichotillomania. A comparison of male and female characteristics. *Psychosomatics* **35**:142–9.

Christenson RC (2004): Olanzapine augmentation of fluoxetine in the treatment of pathological skin picking. *Can J Psychiatry* **49**:788–9.

Cohen JD and Insel TR (2008): Cognitive neuroscience and schizophrenia: translational research in need of a translator. *Biol Psychiatry* **64**:2–3.

De Oliveira SKF (2007): PANDAS: A new disease? *J Pediatr* (Rio J) **83**:201–8.

Deckersbach T, Wilhelm S, Keuthen NJ, Baer L and Jenike MA (2002): Cognitive-behavior therapy for self-injurious skin picking: a case series. *Behav Modif* **26**:361–77.

Denys D, Van Megen HJ and Westenberg HG (2003): Emerging skin-picking after serotonin reuptake inhibitor-treatment in patients with obsessive–compulsive disorder: possible mechanisms and implications for clinical care. *J Psychopharmacol* **17**:127–9.

Diefenbach GJ, Tolin DF, Hannan S, Crocetto J and Worhunsky P (2005): Trichotillomania: impact on psychosocial functioning and quality of life. *Behav Res Ther* **43**:869–84.

Dougherty DD, Loh R, Jenike MA and Keuthen NJ (2006): Single modality versus dual modality treatment for trichotillomania: sertraline, behavioral therapy, or both? *J Clin Psychiatry* **67**:1086–92.

Du Toit PL, van Kradenburg J, Niehaus DJH and Stein DJ (2001): Characteristics and phenomenology of hair-pulling: an exploration of subtypes. *Compr Psychiatry* **42**:247–56.

Ellis MPJ (1986): Oral habits: a behavioural approach. *Br J Psychiatry* **148**:751–2.

Epperson CN, McDougle CJ and Price LH (1996): Intranasal oxytocin in trichotillomania. *Biol Psychiatry* **40**:559–61.

Epperson CN, Fasula D, Wasylink S, Price LH and McDougle CJ (1999): Risperidone addition in serotonin reuptake inhibitor-resistant trichotillomania: three cases. *J Child Adolesc Psychopharmacol* **9**:43–9.

Fineberg NA, Saxena S, Zohar J and Craig KJ (2007): Obsessive–compulsive disorder: boundary issues. *CNS Spectr* **12**:359–64; 367–75.

Flessner CA and Woods DW (2006): Phenomenological characteristics, social problems and the economic impact associated with chronic skin picking. *Behav Modif* **30**(6):944–63.

Flessner CA, Mouton-Odum S, Stocker AJ and Keuthen NJ (2007): StopPicking.com: internet-based treatment for self-injurious skin picking. *Dermatol Online J* **13**:3.

Flessner CA, Busch AM, Heideman PW and Woods DW (2008a): Acceptance-enhanced behaviour therapy (AEBT) for trichotillomania and chronic skin picking: exploring the effects of component sequencing. *Behav Modif* **32**:579–94.

Flessner CA, Woods DW, Franklin ME, Cashin SE and Keuthen NJ, Trichotillomania Learning Center – Scientific Advisory Board (TLC-SAB) (2008b): The Milwaukee Inventory for Subtypes of Trichotillomania – Adult Version (MIST-A): development of an instrument for the assessment of “focused” and “automatic” hair pulling. *J Psychopathol Behav Assess* **30**:20–30.

Friman P, Finney J and Christerpherson E (1987): Behavioral treatment of trichotillomania: an evaluative overview. *Behav Ther* **15**:249–64

Gadde KM, Wagner RH (2nd), Connor KM and Foust MS (2007): Escitalopram treatment of trichotillomania. *Int Clin Psychopharmacol* **22**:39–42.

Garner JP and Mason GJ (2002): Evidence for a relationship between cage stereotypies and behavioural disinhibition in laboratory rodents. *Behav Brain Res* **136**:83–92.

Garner JP, Meehan CL and Mench JA (2003): Stereotypies in caged parrots, schizophrenia and autism: evidence for a common mechanism. *Behav Brain Res* **145**:125–34.

Geller JL, Erlen J and Pinkus RL (1986): A historical appraisal of America’s experience with “pyromania” – a diagnosis in search of a disorder. *Int J Law Psychiatry* **9**:201–29.

Grant JE and Christenson GA (2007): Examination of gender in pathologic grooming behaviors. *Psychiatr Q* **78**:259–67.

Grant JE, Odlaug BL and Won Kim S (2007): Lamotrigine treatment of pathologic skin-picking: an open-label study. *J Clin Psychiatry* **68**:1384–91.

Greer JM and Capecchi MR (2002): Hoxb8 is required for normal grooming behavior in mice. *Neuron* **33**:23–34.

Hajcak G, Franklin ME, Simons RF and Keuthen NJ (2006): Hair pulling and skin picking in relation to affective distress and obsessive compulsive symptoms. *J Psychopathol Behav Assess* **28**:179–87.

Himle JA, Bordnick PS and Thyer BA (1995): A comparison of trichotillomania and obsessive–compulsive disorder. *J Psychopathol Behav Assess* **17**:251–60.

Hollander E (1993): Introduction. In: *The Obsessive–Compulsive Related Disorders*. Washington, DC: American Psychiatric Press.

Hollander E, Stein DJ, DeCaria CM *et al.* (1992): Disorders related to OCD-neurobiology. *Clin Neuropharmacol* **15** Suppl 1 Pt A:259A–60A.

Iancu I, Weizman A, Kindler S, Sasson Y and Zohar J. Serotonergic drugs in trichotillomania. *J Nerv Ment Dis* **184**:641–4.

Jenike MA (1992): Pharmacologic treatment of obsessive compulsive disorders. *Psychiatr Clin North Am* **15**:895–919.

Kalivas J, Kalivas L, Gilman D and Hayden CT (1996): Sertraline in the treatment of neurotic excoriations and related disorders. *Arch Dermatol* **132**:589–90.

Keuthen NJ, Savage CR, O'Sullivan RL *et al.* (1996): Neuropsychological functioning in trichotillomania. *Biol Psychiatry* **39**:747–9.

Keuthen NJ, Jameson M, Loh R *et al.* (2007a): Open-label escitalopram treatment for pathological skin picking. *Int Clin Psychopharmacol* **22**:268–74.

Keuthen NJ, Makris N, Schlerf JE *et al.* (2007b): Evidence for reduced cerebellar volumes in trichotillomania. *Biol Psychiatry* **61**:374–81.

Koran LM, Ringold A and Hewlett W (1992): Fluoxetine for trichotillomania: an open clinical trial. *Psychopharmacol Bull* **28**:145–9.

Leonard HL, Swedo SE, Rapoport JL *et al.* (1989): Treatment of obsessive–compulsive disorder with clomipramine and desipramine in children and adolescents. A double-blind crossover comparison. *Arch Gen Psychiatry* **46**:1088–92.

Leonard HL, Lenane MC, Swedo SE, Rettew DC and Rapoport JL (1991): A double-blind comparison of clomipramine and desipramine treatment of severe onychophagia (nail biting). *Arch Gen Psychiatry* **48**:821–7.

Lewis NDC and Yarnell H (1951): *Pathological Firesetting (Pyromania)*. Nervous and Mental Disease Monograph 82. New York: Coolidge Foundation.

Lochner C and Stein DJ (2006): Does work on obsessive–compulsive spectrum disorders contribute to understanding the heterogeneity of obsessive–compulsive disorder. *Prog Neuropsychopharmacol Biol Psychiatry* **30**:353–61.

Lochner C, du Toit PL, Zungu-Dirwayi N *et al.* (2001): Childhood trauma in obsessive–compulsive disorder, trichotillomania and controls. *Depress Anxiety* **14**:1–3.

Lochner C, Simeon D, Niehaus DJH and Stein DJ (2002): Trichotillomania and skin picking: a phenomenological comparison. *Depress Anxiety* **15**:83–6.

Lochner C, Seedat S, du Toit PL *et al.* (2005): Obsessive–compulsive disorder and trichotillomania: a phenomenological comparison. BMC *Psychiatry* **5**:2.

Lochner C, Seedat S, Niehaus DJ and Stein DJ (2006): Topiramate in the treatment of trichotillomania: an open-label pilot study. *Int Clin Psychopharmacol* **21**:255–9.

Lyon LS (1983): A behavioral treatment of compulsive lip biting. *J Behav Ther Exp Psychiatry* **14**:275–6.

Malone AJ and Massler M (1952): Index of nailbiting in children. *J Abnorm Soc Psychol* **47**:193–202.

McElroy SL, Hudson JI, Pope Jr, HG, Keck PE and Aizley HG (1992). The DSM-III-R impulse control disorders not elsewhere classified: clinical characteristics and relationship to other psychiatric disorders. *Am J Psychiatry* **149**:318–27.

McElroy SL, Phillips KA and Keck PE Jr (1994): Obsessive compulsive spectrum disorders. *J Clin Psychiatry* **55** Suppl:33–51.

Ninan PT, Knight B, Kirk L *et al.* (1998): A controlled trial of venlaflaxine in trichotillomania: interim phase I results. *Psychopharmacol Bull* **34**:221–4.

Ninan PT, Rothbaum BO, Marsteller FA, Knight BT and Eccard MB (2000): A placebo-controlled trial of cognitive-behavioral therapy and clomipramine in trichotillomania. *J Clin Psychiatry* **61**:47–50.

Odlaug BL and Grant JE (2007a): Childhood-onset pathologic skin picking: clinical characteristics and psychiatric comorbidity. *Compr Psychiatry* **48**:388–93.

Odlaug BL and Grant JE (2007b): N-Acetyl cysteine in the treatment of grooming disorders. *J Clin Psychopharmacol* **27**:227–8.

Odlaug BL and Grant JE (2008a): Clinical characteristics and medical complications of pathologic skin picking. *Gen Hosp Psychiatry* **30**:61–6.

Odlaug BL and Grant JE (2008b): Trichotillomania and pathologic skin picking: clinical comparison with an examination of comorbidity. *Ann Clin Psychiatry* **20**:57–63.

O'Sullivan RL, Keuthen NJ, Jenike MA and Gumley G (1996): Trichotillomania and carpal tunnel syndrome. *J Clin Psychiatry* **57**:174.

O'Sullivan RL, Keuthen NJ, Christenson GA *et al.* (1997a): Trichotillomania: behavioral symptom or clinical syndrome? *Am J Psychiatry* **154**:1442–9.

O'Sullivan RL, Rauch SL, Breiter HC *et al.* (1997b): Reduced basal ganglia volumes in trichotillomania measured via morphometric magnetic resonance imaging. *Biol Psychiatry* **42**:39–45.

Phillips KA, First MB and Pincus HA (2003a): *Advancing the DSM: Dilemmas in Psychiatric Diagnosis*. Washington, DC: American Psychiatric Association.

Phillips KA, Price LH, Greenberg BD and Rasmussen SA (2003b): Should the DSM diagnostic groupings be changed? In: Phillips KA, First MB and Pincus HA, eds., *Advancing the DSM: Dilemmas in Psychiatric Diagnosis*. Washington, DC: American Psychiatric Association, pp. 57–84.

Pigott TA, Pato MT, Bernstein SE *et al.* (1990): Controlled comparisons of clomipramine and fluoxetine in the treatment of obsessive–compulsive disorder. Behavioral and biological results. *Arch Gen Psychiatry* **47**:926–32.

Pollard CA, Ibe IO, Krojanker DN *et al.* (1991): Clomipramine treatment of trichotillomania: a follow-up report on four cases. *J Clin Psychiatry* **52**:128–30.

Pukadan D, Anthony J, Mohandas E *et al.* (2008): Use of escitalopram in psychogenic excoriation. *Aust N Z J Psychiatry* **42**:435–6.

Rasmussen SA and Tsuang MT (1984): The epidemiology of obsessive compulsive disorder. *J Clin Psychiatry* **45**:450–7.

Rauch SL, Wright CI, Savage CR *et al.* (2007): Brain activation during implicit sequence learning in individuals with trichotillomania. *Psychiatry Res: Neuroimaging* **154**:233–40.

Rettew DC, Cheslow DL, Rapoport JL, Leonard HL and Lenane MC (1991): Neuropsychological test performance in trichotillomania: a further link with obsessive–compulsive disorder. *J Anxiety Disord* **5**:225–35.

Rettew DC, Swedo SE, Leonard HL, Lenane MC and Rapoport JL (1992): Obsessions and compulsions across time in 79 children and adolescents with obsessive–compulsive disorder. *J Am Acad Child Adolesc Psychiatry* **31**:1050–6.

Rosenbaum MS and Ayllon T (1981): The behavioral treatment of neurodermatitis through habit-reversal. *Behav Res Therapy* **19**:313–18.

Rothbaum BO, Shaw L, Morris R and Ninan PT (1993): Prevalence of trichotillomania in a college freshman population. *J Clin Psychiatry* **54**:72–3.

Sasso DA, Kalanithi PS, Trueblood KV *et al.* (2006): Beneficial effects of the glutamate-modulating agent riluzole on disordered eating and pathologic skin-picking behaviors. *J Clin Psychopharmacol* **26**:685–7.

Seedat S, Kesler S, Niehaus DJ and Stein DJ (2000): Pathological gambling behaviour: emergence secondary to treatment of Parkinson's disease with dopaminergic agents. *Depress Anxiety* **11**:185–6.

Seedat S, Stein DJ and Harvey BH (2001): Inositol in the treatment of trichotillomania and compulsive skin picking. *J Clin Psychiatry* **62**:60–1.

Simeon D, Stein DJ, Gross S *et al.* (1997): A double-blind trial of fluoxetine in pathologic skin picking. *J Clin Psychiatry* **58**:341–7.

Stanley MA, Bowers TC, Swann AC and Taylor DJ (1991): Treatment of trichotillomania with fluoxetine. *J Clin Psychiatry* **52**:282.

Stanley MA, Swann AC, Bowers TC, Davis ML and Taylor DJ (1992): A comparison of clinical features in trichotillomania and obsessive–compulsive disorder. *Behav Res Ther* **30**:39–44.

Stanley MA, Breckenridge JK, Swann AC, Freeman EB and Reich L (1997): Fluvoxamine treatment of trichotillomania. *J Clin Psychopharmacol* **17**:278–83.

Stein DJ (2008): Is disorder X in category or spectrum Y? General considerations and application to the relationship between obsessive–compulsive disorder and anxiety disorders. *Depress Anxiety* **25**:330–5.

Stein DJ, Simeon D, Cohen LJ and Hollander E (1995): Trichotillomania and obsessive–compulsive disorder. *J Clin Psychiatry* **56** Suppl 4:28–34.

Stein DJ, Van Heerden B, Hugo C *et al.* (2002): Functional brain imaging and pharmacotherapy in trichotillomania: single photon emission computed tomography before and after treatment with the selective serotonin reuptake inhibitor citalopram. *Prog Neuropsychopharmacol Biol Psychiatry* **26**:885–90.

Stein DJ, Garner JP, Keuthen NJ *et al.* (2007): Trichotillomania, stereotypic movement disorder, and related disorders. *Curr Psychiatry Rep* **9**:301–2.

Stewart RS and Nejtek VA (2003): An open-label, flexible dose study of olanzapine in the treatment of trichotillomania. *J Clin Psychiatry* **64**:49–52.

Streichenwein SM and Thornby JI (1995): A long-term, double-blind, placebo-controlled crossover trial of the efficacy of fluoxetine for trichotillomania. *Am J Psychiatry* **152**:1192–6.

Swedo S (1993): Trichotillomania. In: Hollander E, ed., *The Obsessive–Compulsive Related Disorders*. Washington, DC: American Psychiatric Press.

Swedo SE and Leonard HL (1992): Trichotillomania: an obsessive–compulsive disorder? *Psychiatr Clin North Am* **15**:777–90.

Swedo SE, Leonard HL, Rapoport JL *et al.* (1989): A double-blind comparison of clomipramine and desipramine in the treatment of trichotillomania (hair pulling). *N Engl J Med* **321**:497–501.

Swedo SE, Rapoport JL, Leonard HL *et al.* (1991): Regional cerebral metabolism of women with trichotillomania. *Arch Gen Psychiatry* **48**:828–33.

Teng EJ, Woods DW, Twohig MP and Marcks BA (2002): Body-focused repetitive behaviour problems: prevalence in a non-referred population and differences in perceived somatic activity. *Behav Modif* **26**:340–60.

Teng EJ, Woods DW and Twohig MP (2006): Habit reversal treatment for chronic skin picking: a pilot investigation. *Behav Modif* **30**:411–22.

Tukel R, Keser V, Karali N, Olgun T and Calikusu C (2001): Comparison of clinical characteristics in trichotillomania and obsessive–compulsive disorder. *J Anxiety Disord* **15**:433–41.

Turley PK and Henson JL (1983): Self-injurious lip-biting: etiology and management. *J Pedontics* **7**:209–20.

Twohig MP and Woods DW (2001): Habit reversal as a treatment for chronic skin picking in typically developing male siblings. *J Appl Behav Anal* **34**:217–20.

Twohig MP, Woods DW, Marcks BA and Teng EJ (2003): Evaluating the efficacy of habit reversal: comparison with a placebo control. *J Clin Psychiatry* **64**:40–8.

Twohig MP, Hayes SC and Masuda A (2006): A preliminary investigation of acceptance and commitment therapy as a treatment for chronic skin picking. *Behav Res Ther* **44**:1513–22.

Van Ameringen M, Mancini C, Oakman JM and Farvolden P (1999): The potential role of haloperidol in the treatment of trichotillomania. *J Affect Disord* **56**: 219–26.

Van Minnen A, Hoogduin KAL, Keijsers GPJ, Hellenbrand I and Hendriks GJ (2003): Treatment of trichotillomania with behavioral therapy or fluoxetine. *Arch Gen Psychiatry* **60**:517–22.

Weissman MM, Bland RC, Canino GJ *et al.* (1994): The cross national epidemiology of obsessive compulsive disorder. The Cross National Collaborative Group. *J Clin Psychiatry* **55** Suppl:5–10.

Welch JM, Lu J, Rodriguiz RM *et al.* (2007): Cortico-striatal synaptic defects and OCD-like behaviours in sapap3-mutant mice. *Nature* **448**:894–901.

Wilhelm A, Keuthen NJ, Deckersbach T *et al.* (1999): Self-injurious skin picking: clinical characteristics and comorbidity. *J Clin Psychiatry* **60**:454–9.

Winchel RM, Jones JS, Stanley B, Molcho A and Stanley M (1992): Clinical characteristics of trichotillomania and its response to fluoxetine. *J Clin Psychiatry* **53**:304–8.

Woods DW, Miltenberger RG and Flach AD (1996): Habits, tics and stuttering: prevalence and relation to anxiety and somatic awareness. *Behav Modif* **20**: 216–25.

Woods DW, Flessner CA, Franklin ME *et al.* (2006a): The Trichotillomania Impact Project (TIP): exploring phenomenology, functional impairment and treatment utilization. *J Clin Psychiatry* **67**:1877–88.

Woods DW, Wetterneck CT and Flessner CA (2006b): A controlled evaluation of acceptance and commitment therapy plus habit reversal for trichotillomania. *Behav Res Ther* **44**:639–56.

World Health Organization (1992): *International Statistical Classification of Diseases and Health Related Problems*, 10th Revision, Vol. 1. Geneva, Switzerland: WHO.

Zohar J and Insel TR (1987): Obsessive–compulsive disorder: psychobiological approaches to diagnosis, treatment, and pathophysiology. *Biol Psychiatry* **22**:667–87.

12

Neurobiology of trichotillomania

SRINIVAS SINGISETTI, SAM R. CHAMBERLAIN AND
NAOMI A. FINEBERG

Summary

Trichotillomania (TTM) is a common debilitating impulse control disorder, which is under-recognized in clinical practice. New research shows interesting similarities between TTM, other impulse control disorders, and obsessive–compulsive disorder (OCD), while also revealing important differences in some endophenotypic measures. In this chapter we review new advances in genetic, family, neurocognitive, neuroimaging, and neuropharmacological studies. Neural abnormalities in the amygdalo–hippocampal formation and frontal–subcortical circuits are discussed. Animal models of hair pulling are also outlined and may prove a fruitful avenue for future research.

Introduction

Trichotillomania is a neuropsychiatric disorder characterized by noticeable hair loss due to a recurrent failure to resist impulses to pull out hairs. The hair pulling is usually preceded by mounting tension and followed by a sense of relief or gratification (WHO 2002). It predominantly affects females (Swedo and Leonard 1992), and its onset is usually in late childhood and adolescence (Walsh and McDougle 2001). A subgroup with very early onset of hair pulling in children under the age of six may be more benign and self-limiting (Keren *et al.* 2006). Often accompanied by shame and distress, TTM is under-recognized in clinical practice and its prevalence is likely to be greater than currently understood (Bohne *et al.*

Neurobiology of Grooming Behavior, eds. Allan V. Kalueff, Justin L. LaPorte, and Carisa L. Bergner. Published by Cambridge University Press.

2005b). There have been no population-based epidemiological studies of TTM. In a sample of 2579 college students in the United States, a lifetime prevalence of TTM was seen in 0.6%, though subthreshold symptoms not reaching diagnostic criteria were identified in 1.5% of males and 3.4% of females (Christenson *et al.* 1991b).

In practice, TTM is often comorbid with other DSM-IV mental disorders (APA 2000) including major depression (39%–65%), generalized anxiety disorder (27%–32%), OCD (13%–27%), and substance abuse (15%–20%) (Christenson and Mansueto 1999). Using a semi-structured interview, one study in patients with TTM found a comorbidity rate of 82%, when past and current Axis I psychiatric diagnoses were included (Christenson *et al.* 1991a).

Nosological status of trichotillomania

Psychiatric classificatory systems including the International Classification of Diseases (ICD-10), and the DSM-IV categorize TTM as an impulse control disorder (WHO 2002; APA 2000). The essential feature of impulse control disorders is the recurrent failure to resist an impulse, drive, or temptation to perform an act that is ultimately harmful to the person or to others. Usually the individual feels an increasing sense of tension or arousal before committing the act and then experiences short-lived pleasure, gratification, or relief at the time of committing the act (APA 2000).

Other impulse control disorders currently included in the major classificatory systems are pathological gambling, pyromania (pathological fire setting), kleptomania (pathological stealing), and intermittent explosive disorder. Disorders of phenomenological and possible biological similarity that have been described in the literature and proposed for future inclusion in this category include skin picking, compulsive shopping, compulsive internet use, and nonparaphilic compulsive sexual behavior (Grant and Potenza 2004, 2006). In view of the repetitive nature of the behaviors and tendency towards habit formation, as well as the involvement of negative and positive reinforcers (such as sense of relief and pleasure) in maintaining the behavior, one may view these disorders as behavioral addictions, akin to substance use addictions (Brewer and Potenza 2008). The extent to which these impulse control disorders share clinical, genetic, phenomenological, and biological features is incompletely understood. Although relatively understudied, research in these disorders has increased recently, and especially so with TTM.

While TTM is classified along with other impulse control disorders, the repetitive motor behavior of hair pulling with perceived diminished control also bears some resemblance to the repetitive compulsive rituals in OCD and the repetitive motor tics in Tourette's syndrome, and arguably justifies its classification as an obsessive–compulsive spectrum disorder (OCSD) (Hollander *et al.* 2007). But in contrast to OCD, in which compulsions occur in a variety of situations, individuals who have TTM tend to pull most often when engaged in sedentary activities. Although the hair pulling in TTM decreases anxiety, as do compulsions in OCD, it may also produce feelings of pleasure, whereas OCD compulsions typically do not. Also, hair-pulling acts in TTM are not preceded by ruminations unlike the compulsions of OCD, which are usually preceded by obsessive thoughts. The relationship between compulsive and impulsive disorders is complex. While some authors have suggested that compulsivity and impulsivity are orthogonal dimensions and that patients can have both compulsive and impulsive symptoms and neurocognitive features (Chamberlain *et al.* 2006a), others have argued that compulsive and impulsive disorders lie at opposite ends of a spectrum (McElroy *et al.* 1994; Stein 2006; Stein and Lochner 2006).

In this chapter, we attempt to identify biological markers (candidate endophenotypes) of TTM by focusing on family and genetic studies, tests of neurocognitive function, structural and functional neuroimaging, and studies on neurochemical substrates. Unlike OCD, where a growing body of research is driving theoretical models of the disorder, research into the etiology and neurobiology of TTM is still embryonic. Where relevant, we aim to compare these emerging results with findings for OCD, other impulse control disorders, and attention-deficit/hyperactivity disorder (ADHD), in order to highlight areas of similarity and difference. We include animal studies, which emphasize the evolutionary significance of TTM and other grooming disorders.

Explanatory models of trichotillomania

Trichotillomania may be conceptualized as a disorder of impulse control and inappropriate habit formation. Explanatory models have proposed dysregulation in Affective state, Behavioral addiction, and disordered Cognitive control (A-B-C approach: Stein *et al.* 2006) as key contributing factors. Such models provide a useful heuristic for neurobiological research into etiological factors underpinning the disorder.

In support of the "affect regulation" model for TTM, one study showed hair pulling was influenced by negative affective states, and by sedentary activities and contemplative attitudes (Christenson *et al.* 1993). Another study used a retrospective self-report measure to show that hair-pulling activity was initially associated with a decrease in feelings of boredom, sadness, anger, and tension and increased pleasure; while there was a subsequent increase in guilt, sadness, and anger after the activity (Diefenbach *et al.* 2008). Studies have consistently reported individuals with TTM to have elevated rates of depressive and anxiety disorders (52%, 27%, respectively) and even higher prevalence of anxiety or depressed mood (66%–68%) (Christenson 1995; Christenson and Mansueto 1999), and suggest these affective states could act as a trigger for hair pulling, which, through repetition, converts into a maladaptive habit. Thus, depressed or anxious individuals may engage in hair pulling to distract themselves from life stressors. Such a model implicates brain regions involved in the mediation of anxiety and negative affect including the amygdala and hippocampus. A recent whole-brain imaging study showed increased gray matter densities in the left amygdalo–hippocampal formation, among other cortical areas, in patients with TTM (Chamberlain *et al.* 2008).

Akin to substance addiction, hair pulling may be maintained by initial reward in the form of pleasure or a sense of accomplishment from the act of pulling (positive reinforcement). One study examining OCSD within OCD subjects found that TTM clustered with pathological gambling and hypersexual disorder within a proposed "reward deficiency" group (Lochner *et al.* 2005a). Anticipation of reward and outcomes are known to activate foci in the ventral striatum and differentially recruit distinct regions that lie along the ascending dopamine projections (Knutson *et al.* 2001).

The basal ganglia are involved in the control of cognitive as well as motor output (Graybiel 1997), and have been shown to contribute to habit and stimulus–response learning. These forms of learning have the property of slow acquisition and can occur without conscious awareness (Graybiel 1998). Striatal damage in rats has been shown to disrupt the ability to perform choreographed grooming sequences (Cromwell and Berridge 1996). During normal human development the frontal cortex is believed to assume control over the formation and execution of habits through inhibitory actions on the basal ganglia. The neurocognitive model of TTM holds that the pathological symptoms stem from overzealous engagement of the basal ganglia in manifesting habits and/or failures of top–down cortical inhibitory processing (Chamberlain *et al.* 2005a; Stein *et al.* 2006). The high rates of co-occurrence of TTM and other putative habit disorders such as compulsive skin picking and nail biting (Bienvenue *et al.* 2000) are consistent with such a model.

Genetics in trichotillomania

Genetic studies in TTM attempting to link the illness to polymorphisms in specific genes are few and largely inconclusive. In one study, slit and trk like 1 (*SLITRK1*) gene was identified in two patients with TTM while it was absent in 2000 controls (Zuchner *et al.* 2006). The SLITRK1 protein may help guide the growth of specialized extensions (axons and dendrites) that allow each nerve cell to communicate with nearby cells. The gene is associated with abnormal axonal-dendritic development in embryonic mouse cells (Grados and Walkup 2006). Mutations in this gene have been identified in Tourette's syndrome (Abelson *et al.* 2005) and suggest a possible common etiological basis for the two disorders.

Variations in the gene for the serotonin (5HT2A) receptor have also been implicated in TTM. In a study by Hemmings *et al.* (2006), both the genotypic and allelic distribution of the 5HT2A receptor T102C variant were found to be significantly different between groups with TTM and controls. The difference was near significant between TTM and OCD, with the variant conferring susceptibility to the development of TTM. Nomura *et al.* (2006) showed involvement of the A1438A polymorphism of the 5HT2A receptor gene in aspects of impulsive behavior, as measured on a "go/no-go" task (a test for motor inhibition). Variations in serotonin receptor and transporter genes have also been associated with more substantial changes to structure and functioning of particular areas of the brain including anatomical and functional variation in cingulate–amygdala circuitry (Pezawas *et al.* 2005) as well as differential activation of negative, positive, and neutral stimuli in cortical, striatal, and limbic regions (Canli *et al.* 2005). These studies implicate serotonin dysregulation as one of the neurochemical substrates mediating emotional contributions to TTM.

Family studies showing association with other known illnesses also provide information on the genetic basis of TTM. Not surprisingly, TTM is often seen with an increased incidence in probands with OCD and their close relatives, and vice versa. Rates of TTM among individuals with OCD are inconsistent across studies and three studies of small samples of OCD subjects have reported rates ranging from 4.6% to 7.1% (Grant and Potenza 2006). Fontenelle *et al.* (2005) looked at various disorders with impulse dyscontrol (including not only the impulse control disorders, but also other disorders in which impulse control is a prominent feature such as alcohol and drug dependence, paraphilias, and bulimia nervosa/binge eating disorder) and found that these disorders were comorbid in up to 35% of OCD patients. Apart from their presence in patients with OCD, grooming disorders including TTM, skin picking, and nail biting also occur more frequently in their unaffected relatives (Bienvenu *et al.* 2000).

The converse is also true and rates of OCD comorbid in patients with TTM (13%–27%) (Christenson and Mansueto 1999) are significantly higher than its community prevalence (1%–3%). Importantly, higher rates of OCD also extend to relatives of TTM probands unaffected by TTM (2.9%) (Lenane *et al.* 1990; Schlosser *et al.* 1994), implying a shared vulnerability within families for both disorders and the possibility of common neurobiological pathways.

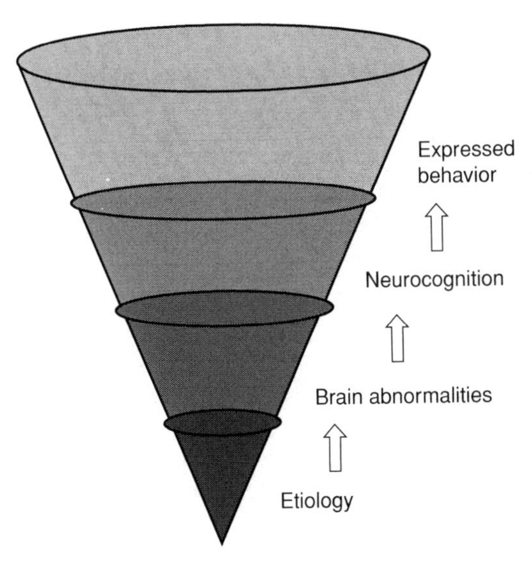

Figure 12.1 Pictorial representation of the position of endophenotypes (e.g., neurocognition and brain abnormalities) on the pathway between disease (expressed behavior) and genetic etiology.

Endophenotype approach

The phenotype represents the observable characteristics of an organism, which are the joint product of both genotypic and environmental influences. In diseases with classic or Mendelian genetics, genotypes are usually indicative of phenotypes. However, this degree of genetic certainty does not exist for diseases with complex genetics such as diabetes, hypertension, and most psychiatric disorders. In these disorders, multiple genes interacting at various levels bring about changes in neuronal circuits and subsequent expression of observable symptoms (McGuffin *et al.* 2002). Indeed, most neuropsychiatric disorders are heterogeneous diseases with multiple different causes and are often highly heritable. But efforts to discover the etiological genes are yet to provide definitive results. Gottesman and Gould (2003) described endophenotypes as "measurable components unseen by the unaided eye along the pathway between disease and distal genotype." These quantitative measures may be neuroanatomical, neurophysiological, biochemical, endocrine, cognitive, or neuropsychological. Endophenotypes are "trait" not "state" dependent (though may require provocation or challenge). They are evident in unaffected relatives at a higher rate than in the background population. In this context, endophenotypes offer an attractive and measurable strategy for delineation of the etiological genes, as they lie closer to them than the clinical phenotypes (Figure 12.1).

Recently the usefulness of endophenotypes has been verified in the study of OCD (Chamberlain *et al.* 2005a), ADHD (Castellanos and Tannock 2002), depression (Niculescu and Akiskal 2001), bipolar affective disorder, and schizophrenia (Gottesman and Gould 2003). In this chapter, we propose exploration of the endophenotype approach to delineate the neurobiological basis of TTM, focusing on neurocognitive functioning, structural and functional brain imaging, and pharmacological substrate changes.

Neurocognitive functioning

While neurocognitive studies in OCD have identified deficits in domains traditionally subserved by cortico-striatal circuitry (Stein *et al.* 2006), studies in TTM have been few and have often used OCD patients as controls (Chamberlain *et al.* 2007a). The traditional notion of TTM being an OCSD has been argued recently (Ferrão *et al.* 2006). Obsessive–compulsive symptoms and spectrum disorders themselves have also been noted by some as heterogeneous constructs. Cluster analysis by one study showed that the symptoms/disorders lie on different dimensions, such as reward deficiency (TTM, pathological gambling, hypersexual disorder, and Tourette's disorder), impulsivity (compulsive shopping, kleptomania, eating disorders, self-injury, and intermittent explosive disorder), and somatic (body dysmorphic disorder and hypochondriasis) (Lochner and Stein 2006; Lochner *et al.* 2005b).

Impulsivity in trichotillomania

Similarities as well as differences have been identified in the neurocognitive functioning deficits in TTM versus OCD, especially in relation to tasks dependent on the frontal–subcortical circuits. Three principal frontal–subcortical circuits originate in the dorsolateral prefrontal cortex, orbitofrontal cortex (OFC), and anterior cingulate (part of the medial frontal) cortex respectively. Circuit-specific marker behaviors associated with each circuit tentatively are: executive dysfunction (dorsolateral prefrontal–subcortical circuit), disinhibition and obsessive–compulsive symptoms (orbitofrontal–subcortical circuit), and apathy (medial frontal–subcortical circuit) (Cummings 1995; Figure 12.2). However, it is now recognized that there are important functional overlaps between these circuits. Impulsivity, a heterogeneous term, is a cardinal and often-studied aspect of TTM and other impulse control disorders. Impulsivity has been described as having three dimensions: (1) an inability to use available information to reflect on the consequences of actions, or poor decision making (simply stated); (2) an inability to forego an immediate small reward in favor of a delayed larger reward; (3) a deficit

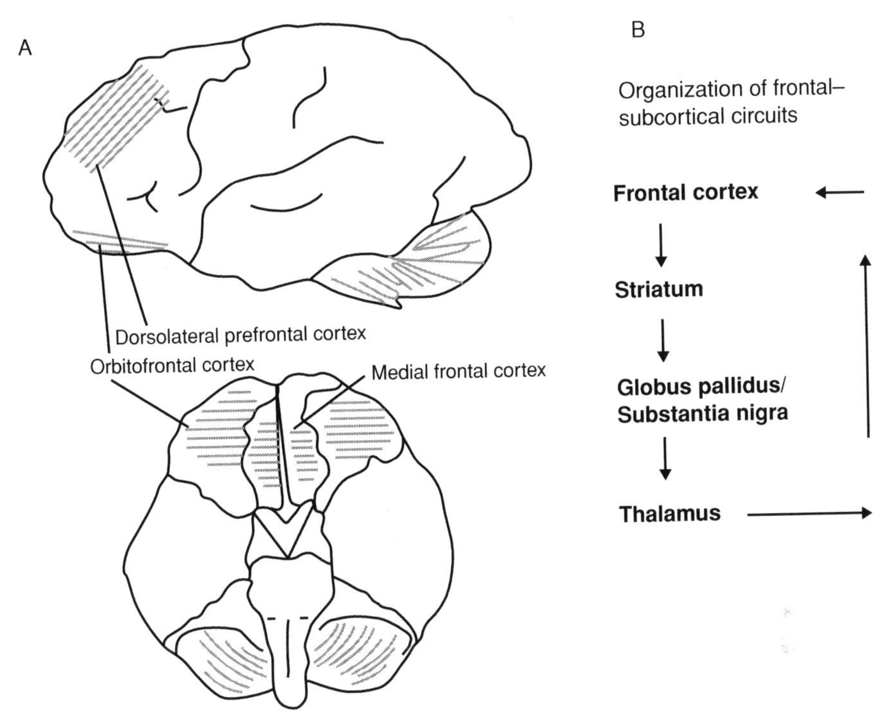

Figure 12.2 (A) Ventral and lateral surfaces of the brain. (B) A simple layout of the frontal–subcortical circuits.

in suppressing prepotent motor responses (response inhibition) (Chamberlain and Sahakian 2007; Torregrossa *et al.* 2008).

Response inhibition is the most often studied dimension of impulsivity and can be assessed clinically by go/no-go and stop-signal tasks. In the former, participants are asked to make speeded motor responses in the "go" tasks (e.g., horizontal lines appearing on a screen), but to withhold on the "no-go" tasks (e.g., vertical lines appearing on a screen). The number of inappropriate motor responses (motor impulsivity) to the no-go stimuli is an indication of response inhibition. Stop-signal tasks are behaviorally more sensitive, and measure the ability to inhibit already activated motor activities, indexed in terms of "stop-signal reaction times" (SSRT), a measure of the time taken by the brain to suppress an inappropriate response (Chamberlain and Sahakian 2007).

Deficits in response inhibition have been identified in various neuropsychiatric disorders including TTM, ADHD, OCD, and chronic substance abuse. Such deficits have also been noted in the first-degree relatives of patients with ADHD and OCD (Chamberlain and Sahakian 2007). The OFC is implicated in the inhibition of prepotent motor responses (response inhibition) in human and animal experiments

(Chamberlain and Sahakian 2007; Eagle *et al.* 2008). Functional magnetic resonance imaging (fMRI) and transcranial stimulation studies on patients with ADHD and normal subjects have implicated the right inferior frontal gyrus in response inhibition (Chamberlain and Sahakian 2007). Abnormal fronto–striatal brain activation has been confirmed in OCD by a recent fMRI study, which studied a sample of ten adolescent boys with OCD and nine matched controls, during three different tasks of inhibitory control – stop-signal, motor stroop, and switch tasks (Woolley *et al.* 2008).

Bohne *et al.* (2008) investigated motor response inhibition by go/no-go experiment in 25 subjects with TTM, 21 subjects with OCD, and 26 healthy participants. There was no evidence of response inhibition deficits in OCD, whereas a subgroup of TTM subjects with earlier age of onset performed "fast and inaccurate" indicating poor response inhibition. In another study, Bohne *et al.* (2005a) studied 21 patients with TTM, 21 patients with OCD, and 26 healthy controls on a block-cued directed forgetting task (which measures the ability to inhibit irrelevant information, where participants are instructed to intentionally forget one set of words and to remember another word list). They noted that cognitive inhibition deficits were specific to OCD. In an earlier study, Stanley *et al.* (1997) compared the performance of 21 patients with TTM and 17 healthy control participants on a broad battery of tests and demonstrated poorer performance on all measures of divided, but not focused, attention.

In contrast to impulsivity, the compulsive symptoms in OCD are mediated by difficulty in shifting attention, called cognitive flexibility (Chamberlain *et al.* 2005a). In a study looking at both impulsivity and compulsivity, Chamberlain *et al.* (2006a) studied differences in the stop-signal task (response inhibition) and the intradimensional/extradimensional shift task (cognitive flexibility) in 20 patients with OCD, 17 patients with TTM, and 20 healthy comparison subjects. Impaired SSRT was found in both OCD and TTM patients, while only OCD patients showed cognitive inflexibility. Between patients with OCD and TTM, motor response inhibition was much greater in the TTM group and correlated with symptom severity. The authors hypothesized response inhibition (impulsivity) to be common to both the illnesses, with cognitive inflexibility limited to OCD.

The ability to use spontaneous strategies to aid performance (strategy implementation, seen to be dependent on fronto–striatal circuits, and necessary in cognitive flexibility) was found deficient in patients with OCD but not in TTM. In a study that used a computer interface to generate novel visuospatial sequences after strategy training in 20 OCD patients, 17 TTM patients, and 20 age-, education-, and IQ-matched controls, performance of TTM patients improved significantly after strategy training to the same extent as controls, while OCD patients failed to improve (Chamberlain *et al.* 2005b).

Thus, more extensive neuropsychological impairment including executive malfunctioning is implicated in OCD, whereas the deficits in TTM seem, on the basis of evidence to date, limited to response inhibition. Therefore it appears that TTM patients may have only limited deficiencies of the cortico–subcortical neural circuits. In OCD, impaired cognitive flexibility and response inhibition extends to unaffected first-degree relatives (Chamberlain *et al.* 2007b) and therefore serves as a cognitive brain marker (endophenotype) in OCD, and possibly OCSD. The neural systems regulating impulsive, compulsive, and habitual behaviors are likely to differ, yet there may well be additional overlapping neurobiology (along the frontal–subcortical circuitry) that explains why several psychiatric disorders have comorbid impulsive and compulsive features (Torregrossa *et al.* 2008). Poor functioning on the stop-signal test by patients with TTM, implicating the right inferior frontal gyrus, mandates further neuroimaging focused on that brain region. Patients with ADHD also perform poorly on the stop-signal test and one may envision a role for cognitive enhancers in the treatment of TTM (Chamberlain *et al.* 2007c).

Structural and functional abnormalities of the brain in trichotillomania

The neuroimaging findings in OCD are relatively robust, compared to findings in other DSM Axis-I disorders, with structural and functional abnormalities frequently reported in the orbitofrontal cortex, anterior cingulate cortex, and caudate nucleus (Chamberlain *et al.* 2005a; Menzies *et al.* 2008). In contrast, imaging studies in TTM are few and have involved only small patient numbers. Comorbidity was often not excluded, and most studies employed a region-of-interest design. Given the phenomenological and familial similarities of TTM with OCD, studies in TTM have hypothesized abnormalities on the premise of OCD. For example, functional imaging studies of OCD using PET and SPECT have reported increased metabolism in the frontal cortex in both the resting state and during symptom activation, which was reduced following successful treatment with a selective serotonin reuptake inhibitor (SSRI) (Stein *et al.* 1999). Studying the effect of 12 weeks of citalopram on ten female TTM patients, Stein *et al.* (2002) noted a significantly reduced activity in inferior–posterior and other frontal–cortical regions compared with pretreatment activity using SPECT. These results imply similarities in the mechanism of action of SSRIs in OCD and TTM in reducing frontal activation. Larger studies are required to replicate this effect and examine the association between radiological changes and symptom response. A single PET study of 10 TTM patients and 20 age-matched female controls showed significantly increased global and normalized right and left cerebellar and right superior parietal (but not frontal) metabolism in TTM (Swedo *et al.* 1991).

Grachev (1997) studied volumetric changes of the neocortex between ten female TTM subjects and ten normal female controls using MRI. There were no significant volumetric changes of the precentral gyrus, postcentral gyrus, supplementary motor cortex, or opercular cortex in TTM patients compared with control subjects, though a post-hoc analysis showed that TTM subjects exhibited significantly reduced left inferior frontal gyrus volume of 27% ($p = 0.04$) and enlarged right cuneal cortex volume of 40% ($p = 0.03$). Another MRI study comparing ten female subjects with TTM versus ten sex-, age-, handedness-, and education-matched normal controls found the left putamen volume to be significantly smaller in TTM patients (O'Sullivan *et al.* 1997). In contrast, in the MRI study by Stein *et al.* (1997b), which compared 13 women with OCD, 17 women with TTM, and 12 healthy controls, no significant differences were found between the groups on caudate volume or ventricular brain ratio. Due to the role of the cerebellum in complex, coordinated motor sequences, Keuthen *et al.* (2007) studied the MRI findings in 14 female subjects with TTM and 12 age-, education-, and gender-matched normal controls. The TTM group showed significantly smaller cerebellar volumes ($p = 0.008$), which remained significant after correcting for the total brain volume and head circumference ($p = 0.037$). Subterritory cerebellar volumes also revealed significant differences though the same did not remain after correction for the total cerebellar volumes.

Cortico–striatal dysfunction is implicated in OCD and habit disorders. A study using fMRI and a serial reaction-time task to assess striatal and hippocampal activation during implicit sequence learning compared ten subjects with TTM with ten age- and education-matched normal controls. No significant differences in implicit learning or in activation within the striatum or hippocampus were noted (Rauch *et al.* 2007). However, the result may have been compromised by lack of statistical power owing to the small sample size.

Given the inconsistencies in outcomes of the small number of existing brain-imaging studies that employed an a-priori volumetric analysis focusing on specific regions of interest, future structural imaging studies in TTM may benefit from an unbiased whole-brain analysis approach (Chamberlain *et al.* 2007c). In the first whole-brain structural MRI study in TTM, Chamberlain *et al.* (2008) studied 20 unmedicated patients and 19 healthy controls. In comparison to the controls, TTM patients showed increased gray matter density in the left amygdalo–hippocampal formation, left putamen, bilateral cingulate cortices, and frontal (especially right) cortical regions. As previously discussed, these brain areas are implicated in affect regulation, habit formation, and behavioral inhibition. The demonstration of structural abnormalities in these key areas is strikingly consistent with the A-B-C model for TTM and suggests dysfunction in affect regulation as well as habit formation and cortical inhibition.

Neuropharmacological substrates

Results from studies on pharmacological treatments of TTM are inconsistent (Bloch *et al.* 2007; Chamberlain *et al.* 2007c; Stein *et al.* 1997a). Though, as in OCD, the serotonergic tricyclic clomipramine showed greater efficacy than the noradrenergic tricyclic desipramine in a double-blind cross-over trial (Swedo *et al.* 1989), the results for SSRIs in a small number of TTM trials has been disappointing. Serotonergic agonists such as the nonselective agent m-chlorophenylpiperazine (mCPP) and the more selective 5-HT1D agonist sumatriptan have been shown to exacerbate symptoms of OCD (Zohar *et al.* 1987) while administration of the same have not resulted in any worsening of symptoms in TTM (Stein *et al.* 1995), suggesting 5-HT1C/D receptors may not be relevant in TTM.

Serotonin has been traditionally seen as a substrate for impulsivity. Patients with TTM showed impulsivity on laboratory-based measures of motor response inhibition such as the SSRT (Chamberlain *et al.* 2006a). It is therefore intriguing that studies involving depletion of serotonin in the central nervous system and those involving administration of serotonin receptor agonists (buspirone) or serotonin reuptake inhibitors (citalopram) in healthy volunteers have not shown any major impact on response inhibition, measured by the SSRT, whereas noradrenergic agents (e.g., atomoxetine, a selective noradrenaline reuptake inhibitor used in treatment of ADHD) improved this function (Chamberlain *et al.* 2006b). It is hypothesized that serotonin is involved in emotional aspects of impulsivity, where the acts are conditioned by rewards and feedback (Cools *et al.* 2005), rather than motor impulsivity (as measured by the SSRT) per se. While the role of SSRIs in the treatment of TTM needs to be reviewed, agents acting upon noradrenergic neurotransmission such as atomoxetine, currently used in the treatment of impulsivity in ADHD, may serve as novel (and as yet untested) options. Finally, it is relevant to consider that TTM, as currently defined, may represent a heterogeneous disorder with different neurobiological determinants prevailing in different cases (Grant and Potenza 2006). Thus, a differential response to pharmacological treatments and probes may be expected. The role of pharmacogenomics in predicting response indicators in TTM is a challenge for future research.

Animal models

Trichotillomania, skin picking, and nail biting are often grouped under a set of conditions called "grooming disorders," which share similarities in phenomenology, female predisposition, and serotonin reuptake inhibitor response with their counterpart disorders in animals (Garner *et al.* 2004; Grant and

Christenson 2007; Hugo *et al.* 2003; Stein *et al.* 1999). Grooming behavior is conserved in a variety of species and hair pulling in TTM may represent a residual evolutionary habit. Along with humans, hair pulling has been reported in six different nonhuman species including mice, guinea pigs, rabbits, sheep and musk-ox, dogs and cats, with some seen in artificial conditions only (Reinhardt 2005). Animal models of hair pulling are important for translational research into TTM, as they provide a substrate for neurobiological methods that are applicable in humans.

"Barbering," a whisker- and fur-plucking behavior in mice that is predominantly female biased and has its onset at puberty, has face validity as a model for TTM (Garner *et al.* 2004). However, unlike TTM, in barbering, two agents are usually needed: the dominant barber and the passive recipient from which the hair is pulled (Kurien *et al.* 2005). Greer and Capecchi (2002) showed that disruption of the *Hoxb8* gene in mice results in grooming behavior with 100% penetrance, leading to removal of hair and lesions from the animal's own body as well as its cage-mates. The *Hoxb8* mutant mice did not differ from normal mice in most other behaviors. The *Hoxb8* gene is expressed in multiple brain regions referred to as the "OCD-circuit," including the putamen/caudate and anterior cingulate. Similarly, mice with genetic deletion of *Sapap3*, a postsynaptic scaffolding protein found at excitatory synapses and highly expressed in the striatum, exhibited increased anxiety and compulsive grooming behavior leading to facial hair loss and skin lesions; both behaviors being alleviated by SSRIs (Welch *et al.* 2007). Furthermore, lentiviral-mediated selective expression of *Sapap3* in the striatum rescued the synaptic and behavioral defects of *Sapap3* mutant mice, thereby implicating the cortico–striatal circuitry in grooming behaviors (Welch *et al.* 2007).

The process by which birds carefully and regularly groom themselves to oil their feathers and meticulously remove sheaths is called preening. In certain hormonal and other conditions, preening becomes excessive and damaging and leads to pulling out of feathers, which is referred to as feather picking. Trichotillomania shares similarities with avian feather picking with respect to treatment response (Bordnick *et al.* 1994). A double-blind, placebo-controlled trial demonstrated efficacy of clomipramine in the treatment of feather picking, while having no effect on normal preening behaviors (Seibert *et al.* 2004). Feather picking has also been described as a displacement activity in response to conflict and stress (Moon-Fanelli *et al.* 1999), and may explain its more frequent presence in captive birds than birds in the wild. Humans also demonstrate various normal displacement activities in response to conflict and tension, e.g., wiping, scraping, and scratching movements. Therefore grooming may be an innate attribute in some species, with stress and tension leading to abnormal manifestations.

Conclusions

This chapter has reviewed what is known of the neurobiology of TTM, with special focus on recent research using such tools as neurocognitive tests, neuroimaging, and pharmacological manipulations. Trichotillomania remains an under-researched, frequently hidden, debilitating disorder, which affects substantial numbers of people worldwide and adversely affects quality of life and day-to-day functioning (Bohne *et al.* 2005b; Christenson *et al.* 1991b). Several factors have been implicated in the genesis of pathological hair pulling, including affect dysregulation, behavioral addiction, and cognitive failures relating to impulse control (the A-B-C approach) (Chamberlain *et al.* 2005a; Stein *et al.* 2006). To some extent concordant with this view, abnormalities in neural regions such as the amygdalo–hippocampal formation, putamen, and frontal cortex have been identified in patients with the disorder. Comparisons with OCD have suggested overlapping features (such as impaired impulse control and involvement of cortico–subcortical circuitry) but also differences (Chamberlain *et al.* 2006a). In terms of the underlying genetic factors, little is known. Abnormalities in genes relating to brain development and serotonin transmission have thus far been tentatively implicated. It is hoped that the further application of techniques from the neurosciences, both in humans and animal models of the disorder, will extend our understanding of the neurobiology, contributing to improved diagnostic classification systems, and yielding novel treatment directions for this distressing condition.

References

Abelson JF, Kwan KY, O'Roak BJ *et al.* (2005): Sequence variants in SLITRK1 are associated with Tourette's syndrome. *Science* **310**:317–20.

American Psychiatric Association (2000): *Diagnostic and Statistical Manual of Mental Disorders*, Text Revised, 4th edn. Washington, DC: APA.

Bienvenu OJ, Samuels JF, Riddle MA *et al.* (2000): The relationship of obsessive-compulsive disorder to possible spectrum disorders: results from a family study. *Biol Psychiatry* **48**:287–93.

Bloch MD, Landeros-Weisenberger A, Dombrowski P *et al.* (2007): Systemic review: pharmacological and behavioral treatment for trichotillomania. *Biol Psychiatry* **62**:839–46.

Bohne A, Keuthen NJ, Tuschen-Caffier B and Wilhelm S (2005a): Cognitive inhibition in trichotillomania and obsessive–compulsive disorder. *Behav Res Ther* **43**:923–42.

Bohne A, Keuthen N and Wilhelm S (2005b): Pathologic hairpulling, skin picking, and nail biting. *Ann Clin Psychiatry* **17**:227–32

Bohne A, Savage CR, Deckersbach T, Keuthen NJ and Wilhelm S (2008): Motor inhibition in trichotillomania and obsessive–compulsive disorder. *J Psychiatr Res* **42**:141–50.

Bordnick PS, Thyer BA and Ritchie BW (1994): Feather picking disorder and trichotillomania: an avian model of human psychopathology. *J Behav Ther Exp Psychiatry* **25**:189–96.

Brewer JA and Potenza MN (2008): The neurobiology and genetics of impulse control disorder: relationship to drug addictions. *Biochem Pharmacol* **75**:63–75.

Canli T, Omura K, Haas BW *et al.* (2005): Beyond affect: a role for genetic variation of the serotonin transporter in neural activation during a cognitive attention task. *Proc Natl Acad Sci USA* **102**:12224–9.

Castellanos FX and Tannock R (2002): Neuroscience of attention-deficit/hyperactivity disorder: the search for endophenotypes. *Nat Rev Neurosci* **3**:617–28.

Chamberlain SR and Sahakian BJ (2007): The neuropsychiatry of impulsivity. *Curr Opin Psychiatry* **20**:255–61.

Chamberlain SR, Blackwell AD, Fineberg NA, Robbins TW and Sahakian BJ (2005a): The neuropsychology of obsessive compulsive disorder: the importance of failures in cognitive and behavioural inhibition as candidate endophenotypic markers. *Neurosci Biobehav Rev* **29**:399–419.

Chamberlain SR, Blackwell AD, Fineberg NA, Robbins TW and Sahakian BJ (2005b): Strategy implementation in obsessive-compulsive disorder and trichotillomania. *Psychol Med* **36**:91–7.

Chamberlain SR, Fineberg NA, Blackwell AD, Robbins TW and Sahakian BJ (2006a): Motor inhibition and cognitive flexibility in obsessive–compulsive disorder and trichotillomania. *Am J Psychiatry* **163**:1282–4.

Chamberlain SR, Muller U, Robbins TW and Sahakian BJ (2006b): Neuropharmacological modulation of cognition. *Curr Opin Neurol* **19**:607–12.

Chamberlain SR, Fineberg NA, Blackwell AD *et al.* (2007a): A neuropsychological comparison of obsessive–compulsive disorder and trichotillomania. *Neuropsychologia* **45**:654–62.

Chamberlain SR, Fineberg NA, Menzies LA *et al.* (2007b): Impaired cognitive flexibility and motor inhibition in unaffected first-degree relatives of patients with obsessive–compulsive disorder. *Am J Psychiatry* **164**:335–7.

Chamberlain SR, Menzies L, Sahakian BJ and Fineberg NA (2007c): Lifting the veil on trichotillomania. *Am J Psychiatry* **164**:568–74.

Chamberlain SR, Menzies L, Fineberg NA *et al.* (2008): Grey matter abnormalities in trichotillomania: morphometric magnetic resonance imaging study. *Br J Psychiatry* **193**:216–21.

Christenson GA (1995): Trichotillomania – from prevalence to comorbidity. *Psychiatr Times* **12**:44–8.

Christenson GA and Mansueto CS (1999): Trichotillomania: descriptive characteristics and phenomenology. In: Stein DJ, Christenson GA and Hollander E, eds., *Trichotillomania*. Washington, DC: American Psychiatric Press, pp. 1–42.

Christenson GA, Mackenzie TB and Mitchell JE (1991a): Characteristics of 60 adult chronic hair pullers. *Am J Psychiatry* **148**:365–70.

Christenson GA, Pyle RL and Mitchell JE (1991b): Estimated lifetime prevalence of trichotillomania in college students. *J Clin Psychiatry* **52**:415–17.

Christenson GA, Ristvedt SL and Mackenzie TB (1993): Identification of trichotillomania cue profiles. *Behav Res Ther* **31**:315–20.

Cools R, Blackwell A, Clark L *et al.* (2005): Tryptophan depletion disrupts the motivational guidance of goal-directed behavior as a function of trait impulsivity. *Neuropsychopharmacology* **30**:1362–73.

Cromwell HC and Berridge KC (1996): Implementation of action sequences by a neostriatal site: a lesion mapping study of grooming syntax. *J Neurosci* **16**:3444–58.

Cummings JL (1995): Anatomic and behavioral aspects of frontal-subcortical circuits. *Ann N Y Acad Sci* **15**:1–13.

Diefenbach GJ, Tolin DF, Meunier S and Worhunsky P (2008): Emotion regulation and trichotillomania: a comparison of clinical and nonclinical hair pulling. *J Behav Ther Exp Psychiatry* **39**:32–41.

Eagle DM, Baunez C, Hutcheson DM *et al.* (2008): Stop-signal reaction-time task performance: role of prefrontal cortex and subthalamic nucleus. *Cereb Cortex* **18**:178–88.

Ferrão YA, Almeida VP, Bedin NR, Rosa R and Busnello ED (2006): Impulsivity and compulsivity in patients with trichotillomania or skin picking compared with patients with obsessive–compulsive disorder. *Compr Psychiatry* **47**:282–8.

Fontenelle LF, Mendlowicz MV and Versiani M (2005): Impulse control disorders in patients with obsessive–compulsive disorder. *Psychiatry Clin Neurosci* **59**:30–7.

Garner JP, Weisker SM, Dufour B and Mench JA (2004): Barbering (fur and whisker trimming) by laboratory mice as a model of human trichotillomania and obsessive–compulsive spectrum disorders. *Comp Med* **54**:216–24.

Gottesman II and Gould TD (2003): The endophenotype concept in psychiatry: etymology and strategic intentions. *Am J Psychiatry* **160**:636–45.

Grachev ID (1997). MRI-based morphometric topographic parcellation of human neocortex in trichotillomania. *Psychiatry Clin Neurosci* **51**:315–21.

Grados MA and Walkup JT (2006): A new gene for Tourette's syndrome: a window into causal mechanisms? *Trends Genet* **22**:291–3.

Grant JE and Christenson GA (2007): Examination of gender in pathologic grooming behaviors. *Psychiatr Q* **78**:251–325.

Grant JE and Potenza MN (2004): Impulse control disorders: clinical characteristics and pharmacological management. *Ann Clin Psychiatry* **16**:27–34.

Grant JE and Potenza MN (2006) Compulsive aspects of impulse-control disorders. *Psychiatr Clin North Am* **29**:539–51.

Graybiel AM (1997): The basal ganglia and cognitive pattern generators. *Schizophr Bull* **23**:459–69.

Graybiel AM (1998): The basal ganglia and chunking of action repertoires. *Neurobiol Learn Mem* **70**:119–36.

Greer JM and Capecchi MR (2002): Hoxb8 is required for normal grooming behavior in mice. *Neuron* **33**:23–34.

Hemmings SM, Kinnear CJ, Lochner C *et al.* (2006): Genetic correlates in trichotillomania – a case-control association study in the South African Caucasian population. *Isr J Psychiatry Relat Sci* **43**:93–101.

Hollander E, Kim S, Khanna S and Pallanti S (2007): Obsessive–compulsive disorder and obsessive–compulsive spectrum disorders: diagnostic and dimensional issues. *CNS Spectr* **12**(2 Suppl 3):5–13.

Hugo C, Seier J, Mdhluli C *et al.* (2003): Fluoxetine decreases stereotypic behavior in primates. *Prog Neuropsychopharmacol Biol Psychiatry* **27**:639–43.

Keren M, Ron-Miara A, Feldman R and Tyano S (2006): Some reflections on infancy-onset trichotillomania. *Psychoanal Study Child* **61**:254–72.

Keuthen NJ, Makris N, Schlerf JE *et al.* (2007): Evidence for reduced cerebellar volumes in trichotillomania. *Biol Psychiatry* **61**:374–81.

Knutson B, Fong GW, Adams CM, Verner JL and Hommer D (2001): Dissociation of reward anticipation and outcome with event related fMRI. *Neuroreport* **12**:3683–7.

Kurien BT, Gross T and Scofield RH (2005): Barbering in mice: a model for trichotillomania. *BMJ* **331**:1503–5

Lenane MC, Swedo SE, Leonard H *et al.* (1990): Psychiatric disorders in first degree relatives of children and adolescents with obsessive compulsive disorder. *J Am Acad Child Adolesc Psychiatry* **29**:407–12

Lochner C and Stein DJ (2006): Does work on obsessive–compulsive spectrum disorders contribute to understanding the heterogeneity of obsessive–compulsive disorder? *Prog Neuropsychopharmacol Biol Psychiatry* **30**:353–61.

Lochner C, Hemmings SM, Kinnear CJ *et al.* (2005a): Cluster analysis of obsessive–compulsive spectrum disorders in patients with obsessive–compulsive disorder: clinical and genetic correlates. *Compr Psychiatry* **46**:14–19.

Lochner C, Seedat S, du Toit PL *et al.* (2005b): Obsessive–compulsive disorder and trichotillomania: a phenomenological comparison. *BMC Psychiatry* **5**:2.

McElroy SL, Phillips KA and Keck PE Jr (1994): Obsessive compulsive spectrum disorder. *J Clin Psychiatry* **55** Suppl:33–51; discussion 52–3.

McGuffin P, Owen MJ and Gottesman II (2002): *Psychiatric Genetics and Genomics*. Oxford, UK: Oxford University Press.

Menzies LA, Chamberlain SR, Laird AR *et al.* (2008): Integrating evidence from neuroimaging and neuropsychological studies of obsessive compulsive disorder: the orbitofronto-striatal model revisited. *Neurosci Biobehav Rev* **32**:525–49.

Moon-Fanelli AA, Dodman NH and O'Sullivan RL (1999): Veterinary models of compulsive self-grooming: parallels with trichotillomania. In: Stein DJ, Christenson GA and Hollander E, eds., *Trichotillomania*. Washington, DC: American Psychiatric Press, pp. 63–92.

Niculescu AB and Akiskal HS (2001): Proposed endophenotypes of dysthymia: evolutionary, clinical and pharmacogenomic considerations. *Mol Psychiatry* **6**:363–6.

Nomura M, Kusumi I, Kaneko M *et al.* (2006): Involvement of a polymorphism in the 5-HT2A receptor gene in impulsive behavior. *Psychopharmacology (Berl)* **187**:30–5.

O'Sullivan RL, Rauch SL, Breiter HC *et al.* (1997): Reduced basal ganglia volumes in trichotillomania measured via morphometric magnetic resonance imaging. *Biol Psychiatry* **42**:39–45.

Pezawas L, Meyer-Lindenberg A, Drabant EM *et al.* (2005): 5-HTTLPR polymorphism impacts human cingulate–amygdala interactions: a genetic susceptibility mechanism for depression. *Nat Neurosci* **8**:828–34.

Rauch SL, Wright CI, Savage CR *et al.* (2007): Brain activation during implicit sequence learning in individuals with trichotillomania. *Psychiatry Res* **154**:233–40.

Reinhardt V (2005): Hair pulling: a review. *Lab Anim* **39**:361–9.

Schlosser S, Black DW, Blum N *et al.* (1994): The demography, phenomenology, and family history of 22 persons with compulsive hair pulling. *Ann Clin Psychiatry* **6**:147–52.

Seibert LM, Crowell-Davis SL, Wilson GH and Ritchie BW (2004): Placebo-controlled clomipramine trial for the treatment of feather picking disorder in cockatoos. *J Am Anim Hosp Assoc* **40**:261–9.

Stanley MA, Hannay HJ and Breckenridge JK (1997): The neuropsychology of trichotillomania. *J Anxiety Disord* **11**:473–88.

Stein DJ (2006): Obsessive–compulsive spectrum disorders: current advances, and future directions. *Psychiatr Clin North Am* **29**:xiii–xv.

Stein DJ and Lochner C (2006): Obsessive–compulsive spectrum disorders: a multidimensional approach. *Psychiatr Clin North Am* **29**:343–51.

Stein DJ, Hollander E, Cohen L, Simeon D and Aronowitz B (1995): Serotonergic responsivity in trichotillomania: neuroendocrine effects of m-chlorophenylpiperazine. *Biol Psychiatry* **37**:414–16.

Stein DJ, Bouwer C and Maud CM (1997a): Use of the selective serotonin reuptake inhibitor citalopram in treatment of trichotillomania. *Eur Arch Psychiatry Clin Neurosci* **247**:234–6.

Stein DJ, Coetzer R, Lee M, Davids B and Bouwer C (1997b): Magnetic resonance brain imaging in women with obsessive–compulsive disorder and trichotillomania. *Psychiatry Res* **74**:177–82.

Stein DJ, Van Heerden B, Wessels CJ *et al.* (1999): Single photon emission computed tomography of the brain with Tc-99m HMPAO during sumatriptan challenge in obsessive–compulsive disorder: investigating the functional role of the serotonin auto-receptor. *Prog Neuropsychopharmacol Biol Psychiatry* **23**:1079–99.

Stein DJ, van Heerden B, Hugo C *et al.* (2002): Functional brain imaging and pharmacotherapy in trichotillomania. Single photon emission computed tomography before and after treatment with the selective serotonin reuptake inhibitor citalopram. *Prog Neuropsychopharmacol Biol Psychiatry* **26**:885–90.

Stein DJ, Chamberlain SR and Fineberg N (2006): An A-B-C model of habit disorders: hair-pulling, skin-picking, and other stereotypic conditions. *CNS Spectr* **11**:824–7.

Swedo SE and Leonard HL (1992): Trichotillomania: an obsessive compulsive spectrum disorder? *Psychiatr Clin North Am* **15**:777–90.

Swedo SE, Leonard HL, Rapoport JL *et al.* (1989): A double-blind comparison of clomipramine and desipramine in the treatment of trichotillomania. *N Engl J Med* **321**:497–501.

Swedo SE, Rapoport JL, Leonard HL *et al.* (1991): Regional cerebral glucose metabolism of women with trichotillomania. *Arch Gen Psychiatry* **48**:828–33.

Torregrossa MM, Quinn JJ and Taylor JR (2008): Impulsivity, compulsivity, and habit: the role of orbitofrontal cortex revisited. *Biol Psychiatry* **63**:253–5.

Walsh KH and McDougle CJ (2001): Trichotillomania. Presentation, etiology, diagnosis and therapy. *Am J Clin Dermatol* **2**:327–33.

Welch JM, Lu J, Rodriguiz RM *et al.* (2007): Cortico-striatal synaptic defects and OCD-like behaviours in Sapap3-mutant mice. *Nature* **448**:894–900.

Woolley J, Heyman I, Brammer M *et al.* (2008): Brain activation in paediatric obsessive–compulsive disorder during tasks of inhibitory control. *Br J Psychiatry* **192**:25–31.

World Health Organization (2002): *The ICD10 Classification of Mental Behavioural Disorders: Clinical Description and Diagnostic Guidelines.* Geneva, Switzerland: WHO.

Zohar J, Mueller EA, Insel TR, Zohar-Kadouch RC and Murphy DL (1987): Serotonergic responsivity in obsessive–compulsive disorder. Comparison of patients and healthy controls. *Arch Gen Psychiatry* **44**:946–51.

Zuchner S, Cuccaro ML, Tran-Viet KN *et al.* (2006): SLITRK1 mutations in trichotillomania. *Mol Psychiatry* **11**:887–9.

Index